CONCEPTS, PROBLEMS AND SOLUTIONS IN GENERAL PHYSICS

VOLUME ONE

A STUDY GUIDE FOR STUDENTS OF ENGINEERING AND SCIENCE

RAYMOND A. SERWAY
CLARKSON COLLEGE OF TECHNOLOGY

W. B. SAUNDERS COMPANY · Philadelphia · London · Toronto

PREFACE

The purpose of this book is to provide students of science and engineering with a review of basic concepts and fundamental principles in general physics, together with a collection of illustrative examples and programmed exercises. It is a widely accepted fact that physics is best understood through the experience of application of fundamental concepts to a variety of physical situations. Therefore, methods and techniques of problem solving are emphasized which, when mastered, should provide the student with the background and confidence required for more complex situations.

Each chapter is divided into several parts. The first part is a presentation of basic concepts and definitions, together with a collection of illustrative examples. The second part is a set of programmed exercises, which serve as a review of the fundamental principles in that chapter, as well as a self-test of techniques in problem solving. The last part contains a summary of important definitions and equations, and a collection of unsolved problems; most of these problems are closely related to the "theory" portion of that chapter. Problems and examples which involve calculus are labeled with a dagger (†), so that they can be avoided in the event that the book is used in a non-calculus physics course. In general, problems range in difficulty from those of the confidence building variety to complex problems involving several thought processes. The more complex problems are labeled with an asterisk (*). Hints to solving these complex problems are usually given. Answers to all problems are given at the end of the book, together with a collection of useful tables and an index.

This book should be most useful in an introductory physics course where problem solving and concepts are emphasized. The added feature of the programmed exercises will hopefully provide some guidelines of "self-learning" for the student. A cross-index is provided in the event that the book is used as a supplement to other texts such as *Physics for Scientists and Engineers* by Melissinos and Lobkowicz, *Fundamentals of Physics* by Halliday and Resnick, *University Physics* by Sears and Zemansky, *Introduction to Physics for Scientists and Engineers* by Bueche and *Elementary Classical Physics* by Weidner and Sells. Since the order of topics is somewhat standard, the book can be used in conjunction with other introductory physics texts.

I would be grateful to the users of this book if they would point out errors by writing to me directly.

RAYMOND A. SERWAY
Clarkson College of Technology
Potsdam, New York

TO THE STUDENT

It is difficult to teach good study habits to students. However, my experience with incoming students has shown that the major difficulty with learning physics is the "wrong" approach, namely, memorization of textbook and lecture material. Memorizing sections of a text, including definitions, derivations, and the like *does not* necessarily mean the student *understands* the material. Understanding the basic concepts and scientific argument is possible only through efficient study habits, and many hours of problem solving, discussions with other students and instructors, and so on. Therefore, my first word of advice is *reduce memorization* of material (including basic equations and definitions) to a *minimum.* (It has been my own policy to list basic equations and constants on the front of exams to encourage students to "learn" physics through "thinking" rather than memorizing.) It is very important, in fact, essential, that you understand basic concepts and principles *first,* before attempting to solve assigned problems.

Second, try to solve as many problems at the end of the chapter as possible. This can only be done after carefully reading through the text material, examples and exercises. Keep in mind that very few people are able to absorb the full meaning of scientific writing after one reading. Lectures should provide some clarification of troublesome points, but several readings of notes and text material are usually necessary. When solving problems, try to find alternate solutions to the same problem. For example, many problems in mechanics can be treated by solving the equations of motion or by the more direct energy method. This book provides a variety of examples and step-by-step problems (programmed exercises) which should be of value in this regard.

The method of solving certain problems, especially those requiring the use of several concepts, should be *carefully* planned. Always read the problem through a few times until you are confident you understand what is being asked. Next, read the problem through with more thought, with special attention to the available information. Finally, write down the basic structure of the method (or methods) you feel is applicable to the problem, and proceed with the solution. Be careful not to misinterpret the problem. The ability to interpret properly what is being asked is an integral part of solving the problem.

Finally, it is always useful to supplement the study of material through models and experiments. Whenever it is possible, try to set up simple experiments at home or in the laboratory to substantiate ideas and models discussed in class or in the text. For example, the common "slinky" toy is invaluable for demonstrating traveling waves; an old pair of Polaroid sunglasses and some discarded lenses and magnifying glass are central components of various experiments in optics; collisions between billiard balls can be conveniently studied in the pool room, with the addition of a paper covered table to provide a permanent record of the collisions. The list is endless. When physical models are not available, try to develop "mental" models, and devise thought-provoking experiments to improve your understanding of the concepts or the situation at hand.

A few words of advice pertaining to the use of this book are in order. Note that each chapter is divided into three major categories: the first is a discussion of the basic concepts, definitions and principles, together with a collection of solved example problems. The second is a collection of programmed exercises which should be read with care. Some of these exercises serve as a review of the concepts and definitions in that chapter. Others are step-by-step solutions to new problems, some of which involve more than one concept discussed in that chapter. Finally, a summary of important concepts and equations is presented, followed by a collection of unsolved problems, with answers at the end of the text. The student should attempt to solve these problems only after reading through and, hopefully, "understanding" the previous portions of that chapter. Some of the more difficult problems are labeled with an asterisk (*), while those requiring the use of calculus are labeled with a dagger (†). It would be to the student's advantage to use the programmed exercises in an "honest" manner. By this, I mean that the answers to each part of the exercise in the *right column* of the page should be covered up, while the questions in the *left column* are being read. Obviously, the exercises will prove most effective when used in this manner. I suggest blocking the right column with a blank sheet of paper (which you can use for calculations, and so on). After writing what you think is the correct answer to that part of the exercise, check the answer in the right column by sliding your blank paper down one frame. If your answer is correct, go on to the next part of the exercise. If your answer is incorrect, reread the section of the text applicable to that exercise and go over it a second time. Repeat this until you are confident that the answer in the text is correct. If there is still disagreement, check your work with the instructor.

Cross-Index To Other Texts
Given by Chapter Numbers*

ML – S		SZ – S		B – S		WS – S		HR – S	
1	1	1	1, 2	1	1	1	1	1	1
1	2	2	2	2	2	2	2	2	2
2, 3	3	3	9	3	9	3, 4	3	3	3
2, 3	4	4	3	4	3, 4	5, 6, 7	7	4	3, 4
4	5	5	5, 11	5	5	8	5	5	5
6	7	6	3, 4	7	7	9, 10	6	6, 7	6
7	6	7	6	8	6	11, 12	8	8, 9	7
10	8	8	7	9	7	13	11	10, 11	8
13	10	9	8	10	4, 8	14	10	12	9
16	13	11	10	11, 12	8	15	12	13	10
17	14	12, 14	12	13	10	18, 20	13	14	11
18	12	15, 16, 17	13	14	11	19, 21	14	15	12
		18, 19, 20	14	15	14			18, 19	13
				16	13			20, 21	14
				17	14				

**Note:* This index is only a *rough* guide, and the symbols used are as follows:

ML – Melissinos & Lobkowicz – *Physics for Scientists and Engineers (Volume I)*
HR – Halliday & Resnick – *Fundamentals of Physics*
SZ – Sears & Zemansky – *University Physics*
B – Bueche – *Introduction to Physics for Scientists and Engineers*
WS – Weidner & Sells – *Elementary Classical Physics*
S – Serway – (This Text)

ACKNOWLEDGMENTS

I am grateful to a number of people who, in various ways, assisted me in the preparation of the final manuscript. Mrs. Agatha Hollister did an outstanding job of typing the manuscript. Helpful criticisms and suggestions were provided by Professors A. Czanderna, H. Helbig, D. Kaup, D. Larsen, F. Lobkowicz, J. Love, F. Otter, J. Marion, and Messrs. G. Anton, B. Davis, A. Miller and A. Serway. I thank the hundreds of students who used this manuscript in its original, "uncut" version for pointing out various errors and misleading statements.

Finally, I dedicate this book to my wife Elizabeth and my children Mark, Michele and David. They were a constant source of inspiration and eventually learned to cope with my unusual hours.

CONTENTS

1

UNITS AND UNCERTAINTIES

1.1 FUNDAMENTAL UNITS

Physics is a science based upon measurements, and the interpretation of such measurements with a set of rules or laws. Consequently, certain observable quantities must be defined such as velocity, force, momentum, and so forth. Such quantities are expressed in terms of three undefinable quantities called mass (M), length (L) and time (T).

Three systems of units are used in practice: (1) Engineering units, where the units of mass, length and time are the slug, the foot and the second, respectively; (2) Standard International (SI), sometimes referred to as mks, which stands for meters, kilograms and seconds; (3) cgs units, which stands for centimeters, grams and seconds. In mechanics, we will use these units interchangeably, but emphasis will be placed on the SI units. The SI units will be used exclusively in the treatment of electricity and magnetism.

The dimensions of area, volume, velocity and acceleration in the three systems described are given in Table 1-1.

TABLE 1-1 DIMENSIONS OF AREA, VOLUME, VELOCITY AND ACCELERATION*

System	Area	Volume	Velocity	Acceleration
SI	m^2	m^3	m/sec	m/sec^2
cgs	cm^2	cm^3	cm/sec	cm/sec^2
Engineering	ft^2	ft^3	ft/sec	ft/sec^2

*The abbreviations m, cm, ft and sec refer to meters, centimeters, feet and second, respectively. Later, we will use the abbreviations kg and g for kilogram and gram, respectively.

It is often useful to determine whether or not an expression is dimensionally correct. Such an analysis is *especially* useful when the derivation of the expression is uncertain. The student should use this technique as *a standard operating procedure.*

Example 1.1

Let us show that the expression $v^2 = v_0{}^2 + 2ax$ is dimensionally correct, where v and v_0 represent velocities, a the acceleration and x the displacement.

v^2 and $v_0{}^2$ have dimensions of L^2/T^2, a has dimensions of L/T^2 and x has dimensions of L. Therefore, ax has dimensions of $L/T^2 \times L = L^2/T^2$, so the expression is dimensionally correct.

Example 1.2

Determine whether or not the expression $x = v_0 t + \frac{1}{2}at^3$ is dimensionally correct, where t is a time interval.

The left side has units of L. The term $v_0 t$ has units of $L/T \times T = L$; however, the term $\frac{1}{2}at^3$ has units of $L/T^2 \times T^3 = LT$, so the expression is dimensionally *incorrect.* The correct expression is obtained if the last term is replaced by $\frac{1}{2}at^2$.

Example 1.3

Newton's second law of motion states that the acceleration of an object is directly proportional to the resultant force F acting on it, and inversely proportional to its mass m. From this law, determine the dimensions of F.

Since $a = F/m$, the dimensions of F are the same as the dimensions of ma. Therefore, $[F] = [m]\ [a] = ML/T^2$.

1.2 CONVERSION OF UNITS

In many situations, it is necessary to convert from one system of units to another. The technique of conversion will be illustrated with a few examples.

Example 1.4

A cube is measured to be 3.00 ft on each side. The volume of the cube in Engineering units is 27.0 ft^3. Since 1 ft = 0.3048 m, the volume of the cube in SI units is

$$V = 27.0\ \text{ft}^3 \times \left(0.3048\ \frac{\text{m}}{\text{ft}}\right)^3 = 0.765\ \text{m}^3$$

Note: Only three significant figures were given in the measured value of the cube's side; therefore, the volume is only accurate to three significant figures.

Example 1.5

A train is traveling at an average speed of 60 mi/hr. The speed of the train in SI units can be found by noting that 1 mile = 1610 m and 1 hr = 3600 sec; therefore,

$$60\ \frac{\text{mi}}{\text{hr}} = \frac{60\ \text{mi} \times 1610\ \frac{\text{m}}{\text{mi}}}{1\ \text{hr} \times 3600\ \frac{\text{sec}}{\text{hr}}} = 27\ \text{m/sec}$$

1.3 ACCURACY OF MEASUREMENTS AND UNCERTAINTIES

All real measurements have some degree of inaccuracy, such as measurements of mass, length and time. Therefore, the errors encountered in the measurements must be taken into account in order to make meaningful comparisons between experiment and theory. The errors in measurements may arise from several factors such as human error (for example, the limiting resolution of the eye) and instrument inaccuracies (for example, weighing a light object with an insensitive balance). We shall be concerned only with the estimation of uncertainties, since the mathematical details of uncertainty calculations are complex. Keep in mind that any calculated quantity involving one or more measured parameters should have as many significant figures as the *least* accurately known quantity in the calculation. There are several rules for estimating uncertainties of the sum, difference, product or quotient of *two* quantities.

A. If two quantities X and Y are *added* or *subtracted,* and the uncertainties in their measured values are $\pm\Delta X$ and $\pm\Delta Y$, respectively, the sum or difference of X and Y will be uncertain by an amount $\pm(\Delta X + \Delta Y)$.

$$(X \pm \Delta X) \pm (Y \pm \Delta Y) = X \pm Y \pm (\Delta X + \Delta Y)$$

B. When two quantities X and Y are *multiplied,* and the uncertainties in their measured values are $\pm\Delta X$ and $\pm\Delta Y$, respectively, the uncertainty in the product XY can be found from the following consideration:

$$(X \pm \Delta X)(Y \pm \Delta Y) = XY \pm Y\Delta X \pm X\Delta Y \pm \Delta X\Delta Y$$

However, if the uncertainties ΔX and ΔY are *small* compared to X and Y, then the last term $\Delta X\Delta Y$ can be *neglected;* hence, the uncertainty in the product XY reduces to $\pm(Y\Delta X + X\Delta Y)$. The rule for *division* is essentially the same as this. That is, the uncertainty in the quotient X/Y is equivalent to the uncertainty in the product XY^{-1}, which then gives the same result for the uncertainty as XY.

Example 1.6

A student measures a certain rectangular plate to have a length of (12.30 ± 0.04) cm and a width of (4.26 ± 0.03) cm. Find the area of the plate and the uncertainty in the calculated area.

$$\text{Area} = \ell w = (12.30 \pm 0.04)\text{ cm} \times (4.26 \pm 0.03)\text{ cm}$$

$$\cong [12.30 \times 4.26 \pm 4.26 \times 0.04 \pm 12.30 \times 0.03]\text{ cm}^2$$

$$\cong [52.40 \pm 0.54]\text{ cm}^2$$

Example 1.7

The mass of an object is measured to be (2.40 ± 0.01) g and its volume is measured to be (4.35 ± 0.03) cm^3. Find the density of the object (that is, the mass per unit volume), the percentage uncertainty in the density and the absolute uncertainty in the density.

If we let $\pm\ \Delta m$ and $\pm\ \Delta V$ represent the uncertainties in the mass and volume, respectively, we have

$$\rho = \frac{m \pm \Delta m}{V \pm \Delta V} = \frac{m\left(1 \pm \frac{\Delta m}{m}\right)}{V\left(1 \pm \frac{\Delta V}{V}\right)}$$

$$\rho \cong \frac{m}{V}\left(1 \pm \frac{\Delta m}{m}\right)\left(1 \mp \frac{\Delta V}{V}\right) \cong \frac{m}{V}\left(1 \pm \frac{\Delta m}{m} \pm \frac{\Delta V}{V}\right)$$

$$\rho \cong \frac{2.40}{4.35}\left[1 \pm \frac{0.01}{2.40} \pm \frac{0.03}{4.35}\right]\frac{g}{cm^3}$$

$$\rho \cong 0.552\,(1 \pm 0.011)\,\frac{g}{cm^3}$$

Therefore, the percentage uncertainty in ρ is 1.1%, and the *absolute* uncertainty in ρ is $\pm 0.552(0.011) = \pm 0.006\ g/cm^3$.

1.4 POWERS OF TEN AND STANDARD PREFIXES

The student should be familiar with the usage of powers of ten. It is a compact form of writing very large or very small numbers. For example, instead of 10,000, we write 10^4, where the exponent represents the number of zeros; that is, $10^4 = 10 \times 10 \times 10 \times 10 = 10{,}000$. Likewise, a small number like 0.0001 can be expressed as 10^{-4}, where the negative exponent is involved, since we are dealing with a number less than one. Some other examples of the use of powers of ten are given below:

$$1{,}000 = 10^3 \qquad 0.003 = 3 \times 10^{-3}$$
$$85{,}000 = 8.5 \times 10^4 \qquad 0.00085 = 8.5 \times 10^{-4}$$
$$3{,}200{,}000 = 3.2 \times 10^6 \qquad 0.00002 = 2 \times 10^{-5}$$

If numbers written as powers of ten are *multiplied,* we simply *add* the exponents, maintaining their signs. For example,

$$(3 \times 10^3) \times (5 \times 10^4) = 15 \times 10^7 = 1.5 \times 10^8$$

$$(2 \times 10^5) \times (4 \times 10^{-2}) = 8 \times 10^3$$

$$(5.6 \times 10^4) \times (4.3 \times 10^8) = 24 \times 10^{12}$$

Likewise, when numbers written as powers of ten are *divided,* we can bring the power of ten from the denominator to the number by changing its sign. For example,

$$\frac{8 \times 10^5}{2 \times 10^2} = 4 \times 10^5 \times 10^{-2} = 4 \times 10^3$$

$$\frac{12 \times 10^{-4}}{4 \times 10^{-9}} = 3 \times 10^{-4} \times 10^9 = 3 \times 10^5$$

Finally, there are a number of commonly used *prefixes* to replace powers of ten. For example, one millimeter is the equivalent of 10^{-3} meters. We express this as one mm. Also, one kilogram is equivalent to 10^3 g. We therefore use the abbreviation kg for kilogram, the prefix being k for kilo. A list of prefixes is given in Table 1–2. Some of these will be used frequently in the text, especially μ (micro-), m (milli-) and k (kilo-).

TABLE 1-2 PREFIXES FOR POWERS OF TEN

Power of Ten	10^{-12}	10^{-9}	10^{-6}	10^{-3}	10^{-2}	10^3	10^6	10^9	10^{12}
Prefix	pico	nano	micro	milli	centi	kilo	mega	giga	tera
Symbol	p	n	μ	m	c	k	M	g	T

1.5 PROBLEMS

1. The area of a table is measured to be 2.2 m^2. Express this area in ft^2.

2. The equatorial radius of the earth is 3963 miles, and the period of revolution of the earth about its axis is 24 hours. From this information, determine the tangential speed of a point on the equator relative to the center of the earth. Express your answer in SI units.

3. The speed of light in a vacuum is known to be about 3.00×10^8 m/sec. Express this in mi/hr.

4. Standard atmospheric pressure is known to be 14.70 lb/in^2. Determine the standard atmospheric pressure in SI units using the information that one pound of force is equal to a force of 4.448 newtons (N), the unit of force in SI, and 1 in = 2.54×10^{-2} m.

5. The radius of a solid sphere is measured to be (3.50 ± 0.02) cm. Find the volume of the sphere and the uncertainty in the calculated volume.

6. A carpenter uses a 6 ft tape rule to measure the length of a large room. In three consecutive steps, he measures distances of 72.00 in, 72.00 in, and 13.50 in, where the uncertainty in each measurement is estimated to be ±0.13 in. What is the total length of the room and the uncertainty in the length?

7. The moment of inertia of a solid cylinder is given by $\frac{1}{2}MR^2$, where M is its mass and R is its radius. A certain cylinder is measured to have a mass of (3.20 ± 0.01) kg and a radius of (0.350 ± 0.003) m. Determine the moment of inertia of the cylinder and its uncertainty.

8. Express the following in powers of ten: (a) 530,000 (b) 0.00025 (c) 5 million (d) 13 billion (e) 1492.

9. Write the following products and divisions in powers of ten.

(a) $(5.2 \times 10^5) \times (4.0 \times 10^8)$

(b) $(2.0 \times 10^{-7}) \times (6.1 \times 10^2)$

(c) $\dfrac{8.4 \times 10^{-4}}{2.0 \times 10^5}$

(d) $\dfrac{9.9 \times 10^{20}}{3.0 \times 10^8}$

2

VECTORS

2.1 INTRODUCTION

Vectors are any quantities which can only be described with both magnitude and direction. They are not to be confused with scalar quantities which are specified by a number with appropriate units (for example, mass, charge, volume). Some examples of vector quantities are velocity, acceleration, force and displacement.

An arbitrary vector **A** has the following properties. Its magnitude is usually represented by the symbol |**A**| or sometimes simply A. This also indicates the length of an arrow representing the vector on graph paper. When vector **A** is added to vector **B**, a third vector **C** is formed, where the order of the sum is not important. That is, **A** + **B** = **B** + **A** = **C**, or in mathematical terms, vector addition is commutative. Vectors also obey the associative law for addition, that is, (**A** + **B**) + **C** = **A** + (**B** + **C**). It is to be understood in all such manipulations that the vectors *all* have the same units. As an example, you can't add a velocity vector to a displacement vector. If we wish to subtract vector **B** from vector **A**, that is, we wish to know the value of **A** − **B**, the result could be obtained by *adding* the vector (−)**B** to the vector **A**. In other words, **A** − **B** = **A** + (−)**B**. The vector (−)**B** can be interpreted as being a vector whose magnitude is the same as **B** but is in the opposite direction to **B**. Therefore, multiplying any vector by the factor −1 simply represents an operation corresponding to reversing the direction of the vector. If a vector **A** is multiplied by a positive number m, the new vector m**A** is in the same direction as **A**, but m times as long as **A**. It follows that multiplication of **A** by a negative number reverses the direction of the new vector relative to **A**.

2.2 VECTOR SUMS AND DIFFERENCES

Graphical methods for adding or subtracting vectors are sometimes useful. The simplest rule for graphically adding two vectors **A** and **B** is to draw the tail of vector **B** starting from the head of vector **A**. The resultant vector **C** is the vector drawn from the tail of **A** to the head of **B** as shown in Figure 2–1.

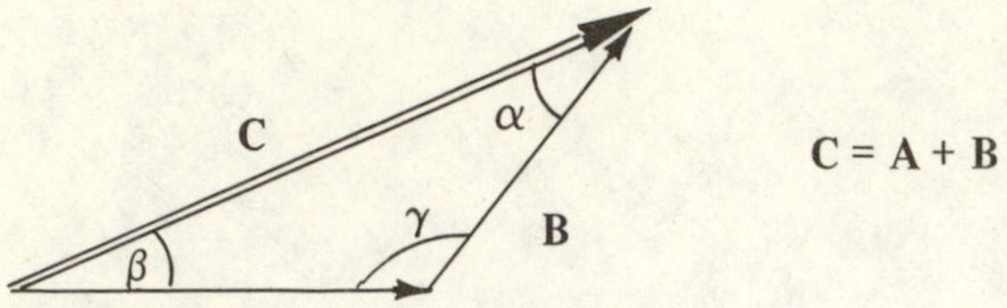

Figure 2-1. Graphical method of adding two vectors.

This can be extended to adding several vectors, say, **A** + **B** + **C** + **D**, where the sum vector **E** is the vector that completes the polygon as illustrated in Figure 2–2.

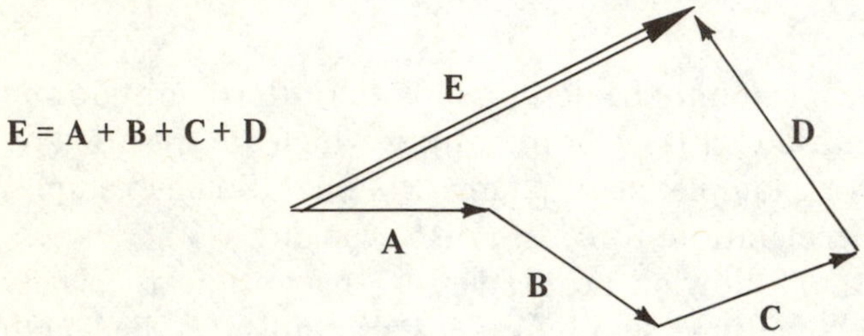

Figure 2-2 Polygon method of adding several vectors.

When the vectors **A** and **B** are added together, the magnitude of the resultant vector **C** can also be computed from the trigonometric relation

$$C = \sqrt{A^2 + B^2 - 2AB \cos \gamma} \tag{2.1}$$

where γ is the angle between vectors **A** and **B** as defined in Figure 2–1. Likewise, the remaining angles in Figure 2–1 may be found from the trigonometric relation

$$\frac{\sin \alpha}{A} = \frac{\sin \beta}{B} = \frac{\sin \gamma}{C} \tag{2.2}$$

The graphical method of subtracting two vectors is illustrated in Figure 2–3.

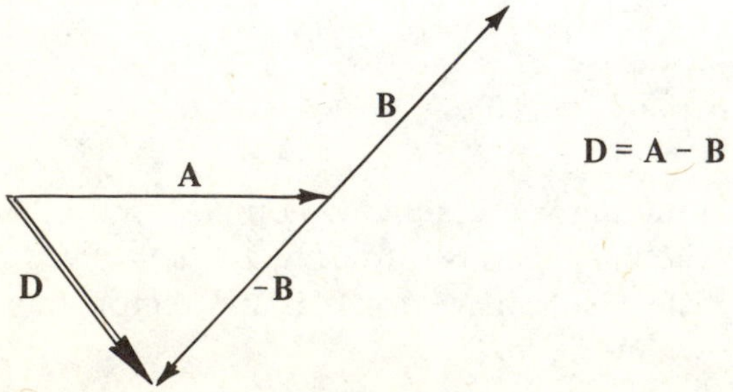

Figure 2-3 Graphical method of subtracting two vectors.

Example 2.1

Two vectors **A** and **B** are given in Figure 2-4. Vector **A** is three units long and points along the positive x axis. Vector **B** is four units long and makes an angle of +30° with the positive x axis. Find the magnitude and direction of the resultant vector **C**.

Solution

The problem can be solved graphically as shown in Figure 2-4, using ruled paper and protractor. The analytical solution for the magnitude of **C** is obtained from the relation

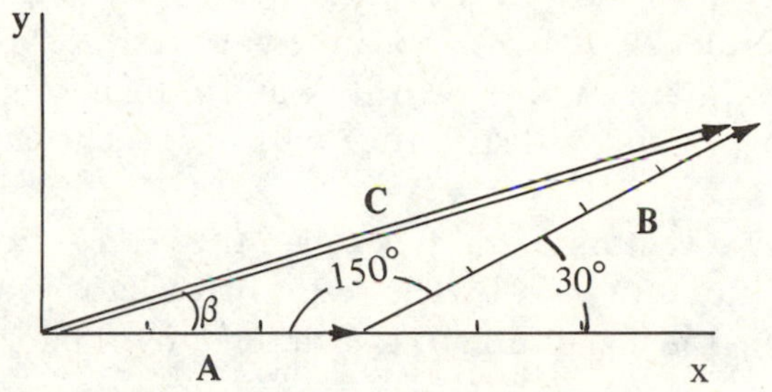

Figure 2-4

$$C = \sqrt{A^2 + B^2 - 2AB \cos \gamma} = \sqrt{9 + 16 - 2(3)(4) \cos (150)}$$

$$C = \sqrt{25 + 24\,(0.866)} \qquad = 6.8$$

and

$$\frac{\sin \beta}{B} = \frac{\sin (150)}{C}$$

$$\sin \beta = \frac{4}{6.8}(0.5) = 0.294$$

$$\beta \cong 17^\circ$$

2.3 SCALAR AND VECTOR PRODUCTS

The rules for multiplying two vectors **A** and **B** are somewhat more complicated compared to the rules of addition. Two kinds of vector multiplication are commonly used. One is called the dot product **A·B** or the scalar product, since the result is a *scalar* quantity. The scalar product **A·B** is defined by the relation

$$\mathbf{A \cdot B} = AB \cos \theta \qquad (2.3)$$

where θ is the smaller angle between **A** and **B** as shown in Figure 2-5.

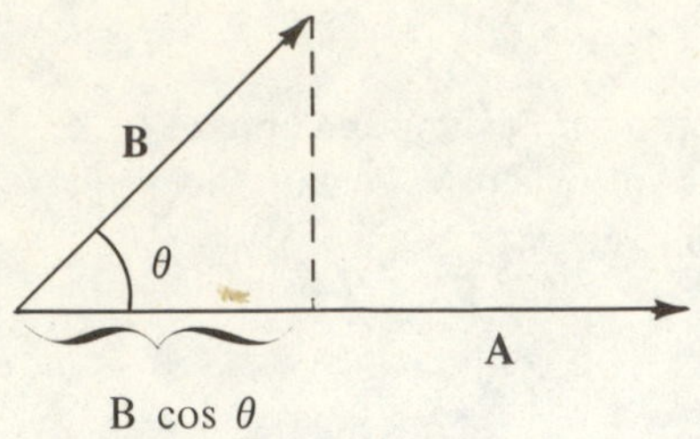

Figure 2-5 Graphical representation of the projection of **B** on **A**.

Notice that B cos θ is the projection of **B** on **A**, therefore the dot product **A·B** is equivalent to the product of the magnitude of **A** times the projection of **B** onto **A**. One property of the scalar product which follows from the definition is the fact that **A·B** = **B·A**. Furthermore, it follows that **A·B** = 0 if **A** is perpendicular to **B** ($\theta = \pi/2$), or, in the more trivial case, if either **A** or **B** is zero. Likewise, **A·B** = AB if **A** and **B** are in the same direction ($\theta = 0$), and **A·B** = −AB if **A** and **B** lie in opposite directions ($\theta = \pi$). Finally, the scalar product is distributive with respect to a sum; that is, **A·(B + C) = A·B + A·C**.

The second kind of product of two vectors **A** and **B** is called the *vector* product or *cross* product, represented by the notation **A × B**. The result of a vector product is a *vector* quantity **C = A × B**, where the magnitude of **C** is defined by the relation

$$C = AB \sin\theta \tag{2.4}$$

and θ is the angle between **A** and **B** as defined in Figure 2-6. The direction of **A × B** is a vector perpendicular to the plane formed by **A** and **B**, where its sense is determined by the direction of advance of a right hand screw.

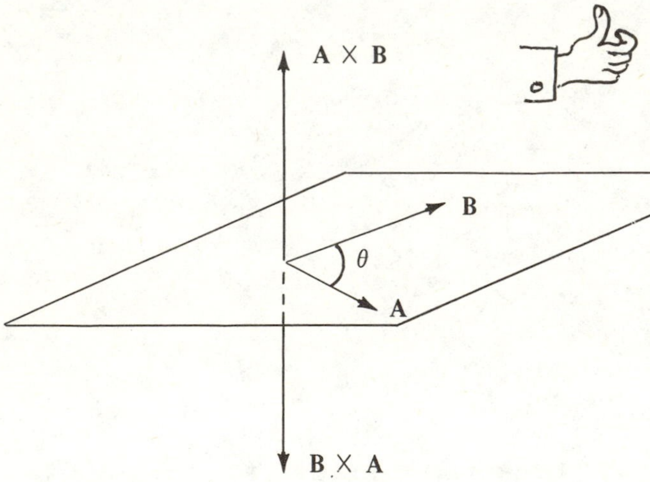

Figure 2-6 Graphical representation of the cross product **A × B**. Note that **A × B = −B × A**.

The right-hand rule is useful to determine the direction of **A × B**. The four fingers of the right hand are pointed along **A**, and then rotated into **B**. The right thumb points in the direction of **A × B**, as in Figure 2-6. One property of the cross product which follows from the definition is the fact that **A × B** = −**B × A**. In addition, if **A** is parallel to **B** ($\theta = 0$ or π), **A × B** = 0, and if **A** is perpendicular to **B**, |**A × B**| = AB. The vector product is also distributive with respect to a sum; that is, **A × (B + C) = A × B + A × C**.

2.4 VECTOR COMPONENTS AND UNIT VECTORS

If a vector **A** lies in the xy plane as shown in Figure 2-7 (*a*), the vector may be written as the sum of two vector components $\mathbf{A_x}$ and $\mathbf{A_y}$. It is useful to write these components as $A_x\mathbf{i}$ and $A_y\mathbf{j}$, respectively, where A_x and A_y are the magnitudes of the vector components and the vectors **i** and **j** are *unit* vectors along the x and y directions. These unit vectors are defined by

$$\mathbf{i} = \frac{\mathbf{A_x}}{A_x},\ \mathbf{j} = \frac{\mathbf{A_y}}{A_y}$$

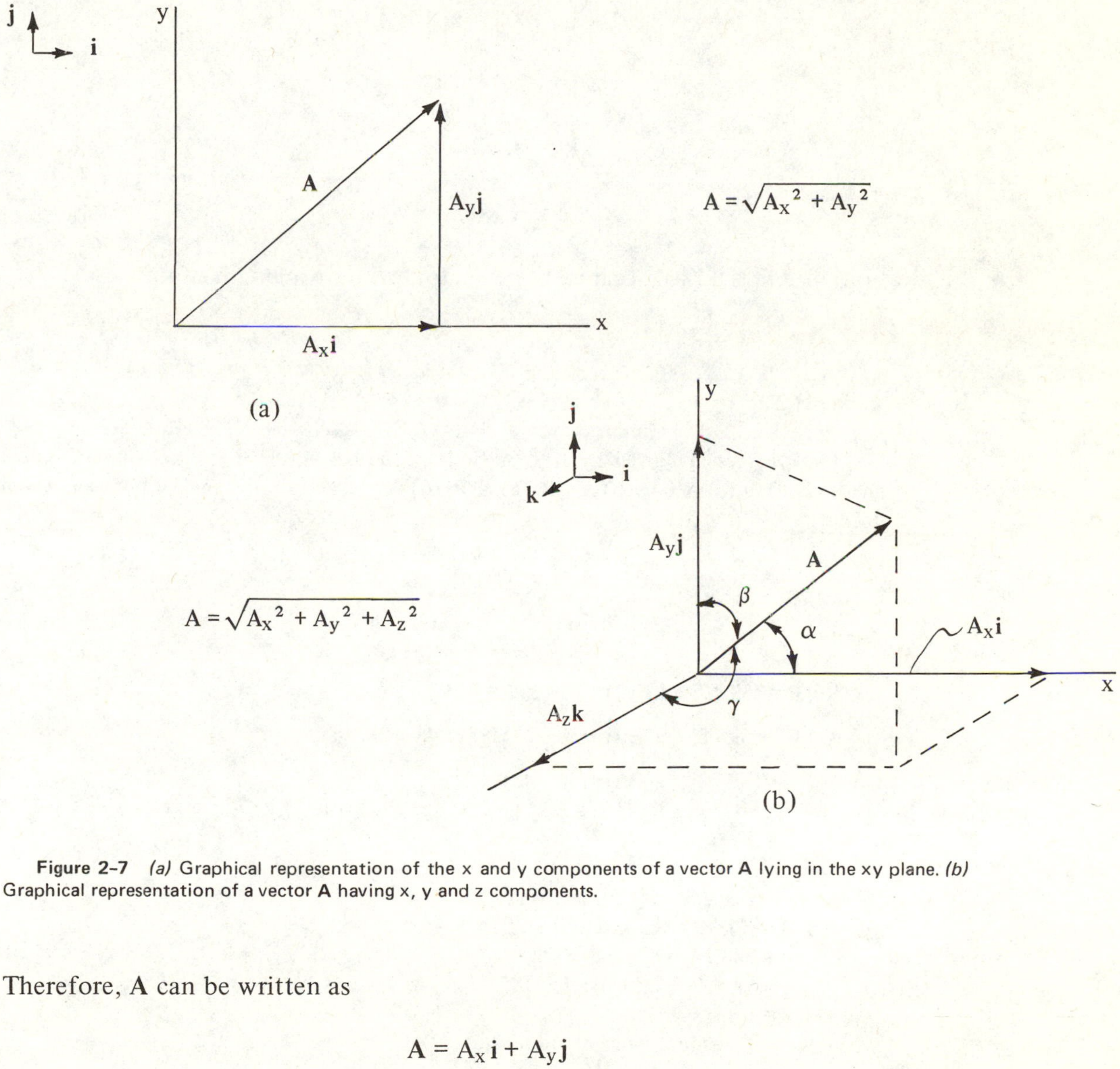

Figure 2-7 *(a)* Graphical representation of the x and y components of a vector **A** lying in the xy plane. *(b)* Graphical representation of a vector **A** having x, y and z components.

Therefore, **A** can be written as

$$\mathbf{A} = A_x\mathbf{i} + A_y\mathbf{j}$$

where

$$A = \sqrt{A_x^2 + A_y^2}$$

If a vector **A** has three components as in Figure 2–7(*b*), we add a third unit vector **k** to describe a unit vector in the z direction.

The unit vectors, **i**, **j** and **k** have some useful properties. Since they are mutually perpendicular, it follows from the definition of the scalar product that

$$\begin{aligned} \mathbf{i}\cdot\mathbf{i} = \mathbf{j}\cdot\mathbf{j} = \mathbf{k}\cdot\mathbf{k} = 1 \\ \mathbf{i}\cdot\mathbf{j} = \mathbf{i}\cdot\mathbf{k} = \mathbf{j}\cdot\mathbf{k} = 0 \end{aligned} \tag{2.5}$$

and from the definition of the vector product it follows that

$$\begin{aligned} \mathbf{i}\times\mathbf{i} = \mathbf{j}\times\mathbf{j} = \mathbf{k}\times\mathbf{k} = 0 \\ \mathbf{i}\times\mathbf{j} = -\mathbf{j}\times\mathbf{i} = \mathbf{k} \\ \mathbf{j}\times\mathbf{k} = -\mathbf{k}\times\mathbf{j} = \mathbf{i} \\ \mathbf{k}\times\mathbf{i} = -\mathbf{i}\times\mathbf{k} = \mathbf{j} \end{aligned} \tag{2.6}$$

In addition, $|\mathbf{i}| = |\mathbf{j}| = |\mathbf{k}| = 1$, that is, these vectors *all* have a length of unity.

Example 2.2

Two vectors are given by $\mathbf{A} = 3\mathbf{i} - 2\mathbf{j} + \mathbf{k}$ and $\mathbf{B} = \mathbf{i} + 3\mathbf{j} - 2\mathbf{k}$. Using analytical methods, find (a) **A** + **B**, (b) **A** − **B**, (c) **A**·**B**, (d) **A** × **B** and (e) the angle θ between **A** and **B**.

Solution

(a) $\mathbf{A} + \mathbf{B} = (3\mathbf{i} - 2\mathbf{j} + \mathbf{k}) + (\mathbf{i} + 3\mathbf{j} - 2\mathbf{k}) = 4\mathbf{i} + \mathbf{j} - \mathbf{k}$

(b) $\mathbf{A} - \mathbf{B} = (3\mathbf{i} - 2\mathbf{j} + \mathbf{k}) - (\mathbf{i} + 3\mathbf{j} - 2\mathbf{k}) = 2\mathbf{i} - 5\mathbf{j} + 3\mathbf{k}$

(c)
$$\begin{aligned} \mathbf{A}\cdot\mathbf{B} &= (3\mathbf{i} - 2\mathbf{j} + \mathbf{k}) \cdot (\mathbf{i} + 3\mathbf{j} - 2\mathbf{k}) \\ &= 3\mathbf{i}\cdot\mathbf{i} + 3\mathbf{i}\cdot 3\mathbf{j} - 3\mathbf{i}\cdot 2\mathbf{k} - 2\mathbf{j}\cdot\mathbf{i} - 2\mathbf{j}\cdot 3\mathbf{j} + 2\mathbf{j}\cdot 2\mathbf{k} + \mathbf{k}\cdot\mathbf{i} + \mathbf{k}\cdot 3\mathbf{j} - \mathbf{k}\cdot 2\mathbf{k} \\ &= 3 - 6 - 2 \\ \mathbf{A}\cdot\mathbf{B} &= -5 \end{aligned}$$

where we have made use of Equation (2.5).

(d)
$$\begin{aligned} \mathbf{A}\times\mathbf{B} &= (3\mathbf{i} - 2\mathbf{j} + \mathbf{k}) \times (\mathbf{i} + 3\mathbf{j} - 2\mathbf{k}) \\ &= 3\mathbf{i}\times\mathbf{i} + 3\mathbf{i}\times 3\mathbf{j} - 3\mathbf{i}\times 2\mathbf{k} - 2\mathbf{j}\times\mathbf{i} - 2\mathbf{j}\times 3\mathbf{j} + 2\mathbf{j}\times 2\mathbf{k} + \mathbf{k}\times\mathbf{i} + \mathbf{k}\times 3\mathbf{j} - \mathbf{k}\times 2\mathbf{k} \\ &= 9\mathbf{k} + 6\mathbf{j} + 2\mathbf{k} + 4\mathbf{i} + \mathbf{j} - 3\mathbf{i} \\ \mathbf{A}\times\mathbf{B} &= \mathbf{i} + 7\mathbf{j} + 11\mathbf{k} \end{aligned}$$

where we have made use of Equation (2.6).

(e) From the definition of the scalar product, Equation (2.3), and the results of (c), we have

$$\mathbf{A}\cdot\mathbf{B} = AB\cos\theta = -5$$

But

$$A = \sqrt{A_x^2 + A_y^2 + A_z^2} = \sqrt{3^2 + (-2)^2 + 1^2} = \sqrt{14}$$

$$B = \sqrt{B_x^2 + B_y^2 + B_z^2} = \sqrt{1^2 + 3^2 + (-2)^2} = \sqrt{14}$$

Therefore,

$$\cos\theta = -\frac{5}{AB} = -\frac{5}{14} = -0.357$$

$$\theta \approx 111^\circ$$

Example 2.3

Three vectors are given by $\mathbf{A} = \mathbf{i} - 3\mathbf{j}$, $\mathbf{B} = -2\mathbf{i} + 4\mathbf{j}$, and $\mathbf{C} = 2\mathbf{i} + \mathbf{j} - 3\mathbf{k}$. (a) Show that $\mathbf{A} \times \mathbf{B} = -\mathbf{B} \times \mathbf{A}$; (b) Find the triple product $\mathbf{A} \cdot (\mathbf{B} \times \mathbf{C})$.

Solution

(a) $\mathbf{A} \times \mathbf{B} = (\mathbf{i} - 3\mathbf{j}) \times (-2\mathbf{i} + 4\mathbf{j}) = -\mathbf{i} \times 2\mathbf{i} + \mathbf{i} \times 4\mathbf{j} + 3\mathbf{j} \times 2\mathbf{i} - 3\mathbf{j} \times 4\mathbf{j}$

$= 4\mathbf{k} - 6\mathbf{k} = -2\mathbf{k}$

$\mathbf{B} \times \mathbf{A} = (-2\mathbf{i} + 4\mathbf{j}) \times (\mathbf{i} - 3\mathbf{j}) = -2\mathbf{i} \times \mathbf{i} + 2\mathbf{i} \times 3\mathbf{j} + 4\mathbf{j} \times \mathbf{i} - 4\mathbf{j} \times 3\mathbf{j}$

$\mathbf{B} \times \mathbf{A} = 6\mathbf{k} - 4\mathbf{k} = 2\mathbf{k}$

$\therefore \mathbf{A} \times \mathbf{B} = -\mathbf{B} \times \mathbf{A}$

(b) In calculating the triple product $\mathbf{A} \cdot (\mathbf{B} \times \mathbf{C})$, we *must first* take the vector product $\mathbf{B} \times \mathbf{C}$, then the result of this is "dotted" into $\mathbf{A}$. The operation $(\mathbf{A} \cdot \mathbf{B}) \times \mathbf{C}$ would not make sense, since $\mathbf{A} \cdot \mathbf{B}$ is a scalar, and the cross product of a scalar with a vector is meaningless.

$$\mathbf{B} \times \mathbf{C} = (-2\mathbf{i} + 4\mathbf{j}) \times (2\mathbf{i} + \mathbf{j} - 3\mathbf{k})$$

$$= -4\mathbf{i} \times \mathbf{i} - 2\mathbf{i} \times \mathbf{j} + 2\mathbf{i} \times 3\mathbf{k} + 4\mathbf{j} \times 2\mathbf{i} + 4\mathbf{j} \times \mathbf{j} - 4\mathbf{j} \times 3\mathbf{k}$$

$$= -2\mathbf{k} - 6\mathbf{j} - 8\mathbf{k} - 12\mathbf{i}$$

$$\mathbf{B} \times \mathbf{C} = -12\mathbf{i} - 6\mathbf{j} - 10\mathbf{k}$$

$$\mathbf{A} \cdot (\mathbf{B} \times \mathbf{C}) = (\mathbf{i} - 3\mathbf{j}) \cdot (-12\mathbf{i} - 6\mathbf{j} - 10\mathbf{k}) = -12\mathbf{i} \cdot \mathbf{i} + 18\mathbf{j} \cdot \mathbf{j}$$

$$\mathbf{A} \cdot (\mathbf{B} \times \mathbf{C}) = 6$$

2.5 PROGRAMMED EXERCISES

1.A

A point A in the xy plane has coordinates given by (−3,2) meters. Write an expression for the position vector $\mathbf{r}_A$ for this point in unit vector notation, and draw a diagram showing the direction of this vector.

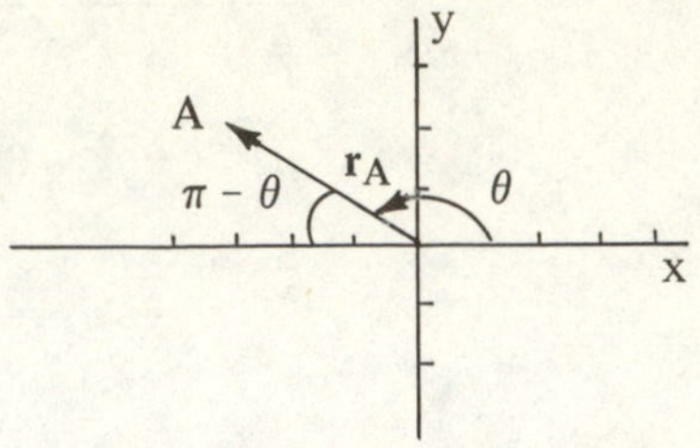

$\mathbf{r}_A = (-3\mathbf{i} + 2\mathbf{j})$ m

1.B

What are the x and y components of $\mathbf{r}_A$?

$r_x = x = -3$ m

$r_y = y = 2$ m

1.C

What is the magnitude of $\mathbf{r}_A$?

$r_A = \sqrt{x^2 + y^2} = \sqrt{(-3)^2 + 2^2} = \sqrt{13}$ m

1.D

Find the angle that $\mathbf{r}_A$ makes with the +x axis.

From the diagram in 1.A, we see that $\cos(\pi - \theta) = -3/\sqrt{13} = -0.832$. Therefore, $\pi - \theta \cong 33°$, or $\theta \cong 147°$.

1.E

A second point B in the xy plane has coordinates (2,4). Write an expression for the position vector $\mathbf{r}_B$, and draw a diagram showing $\mathbf{r}_B$ and $\mathbf{r}_A$.

$\mathbf{r}_B = (2\mathbf{i} + 4\mathbf{j})$m

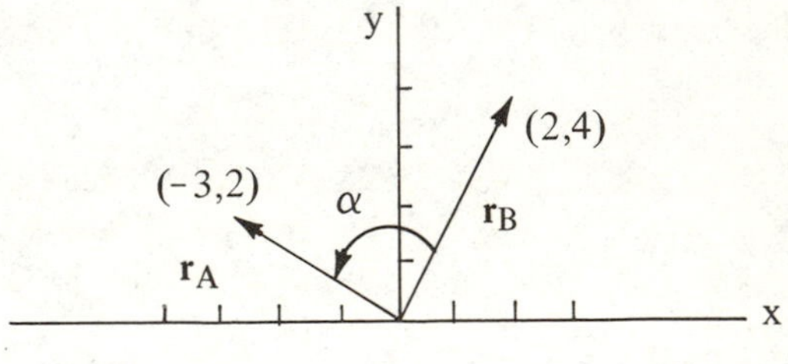

1.F

Find the angle α between $\mathbf{r}_A$ and $\mathbf{r}_B$ using the definition of the dot product.

Note: $r_B = \sqrt{2^2 + 4^2} = \sqrt{20}$ m

$\mathbf{r}_A \cdot \mathbf{r}_B = r_A r_B \cos\alpha = (-3\mathbf{i} + 2\mathbf{j}) \cdot (2\mathbf{i} + 4\mathbf{j})$

$\sqrt{13}\sqrt{20} \cos\alpha = -6 + 8 = 2$

$\cos\alpha = 2/\sqrt{260} = 0.124$

$\alpha \cong 83°$

1.G

Find $\mathbf{r}_A + \mathbf{r}_B$ and $\mathbf{r}_A - \mathbf{r}_B$ and show these in diagrams.

$\mathbf{r}_A + \mathbf{r}_B = (-3\mathbf{i} + 2\mathbf{j}) + (2\mathbf{i} + 4\mathbf{j}) = -\mathbf{i} + 6\mathbf{j}$

$\mathbf{r}_A - \mathbf{r}_B = (-3\mathbf{i} + 2\mathbf{j}) - (2\mathbf{i} + 4\mathbf{j}) = -5\mathbf{i} - 2\mathbf{j}$

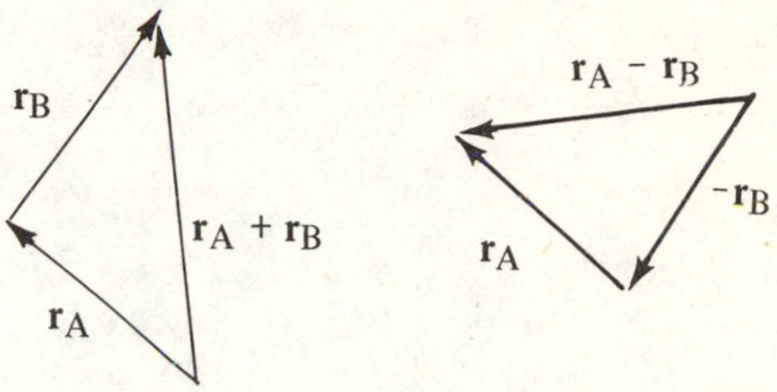

2.A

A vector **A** is represented by $\mathbf{A} = 2\mathbf{i} + 3\mathbf{j} + \mathbf{k}$. What are the vector components of **A**?

$\mathbf{A}_x = 2\mathbf{i}$

$\mathbf{A}_y = 3\mathbf{j}$

$\mathbf{A}_z = \mathbf{k}$

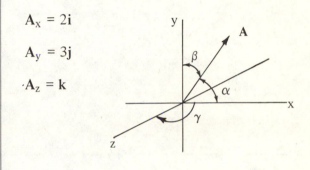

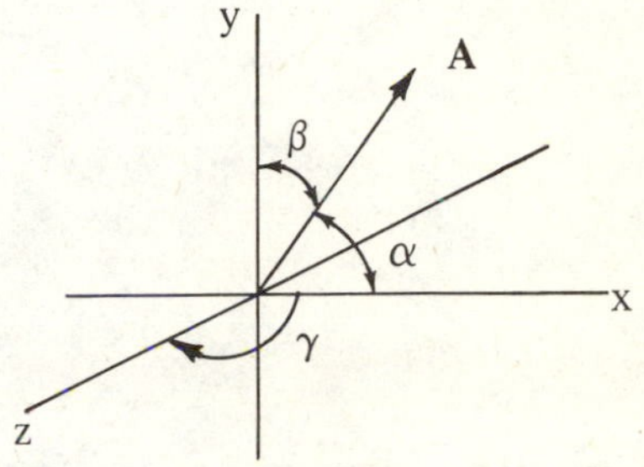

2.B

What is the magnitude of **A**? What does the number represent?

$A = \sqrt{A_x{}^2 + A_y{}^2 + A_z{}^2} = \sqrt{2^2 + 3^2 + 1^2}$

$A = \sqrt{14}$

This is the "length" of **A** and has the same units as **A**.

2.C

Find the result of the scalar product $\mathbf{A} \cdot \mathbf{i}$.

$\mathbf{A} \cdot \mathbf{i} = (2\mathbf{i} + 3\mathbf{j} + \mathbf{k}) \cdot \mathbf{i}$

$\mathbf{A} \cdot \mathbf{i} = 2\mathbf{i} \cdot \mathbf{i} + 3\mathbf{j} \cdot \mathbf{i} + \mathbf{k} \cdot \mathbf{i} = 2$

2.D

Find the angle that **A** makes with the +x axis using the definition of the scalar product and the results of 2.B and 2.C.

$\mathbf{A} \cdot \mathbf{i} = A|\mathbf{i}| \cos\alpha = A \cos\alpha = 2$

$\cos\alpha = 2/\sqrt{14} = 0.535$

$\alpha \cong 58°$

2.E

Determine the scalar products $\mathbf{A \cdot j}$ and $\mathbf{A \cdot k}$.

$$\mathbf{A \cdot j} = (2\mathbf{i} + 3\mathbf{j} + \mathbf{k}) \cdot \mathbf{j} = 3$$

$$\mathbf{A \cdot k} = (2\mathbf{i} + 3\mathbf{j} + \mathbf{k}) \cdot \mathbf{k} = 1$$

since $\mathbf{i \cdot j} = \mathbf{i \cdot k} = \mathbf{k \cdot j} = 0$

2.F

Use the results of 2.B and 2.E to find the angle β that **A** makes with the +y axis.

$$\mathbf{A \cdot j} = A|\mathbf{j}| \cos\beta = A \cos\beta = 3$$

$$\cos\beta = 3/\sqrt{14} = 0.802$$

$$\beta \cong 36°$$

2.G

A second vector is given by $\mathbf{B} = -\mathbf{i} + 5\mathbf{j} - 4\mathbf{k}$. Find an expression for the vector $2\mathbf{A} - \mathbf{B}$.

$$2\mathbf{A} - \mathbf{B} = 2(2\mathbf{i} + 3\mathbf{j} + \mathbf{k}) - (-\mathbf{i} + 5\mathbf{j} - 4\mathbf{k})$$

$$2\mathbf{A} - \mathbf{B} = 4\mathbf{i} + 6\mathbf{j} + 2\mathbf{k} + \mathbf{i} - 5\mathbf{j} + 4\mathbf{k}$$

$$2\mathbf{A} - \mathbf{B} = 5\mathbf{i} + \mathbf{j} + 6\mathbf{k}$$

2.H

Determine the scalar product $\mathbf{A \cdot B}$.

$$\mathbf{A \cdot B} = (2\mathbf{i} + 3\mathbf{j} + \mathbf{k}) \cdot (-\mathbf{i} + 5\mathbf{j} - 4\mathbf{k})$$

$$\mathbf{A \cdot B} = 2\mathbf{i} \cdot (-\mathbf{i}) + 3\mathbf{j} \cdot (5\mathbf{j}) + \mathbf{k} \cdot (-4\mathbf{k})$$

$$\mathbf{A \cdot B} = -2 + 15 - 4 = 9$$

3.A

A boy walks from his hometown to a camp located 3 mi East and 4 mi South of his hometown. The second day he walked from the camp to a village located 1 mi West and 6 mi North of the camp. Write expressions for the position vectors of the camp and the village *relative to the boy's hometown.*

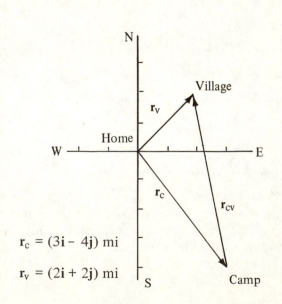

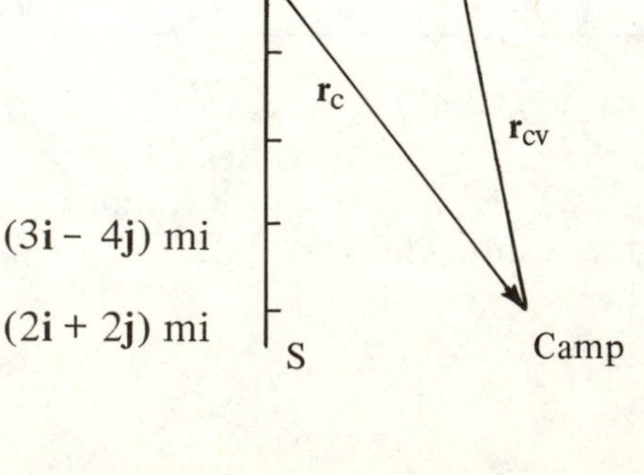

$$\mathbf{r}_C = (3\mathbf{i} - 4\mathbf{j}) \text{ mi}$$

$$\mathbf{r}_V = (2\mathbf{i} + 2\mathbf{j}) \text{ mi}$$

3.B

How far did the boy walk the first day? Assume he walks in a straight path.

$r_c = \sqrt{(3)^2 + (-4)^2} = 5$ mi

3.C

How far is the village from the boy's hometown, and what direction is it in?

$r_v = \sqrt{2^2 + 2^2} = 2\sqrt{2}$ mi

The village is 2 mi East and 2 mi North of the boy's hometown, or in the NE direction.

3.D

Find an expression for the position vector $\mathbf{r}_{cv}$ of the village relative to the camp.

From the vector diagram in 3.A, we see that

$$\mathbf{r}_v = \mathbf{r}_c + \mathbf{r}_{cv}$$

or,

$$\mathbf{r}_{cv} = \mathbf{r}_v - \mathbf{r}_c$$

$\mathbf{r}_{cv} = (2\mathbf{i} + 2\mathbf{j}) - (3\mathbf{i} - 4\mathbf{j})$

$\mathbf{r}_{cv} = (-\mathbf{i} + 6\mathbf{j})$ mi

3.E

What is the distance between the camp and the village?

$r_{cv} = \sqrt{(-1)^2 + 6^2} = \sqrt{37}$ mi

3.F

If the boy returns home on the third day, what is the total distance of his trip?

$r_{total} = r_c + r_{cv} + r_v = 5 + \sqrt{37} + 2\sqrt{2}$

$r_{total} \approx 13.9$ mi

4.A

Two vectors are given by $\mathbf{A} = \mathbf{i} + 3\mathbf{j}$ and $\mathbf{B} = -2\mathbf{i} + \mathbf{j}$. Make a plot of these vectors.

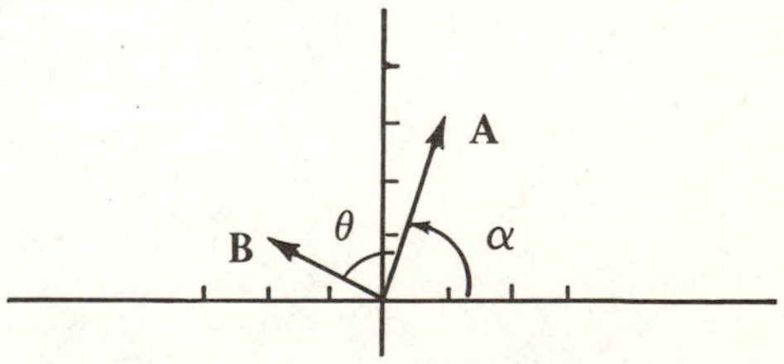

4.B

What are the x and y components of **A** and **B**?

$A_x = 1, A_y = 3$

$B_x = -2, B_y = 1$

4.C

What are the magnitudes of **A** and **B**?

$$A = \sqrt{A_x{}^2 + A_y{}^2} = \sqrt{1^2 + 3^2} = \sqrt{10}$$

$$B = \sqrt{B_x{}^2 + B_y{}^2} = \sqrt{(-2)^2 + 1^2} = \sqrt{5}$$

4.D

Find the angle α between **A** and the +x axis.

$$\tan \alpha = \frac{A_y}{A_x} = \frac{3}{1}$$

$$\alpha = \text{arc tan } 3 \cong 72°$$

4.E

Find the angle θ between **A** and **B** using the definition of the scalar product, and the results of 4.C

$$\mathbf{A}\cdot\mathbf{B} = AB \cos \theta = \sqrt{10}\sqrt{5} \cos \theta$$

$$\mathbf{A}\cdot\mathbf{B} = (\mathbf{i} + 3\mathbf{j})\cdot(-2\mathbf{i} + \mathbf{j}) = -2 + 3 = 1$$

$$\sqrt{10}\sqrt{5} \cos \theta = 1$$

$$\cos \theta = 1/\sqrt{50} = 0.141$$

$$\theta \cong 82°$$

4.F

Determine the vector product **A** × **B**.

$$\mathbf{A} \times \mathbf{B} = (\mathbf{i} + 3\mathbf{j}) \times (-2\mathbf{i} + \mathbf{j})$$

$$\mathbf{A} \times \mathbf{B} = \mathbf{i} \times (-2\mathbf{i}) + \mathbf{i} \times \mathbf{j} + 3\mathbf{j} \times (-2\mathbf{i}) + 3\mathbf{j} \times \mathbf{j}$$

$$\mathbf{A} \times \mathbf{B} = \mathbf{k} + 6\mathbf{k} = 7\mathbf{k}$$

4.G

Use the definition of the cross product and the results of 4.F to find θ.

$$\mathbf{A} \times \mathbf{B} = AB \sin \theta\mathbf{k} = \sqrt{10}\sqrt{5} \sin \theta\mathbf{k}$$

$$\sqrt{10}\sqrt{5} \sin \theta\mathbf{k} = 7\mathbf{k}$$

$$\sin \theta = 7/\sqrt{50} = 0.990$$

$$\theta \approx 82°$$

4.H

What vector **C** must be added to **A** and **B** to make the resultant sum zero?

$$\mathbf{A} + \mathbf{B} + \mathbf{C} = 0$$

$$\mathbf{C} = -(\mathbf{A} + \mathbf{B}) = -(-\mathbf{i} + 4\mathbf{j})$$

$$\mathbf{C} = \mathbf{i} - 4\mathbf{j}$$

2.6 PROBLEMS

1. A vector **A** has a magnitude of 4 units and makes an angle θ with the positive x axis. Find (a) the magnitudes of the x and y components of the vector **A**, and (b) an equation for **A** in unit vector notation.

2. The polar coordinates of a point in the xy plane are given by $r = 20$ m, $\theta = 135°$, where θ is measured from the positive x axis. Find (a) the rectangular coordinates of this point, and (b) an expression for the vector displacement from the origin to this point.

3. Two vectors are given by $\mathbf{A} = 3\mathbf{i} + 4\mathbf{j}$ and $\mathbf{B} = -2\mathbf{i} - 2\mathbf{j}$. Find (a) the vector sum $\mathbf{A} + \mathbf{B}$, (b) the difference $\mathbf{A} - \mathbf{B}$, (c) the magnitude and direction of the vector sum $\mathbf{A} + \mathbf{B}$, (d) $\mathbf{A} \cdot \mathbf{B}$ and (e) $\mathbf{A} \times \mathbf{B}$.

4. The cartesian coordinates of a point are given by $x = -4$ m, $y = -5$ m, $z = 0$. Find the polar coordinates of this point and an expression for the vector **R** from the origin to this point.

5. Show that the cross product A X B can be written as the determinant

$$\mathbf{A} \times \mathbf{B} = \begin{vmatrix} \mathbf{i} & \mathbf{j} & \mathbf{k} \\ A_x & A_y & A_z \\ B_x & B_y & B_z \end{vmatrix}$$

6. Find the angles that the vector $\mathbf{A} = 3\mathbf{i} + 5\mathbf{j} + 2\mathbf{k}$ makes with the x, y and z axes, respectively, using the definition of the scalar product and the angles defined in Figure 2-7(*b*).

7. The cosines of the angles α, β and γ shown in Figure 2-7(*b*) are known as the direction cosines, since they completely describe the direction of an arbitrary vector. Show that these angles are related to each other by the equation $\cos^2\alpha + \cos^2\beta + \cos^2\gamma = 1$ and show that the relation is satisfied for the angles obtained in Problem 6.

8. Show that the vector $\mathbf{A} = 2\mathbf{i} + \mathbf{j} + 2\mathbf{k}$ is perpendicular to the vector $\mathbf{B} = -2\mathbf{i} + 4\mathbf{j} + 4\mathbf{k}$.

9. Show that the two vectors $\mathbf{A} = \mathbf{i} - 3\mathbf{j} + 4\mathbf{k}$ and $\mathbf{B} = -2\mathbf{i} + 6\mathbf{j} - 8\mathbf{k}$ are antiparallel.

10. A truck travels 5 miles eastward, then 8 miles northeastward and finally 10 miles in a direction 37° N of W. Find the magnitude and direction of his resultant displacement.

11. Three forces act on a body located at the origin as shown in Figure 2-8. Find the vector sum of the three forces using an analytical method. (Note: N = newtons.)

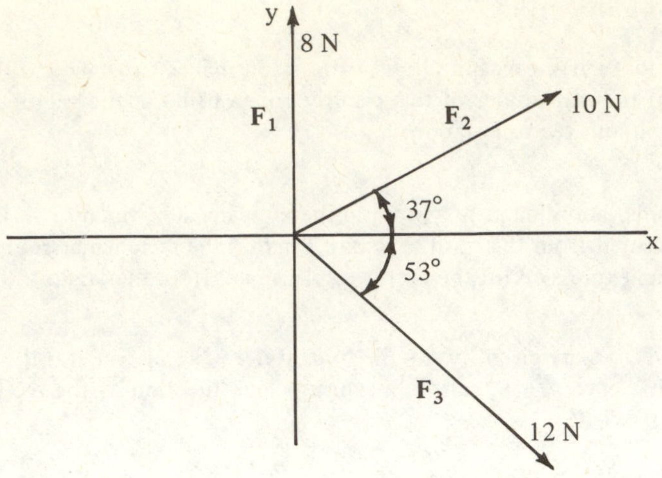

Figure 2-8

12. An arbitrary vector **A** is sometimes specified by its length, a polar angle θ and an azimuthal angle ϕ as shown in Figure 2-9. Show that the components of the vector **A** are given by $A_x = A \sin\theta \cos\phi$, $A_y = A \sin\theta \sin\phi$ and $A_z = A \cos\theta$.

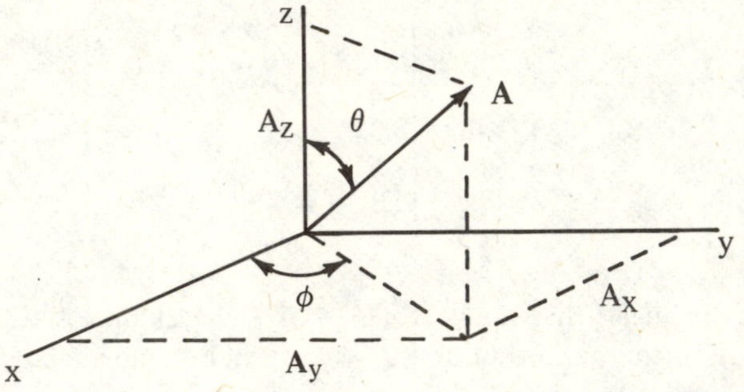

Figure 2-9

13. Two forces of magnitudes $\mathbf{F}_1 = 8$ N and $\mathbf{F}_2 = 12$ N act on an object located at the origin in the directions shown in Figure 2-10. (a) Obtain vector expressions for $\mathbf{F}_1$, $\mathbf{F}_2$, the vector sum and the vector differences $\mathbf{F}_1 + \mathbf{F}_2$ and $\mathbf{F}_1 - \mathbf{F}_2$. (b) What force **F** must be added to the object to make the resultant force zero?

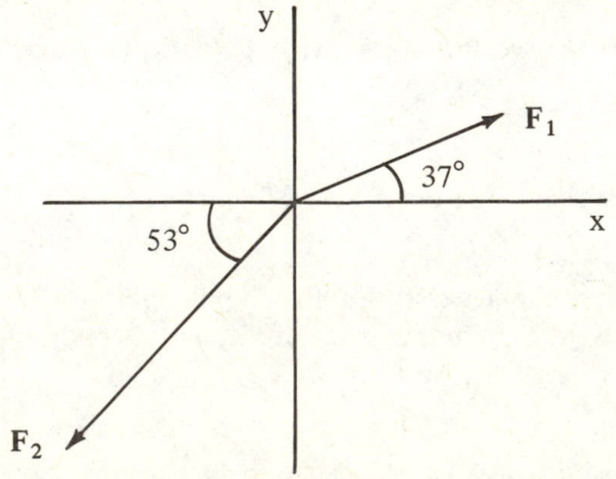

Figure 2-10

14. Consider a cube with a side of length a, with the origin of coordinates at one of the corners as shown in Figure 2-11. (a) Obtain vector expressions for $\mathbf{r}_1$, $\mathbf{r}_2$ and $\mathbf{r}_3$. (b) Determine the angles θ_1, θ_2 and θ_3 from the definition of the scalar product.

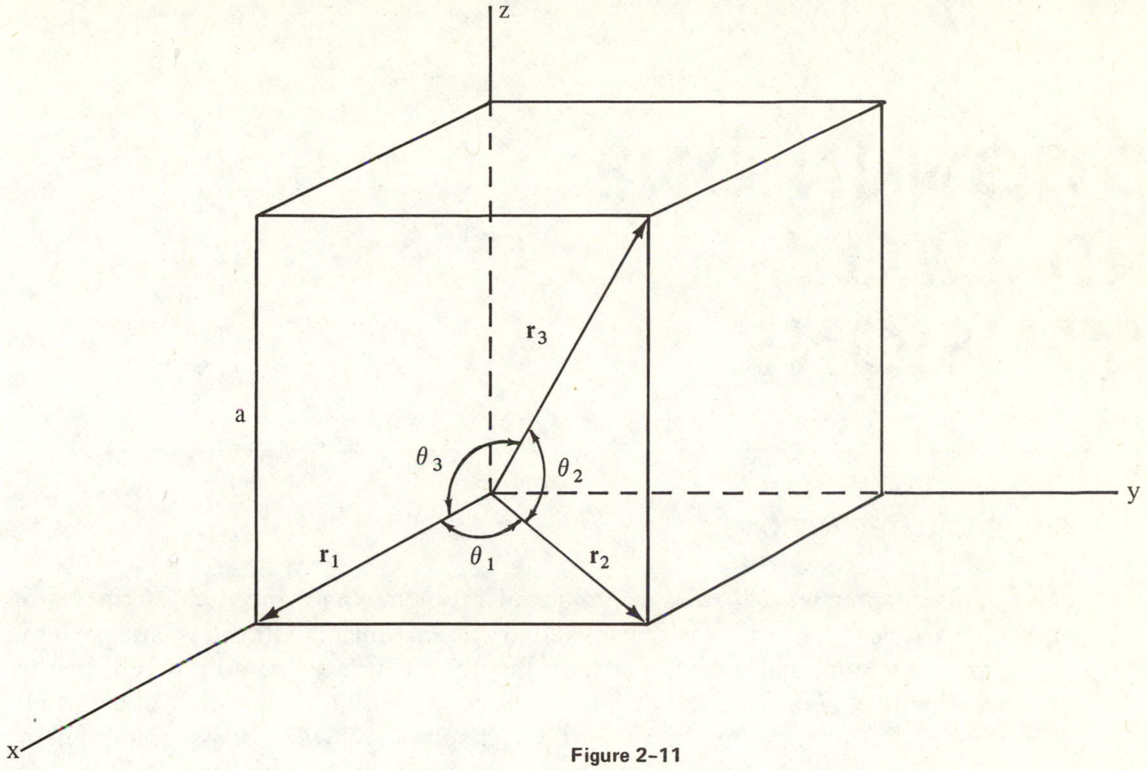

Figure 2-11

15. Show that Equation (2.1) for the law of cosines follows from the fact that three vectors which form a closed triangle are subject to the condition that $\mathbf{A} + \mathbf{B} + \mathbf{C} = 0$.

16. Two vectors **A** and **B** making an angle θ with each other form two sides of a parallelogram. (a) Show that the area of the parallelogram is given by $|\mathbf{A} \times \mathbf{B}|$. (b) If $\mathbf{A} = 2\mathbf{i} + 4\mathbf{j}$ meters and $\mathbf{B} = \mathbf{i} - 2\mathbf{j}$ meters, find the area of the parallelogram.

17. Two *unit* vectors **a** and **b** make angles of θ and ϕ, respectively, with respect to the x axis as in Figure 2-12 [take $\theta > \phi$]. Find algebraic expressions for **a** and **b** in unit vector notation and, from your results, prove that $\cos(\theta - \phi) = \cos\theta \cos\phi + \sin\theta \sin\phi$ and $\sin(\theta - \phi) = \sin\theta \cos\phi - \cos\theta \sin\phi$.

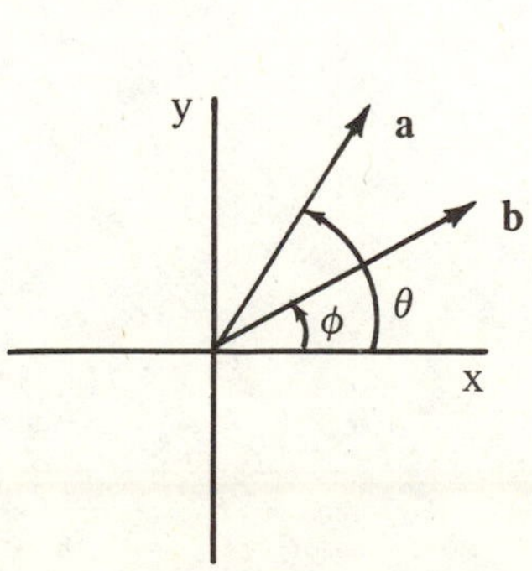

Figure 2-12

3

MOTION IN ONE AND TWO DIMENSIONS

3.1 BASIC DEFINITIONS

In this chapter we will treat a number of problems involving the description of moving particles. A *particle* is considered to be an object without spatial extent, therefore rotational and vibrational motions do not have to be considered. In this first treatment of *dynamics,* we deal only with the translational motion of a particle in one and two dimensions. In subsequent chapters we will discuss more general cases of the motion of rigid bodies. In reality, objects have spatial extent. However, we will see that the approximation of treating an object as a particle, with its mass located at a point in space, greatly simplifies the analysis of motion. Such an approximation can be useful in analyzing such problems as the motions of planetary bodies and atomic constituents.

The fundamental equations that are used in this chapter can be obtained from the definitions of velocity and acceleration. The *velocity* of a particle is defined as the time rate of change of its position in space. The *position* of the particle at some instant of time t can be specified by the vector r, which in Cartesian coordinates is

$$\mathbf{r} = x\mathbf{i} + y\mathbf{j} + z\mathbf{k} \tag{3.1}$$

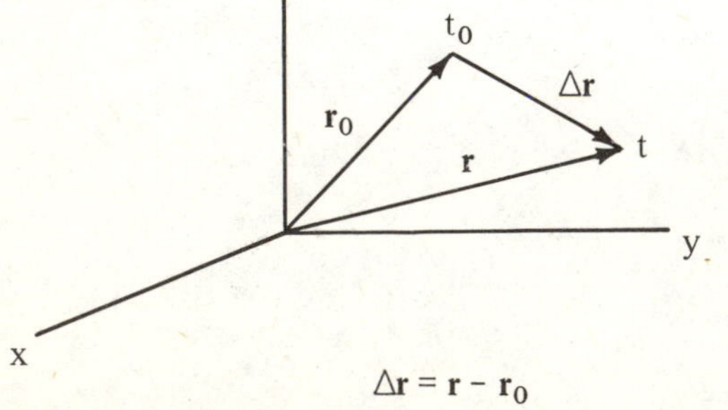

Figure 3-1 The position vectors $\mathbf{r}_0$ and $\mathbf{r}$ of a particle at times t_0 and t, and the corresponding displacement vector $\Delta\mathbf{r}$.

If the position vectors of the particle at times t_0 and t are $\mathbf{r}_0$ and $\mathbf{r}$, respectively (where $t > t_0$), we can denote the change in position by a *displacement* vector $\Delta \mathbf{r}$, where $\Delta \mathbf{r} = \mathbf{r} - \mathbf{r}_0$ as illustrated in Figure 3-1. The *average velocity* of the particle during this displacement is defined as

$$\mathbf{v}_{av} = \frac{\mathbf{r} - \mathbf{r}_0}{t - t_0} = \frac{\Delta \mathbf{r}}{\Delta t} \tag{3.2}$$

where $\Delta t = t - t_0$ is the *time elapsed* for the motion between $\mathbf{r}_0$ and $\mathbf{r}$.

The *instantaneous velocity* of a particle is defined as the velocity of a particle at some instant of time. If we allow the vector displacements $\Delta \mathbf{r}$ to become smaller, the corresponding time intervals Δt become smaller. The limiting value of the ratio $\Delta \mathbf{r}/\Delta t$ as Δt approaches zero is the *instantaneous velocity* $\mathbf{v}$, or

$$\mathbf{v} = \lim_{\Delta t \to 0} \frac{\Delta \mathbf{r}}{\Delta t} = \frac{d\mathbf{r}}{dt} \tag{3.3}$$

The *speed* of the particle is the absolute value of $\mathbf{v}$, and is sometimes written as $|\mathbf{v}|$ or simply v. Average velocity and instantaneous velocity are *not* to be confused with each other. For example, if an airplane makes a round trip flight from New York to Los Angeles and back to New York, its average velocity is *zero* since its net displacement is zero. However, its instantaneous velocity during the flight is obviously not zero.

There are other properties of $\mathbf{v}_{av}$ and $\mathbf{v}$ which should be pointed out. Both are *vector* quantities, where $\mathbf{v}_{av}$ is in the direction of $\Delta \mathbf{r}$ and $\mathbf{v}$ is in the direction tangent to the path of the particle. Both have dimensions of length/time, L/T. However, $\mathbf{v}$ is specified only if the path of the particle is known, while $\mathbf{v}_{av}$ can be specified with simply a knowledge of the positions of the particle at the beginning and end points and the time interval.

The *average acceleration,* $\mathbf{a}_{av}$, of a particle is the ratio of the change in the velocity to the time interval over which this change occurs.

$$\mathbf{a}_{av} = \frac{\Delta \mathbf{v}}{\Delta t} = \frac{\mathbf{v} - \mathbf{v}_0}{t - t_0} \tag{3.4}$$

where $\mathbf{v}_0$ and $\mathbf{v}$ are the velocity vectors at times t_0 and t, respectively, as shown in Figure 3-2.

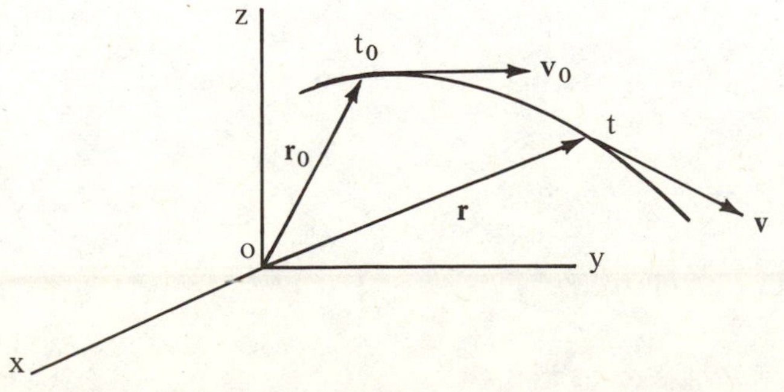

Figure 3-2 The velocity vector as a function of time. **v** is tangent to the path of the particle.

The instantaneous acceleration, **a**, of a particle is defined as the limit of the average acceleration as $\Delta t \to 0$. Therefore, it is the acceleration of the particle at some instant of time or at a distinct point along its path.

$$\mathbf{a} = \lim_{\Delta t \to 0} \frac{\Delta \mathbf{v}}{\Delta t} = \frac{d\mathbf{v}}{dt} \tag{3.5}$$

Both $\mathbf{a}_{av}$ and **a** are vector quantities, where $\mathbf{a}_{av}$ is a vector in the direction of $\Delta\mathbf{v}$. Linear acceleration has dimensions of L/T^2.

3.2 MOTION IN ONE DIMENSION WITH CONSTANT ACCELERATION

Here we will consider the motion of a particle in one dimension, say the x-axis, and for simplicity we will assume that the acceleration is constant. For this special case, the average acceleration *equals* the instantaneous acceleration, that is, both are equal to some constant. In addition, the position, velocity and acceleration vectors reduce to

$$\mathbf{r} = x\mathbf{i}, \qquad \mathbf{v} = v_x\mathbf{i} \qquad \text{and} \qquad \mathbf{a} = a_x\mathbf{i}$$

where $v_x = dx/dt$ and $a_x = dv_x/dt$. Therefore, the average velocity and acceleration in one dimension can be written as

$$v_{av} = \frac{x - x_0}{t - t_0}$$

and

$$a_x = \frac{v_x - v_{x0}}{t - t_0}$$

Note, we have dropped the unit vector notation. However, we will use the convention that a positive sign for v_x corresponds to a vector pointing in the +x direction, while a negative sign for v_x corresponds to a vector in the negative x direction. A similar convention is used for x and a_x. If we take the coordinate and velocity to be x_0 and v_{x0} at $t_0 = 0$, the expressions above reduce to

$$\begin{aligned} x &= x_0 + v_{av} t \\ v_x &= v_{x0} + a_x t \end{aligned} \tag{3.6}$$

However, by definition, $v_{av} = \frac{1}{2}(v_{x0} + v_x)$, so the expression for x becomes

$$x = x_0 + \frac{1}{2}(v_{x0} + v_x)\, t \tag{3.7}$$

Substituting the value for v_x from (3.6) into (3.7) gives

$$x = x_0 + v_{x0} t + \frac{1}{2} a_x t^2 \tag{3.8}$$

Finally, substituting the value of t from (3.6) into (3.7) gives

$$v_x^2 = v_{x0}^2 + 2a_x(x - x_0) \tag{3.9}$$

Equations (3.6) through (3.9) will be used frequently, and perhaps should be memorized. The student should also perform the algebra used in arriving at Equations (3.8) and (3.9).

One common set of problems deals with the motion of a *freely falling body* in the earth's gravitational field. If we neglect air resistance, and assume that the acceleration of gravity is independent of the height of the body above the earth's surface, then the motion is one dimensional, with constant acceleration. Usually, we take the vertical direction, the y axis, so our x subscripts are replaced by y, the acceleration $a_y = -g$ [calling y positive upwards, the acceleration of gravity is in the negative y direction]. Equations (3.6) through (3.9) become

Freely Falling Body

$$v_y = v_{y0} - gt \tag{3.10}$$

$$y = y_0 + v_{y0}t - \frac{1}{2}gt^2 \tag{3.11}$$

$$y = y_0 + \frac{1}{2}[v_{y0} + v_y]\,t \tag{3.12}$$

$$v_y^2 = v_{y0}^2 - 2g[y - y_0] \tag{3.13}$$

The student should note that Equations (3.6) through (3.9) are all dimensionally consistent. This is one technique for determining whether or not you've written a valid expression. The use of these results should become clear with the following examples.

Example 3.1

An airplane traveling in the Easterly direction lands on an airstrip at a speed of 200 miles/hr and slows down uniformly to a speed of 20 miles/hr in a distance of 1200 ft. Find (a) the magnitude and direction of the acceleration and (b) the total time elapsed during this period.

Solution

(a) In this problem, we are given the total displacement $x - x_0 = 1200$ ft (x_0 is the coordinate of the plane when it first lands), the initial and final velocities $v_{x0} = 200$ mi/hr and $v_x = 20$ mi/hr. Calling the direction from west to east the positive x axis, we can apply Equation (3.9) directly. However, we should convert mi/hr to ft/sec to make the calculations dimensionally correct. A useful conversion to remember is 60 mi/hr $\cong$ 88 ft/sec, therefore

$$200 \text{ mi/hr} = 200 \text{ mi/hr}\left(\frac{88}{60}\right)\frac{\text{ft/sec}}{\text{mi/hr}} = 293 \text{ ft/sec}$$

$$20 \text{ mi/hr} = 20\left(\frac{88}{60}\right)\text{ft/sec} = 29.3 \text{ ft/sec}$$

Applying Equation (3.9) and the above conversions gives

$$a_x = \frac{v_x{}^2 - v_{x0}{}^2}{2(x - x_0)} = \frac{(29.3)^2 - (293)^2}{2(1200)\text{ ft}} \text{ ft}^2/\text{sec}^2 = -35.4 \text{ ft/sec}^2$$

The minus sign indicates that the acceleration vector is in the negative x direction, corresponding to a deceleration.

(b) The total time elapsed during the deceleration from 200 mi/hr to 20 mi/hr is obtained using

$$x = x_0 + \frac{1}{2}(v_x + v_{x0})\, t$$

Solving for t gives

$$t = \frac{2(x - x_0)}{v_x + v_{x0}} = \frac{2(1200)\text{ ft}}{(29.3 + 293)\text{ ft/sec}} = 7.45 \text{ sec}$$

Example 3.2

An inquisitive student wishes to make use of the laws of linear motion with constant acceleration. The student drops an apple from the top of a tall building to his girl friend below. Using a stopwatch, he notes that it takes 2.5 sec for the apple to reach the girl. What is the height of the building in meters, and what is the velocity of the apple just before it reaches the girl?

Solution

If we measure the coordinate from the position where the apple is released, and take y positive upwards, we have from Equation (3.11)

$$y = -\frac{1}{2}gt^2 = -\frac{1}{2}\left(9.8 \frac{\text{m}}{\text{sec}^2}\right)(2.5)^2 \text{ sec}^2 = -31 \text{ m}$$

where $y_0 = 0$ and $v_{y0} = 0$ since the apple is released from rest. The negative sign for y indicates that the displacement is in the negative y direction. The velocity of the apple after 2.5 sec is obtained directly from Equation (3.10).

$$v_y = -gt = -9.8 \frac{\text{m}}{\text{sec}^2}(2.5)\text{ sec} = -25 \text{ m/sec}$$

Of course, Equation (3.13) could also be used to get the same result.

Example 3.3

A boy throws a ball vertically upward from the top of a 105 ft building with a velocity of 64 ft/sec. Find (a) the maximum height that the ball rises, (b) the time it takes the ball to reach the level of its starting point, (c) the position of the ball 5 sec after it is thrown, (d) the velocity of the ball just before hitting the ground, and (e) the total elapsed time before the ball hits the ground.

Solution

Let us take the origin of coordinates from the *top* of the building as in Figure 3-3, and assume that the ball is thrown from the top of the building. Calling y positive upwards, we can find the time t_1 it takes the ball to reach the peak (where $v_y = 0$) from Equation (3.10), and substitute this into Equation (3.11) to get the maximum height y_1.

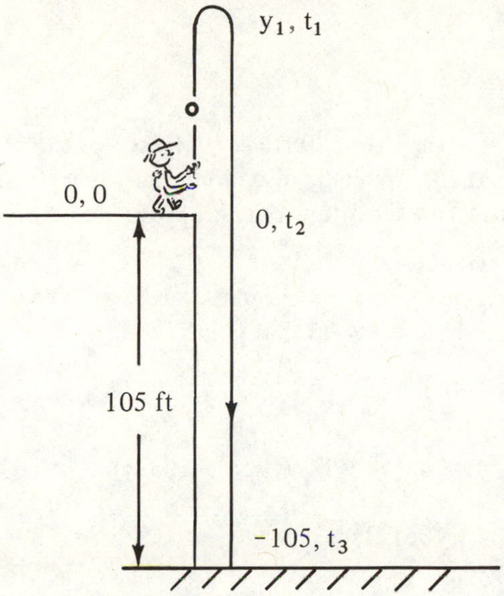

Figure 3-3

(a) At the peak, $v_y = 0$, therefore from Equation (3.10) we have $v_{yo} - gt_1 = 0$. Therefore, $t_1 = \frac{v_{yo}}{g} = \frac{64 \text{ ft/sec}}{32 \text{ ft/sec}^2} = 2 \text{ sec}$

$$y_1 = y_o + v_{yo}t_1 - \frac{1}{2}gt_1{}^2 = (64 \text{ ft/sec})\ 2 \text{ sec} - \frac{1}{2}(32 \text{ ft/sec}^2)\ (2)^2 \text{ sec}^2$$

$$y_1 = 128 \text{ ft} - 64 \text{ ft} = 64 \text{ ft}$$

We could also get this result from Equation (3.12)

$$y_1 = y_o + \frac{1}{2}(v_{yo} + v_y)\ t_1 = \frac{1}{2}\left(64 \frac{\text{ft}}{\text{sec}}\right) 2 \text{ sec} = 64 \text{ ft}$$

(b) The time it takes the ball to reach the level of its starting point is simply $2t_1 = 4$ sec. We could check this result from Equation (3.11), which is quadratic in t and has two solutions. At this point $y = y_o$, therefore Equation (3.11) becomes

$$y_o = y_o + v_{yo}\ t - \frac{1}{2}gt^2$$

$$64t - 16t^2 = 0$$

$$t^2 - 4t = 0$$

$$t(t - 4) = 0$$

Therefore, $t = 0$ corresponding to the starting point,
and $t_2 = 4$ sec corresponding to the time when the ball has a velocity -64 ft/sec.

We could also get this result from Equation (3.10), since $v_y = -64$ ft/sec when the ball is in its downward flight at $y_0 = 0$; therefore, $-gt_2 = -128$ ft/sec, $t_2 = 4$ sec.

(c) The position of the ball at any time t can be obtained directly from Equation (3.11), since v_{y0} is known.

$$y\,(t = 5) = y_0 + v_{y0}t - \frac{1}{2}gt^2 = 64\text{ ft/sec }(5\text{ sec}) - \frac{1}{2}(32\text{ ft/sec}^2)\,(5\text{ sec})^2$$

$$y\,(t = 5) = 320\text{ ft} - 400\text{ ft} = -80\text{ ft}$$

The negative sign simply indicates that the ball is 80 ft *below* the top of the building. This is consistent with our results of part (b). We could also obtain the same result by finding v_y after 5 sec from Equation (3.10), and substitute this into Equation (3.12) to get y at $t = 5$ sec.

(d) When the ball is about to hit the ground, its coordinate is $y = -105$ ft. Therefore, Equation (3.13) can be applied directly to get v_y at that point:

$$v_y{}^2 = v_{y0}{}^2 - 2g(y - y_0)$$

$$v_y{}^2 = (64)^2\text{ ft}^2/\text{sec}^2 - 2(32)\text{ ft/sec}^2(-105\text{ ft})$$

$$v_y{}^2 = 64[64 + 105]\text{ ft}^2/\text{sec}^2$$

$$v_y = \pm\, 104\text{ ft/sec}$$

Of course, since this ball is traveling *downward* at this time, the acceptable solution is $v_y = -104$ ft/sec.

(e) The total time of flight can be obtained using Equation (3.10) and the results of part (d).

$$v_y = v_{y0} - gt$$

$$-104\text{ ft/sec} = 64\text{ ft/sec} - 32\,t_3$$

$$32\,t_3 = 168\text{ ft/sec}$$

$$t_3 = \frac{168\text{ ft/sec}}{32\text{ ft/sec}^2} = 5.3\text{ sec}$$

†3.3 MOTION IN ONE DIMENSION WITH VARIABLE ACCELERATION

Situations may arise where, because of some unusual external forces, the motion of a particle is nonuniform and the acceleration varies in time. In fact, whenever the coordinate of the particle as a function of time involves a polynomial in t, and there is a term, say, bt^n, where $n > 2$, the acceleration of the particle will vary in time. Other examples of variable acceleration arise when the velocity vector remains constant in magnitude, but varies in orientation. This is common in problems dealing with circular motion which will be discussed in the next chapter.

†Example 3.4

The coordinate of a particle moving along the x-axis depends on time according to the expression

$$x = 5\,t^2 - 2\,t^3$$

where x is in meters and t is in seconds. Find (a) the velocity and acceleration of the particle as a function of time, (b) the time it takes the particle to reach its maximum positive x-coordinate, (c) the displacement during the first 2 seconds, (d) the velocity and acceleration of the particle after 2 seconds.

Solution

(a) The velocity and acceleration can be obtained by using Equations (3.3) and (3.5), which in one dimension become $v_x = \frac{dx}{dt}$ and $a_x = \frac{dv_x}{dt}$. Taking first and second derivatives of x given above, we have

$$v_x = \frac{dx}{dt} = \frac{d}{dt}[5\,t^2 - 2\,t^3] = 10\,t - 6\,t^2$$

$$a_x = \frac{dv_x}{dt} = \frac{d}{dt}[10\,t - 6\,t^2] = 10 - 12\,t$$

(b) When the particle has reached its maximum x-coordinate, $v_x = 0$ [that is, it stops and heads back towards the origin]. Therefore, from part (a),

$$10\,t - 6\,t^2 = t(10 - 6\,t) = 0$$

This has two solutions, $t = 0$ corresponding to the beginning of motion where $v_x = 0$, and $t = 5/3$ sec, which is the time in question, or the turning point.

(c) The coordinate of the particle at $t = 2$ seconds is

$$x\,(t = 2) = 5(2)^2 - 2(2)^3 = 4\text{m}$$

Since the particle starts from the origin $x_0 = 0$ at $t_0 = 0$, the displacement of the particle is also 4m. Note, however, that the total distance traveled in 2 seconds is *greater* than 4m. In fact, the total distance traveled in 2 seconds equals $x_{max} + (x_{max} - 4)$, where x_{max} is the maximum positive x coordinate which occurs at $t = 5/3$ sec (see part (b)). The student should show that this total distance traveled in 2 sec is 5.26 m.

(d) Using the results of part (a) we have

$$v_x\,(t = 2) = 10(2) - 6(2)^2 = -4 \text{ m/sec}$$

$$a_x\,(t = 2) = 10 - 12(2) = -14 \text{ m/sec}^2$$

We can see that at some later time, both the velocity and acceleration will be larger in magnitude.

3.4 TWO DIMENSIONAL MOTION WITH CONSTANT ACCELERATION-PROJECTILES

In this section we will deal with the motion of a particle in two dimensions under constant acceleration. A very interesting and useful example of this type of motion is that of a projectile in the earth's gravitational field. For simplicity, we will assume that (a) the curvature of the earth is negligible, (b) the variation of the acceleration of gravity with altitude is negligible and (c) air resistance is negligible. Assumption (a) is reasonable if the horizontal range is small. Assumption (b) is reasonable if the altitude of the particle is always small compared to the earth's radius, and (c) is approximately valid for small initial velocities and for projectiles with small cross-sections. We will choose the motion of the projectile to be in the X-Y plane, with the initial velocity v_0 at an angle θ_0 with respect to the horizontal as shown in Figure 3–4.

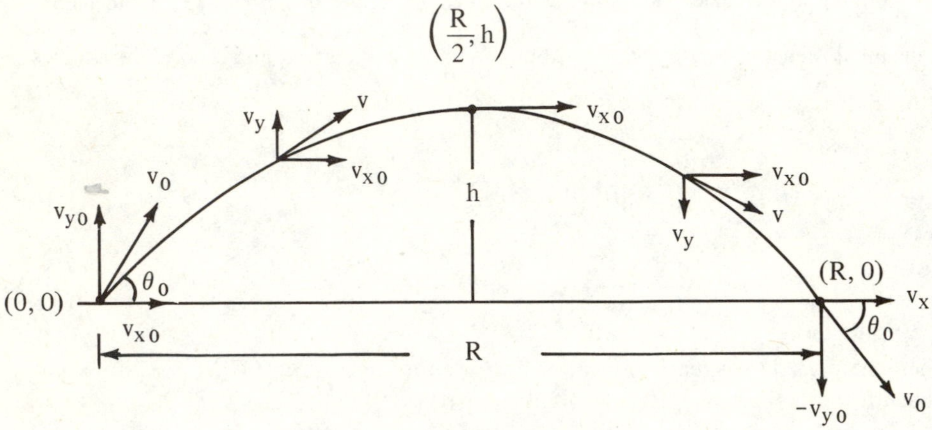

Figure 3-4 Trajectory of a projectile.

If we integrate Equation (3.5), with **a** = constant, we get

$$\mathbf{v} = \mathbf{v}_0 + \mathbf{a}\, t \tag{3.14}$$

where $t_0 = 0$ at the beginning of motion. Substituting Equation (3.3) into (3.14), and integrating once more gives

$$\mathbf{r} = \mathbf{r}_0 + \mathbf{v}_0 t + \frac{1}{2}\mathbf{a}\, t^2 \tag{3.15}$$

When the origin of coordinates is taken to be at the starting point of motion, then $\mathbf{r}_0 = 0$ at $t_0 = 0$.

Because of assumptions (b) and (c), the acceleration **a** has *no* x component, and can be written as $\mathbf{a} = -g\mathbf{j}$. That is, the acceleration of the projectile is the *same* as that of a freely falling body. Since $\mathbf{v} = v_x \mathbf{i} + v_y \mathbf{j}$, Equation (3.14) can be written in component form as

$$v_x = v_{xo} \tag{3.16}$$

$$v_y = v_{yo} - gt \tag{3.17}$$

Likewise, since $\mathbf{r} = x\mathbf{i} + y\mathbf{j}$, and taking $\mathbf{r}_0 = 0$, the component form of Equation (3.15) can be written as

$$x = v_{xo} t \tag{3.18}$$

$$y = v_{yo} t - \frac{1}{2} g t^2 \tag{3.19}$$

Therefore, we can think of the motion as the superposition of two one-dimensional uniform motions, one in the x-direction with zero acceleration, the other in the y-direction with constant acceleration equal to the acceleration of gravity. There is nothing about this situation that is physically different from the previous examples, other than the two-dimensional nature of the situation. Motion in the x-direction is particularly simple since the horizontal component of velocity is constant during the *entire* motion, that is, $v_x = v_{xo}$ = constant. It is convenient to work with the component forms of **v** and **r**, and then find their magnitudes using the expressions

$$v = \sqrt{v_x^{\,2} + v_y^{\,2}}$$

$$r = \sqrt{x^2 + y^2}$$

It is also useful to write $v_{xo} = v_0 \cos \theta_0$ and $v_{yo} = \sin \theta_0$, which follow from Figure 3–4. We can obtain the time it takes to reach the peak, t_1, by noting that $v_y = 0$ at this point, so that Equation (3.17) gives

$$t_1 = \frac{v_{yo}}{g} = \frac{v_0 \sin \theta_0}{g} \tag{3.20}$$

If we let $y = h$ when $t = t_1$, then substitution of Equation (3.20) into (3.19) gives

$$h = \frac{v_0^{\,2} \sin^2 \theta_0}{2g} \qquad \text{Maximum Height} \tag{3.21}$$

The *total* time of flight is $2t_1$. This can be shown by setting $y = 0$ in Equation (3.19). The solutions of this resulting expression are $t = 0$ and $t = 2\,v_{yo}/g = 2t_1$. We could have guessed this result from the symmetry in the trajectory. The *horizontal range* R can be obtained directly from Equation (3.18), with $t = 2t_1$. This gives

$$R = v_{xo}\, 2t_1 = v_0 \cos \theta_0 \frac{2v_0 \sin \theta_0}{g} = \frac{2v_0^{\,2} \sin \theta_0 \cos \theta_0}{g} \tag{3.22}$$

Since $\sin(2\alpha) = 2 \sin \alpha \cos \alpha$, Equation (3.22) can be written as

$$R = \frac{v_0^{\,2} \sin(2\theta_0)}{g} \qquad \text{Horizontal Range} \tag{3.23}$$

Note that Equations (3.21) and (3.23) are very specific formulas which can only be used to obtain the maximum height and horizontal range. The general expressions given by Equations (3.16) through (3.19) are the *most important* results, since they give the coordinates and velocity components of the projectile as a function of time.

Example 3.5

A stone is thrown from the top of a cliff at an angle of 37° from the horizontal as in Figure 3-5. The cliff is 100 ft above the water level and the stone hits the water a distance of 200 ft measured horizontally from the cliff. Find (a) the total time of flight for the stone, (b) the initial speed of the stone, and (c) the maximum altitude reached by the stone.

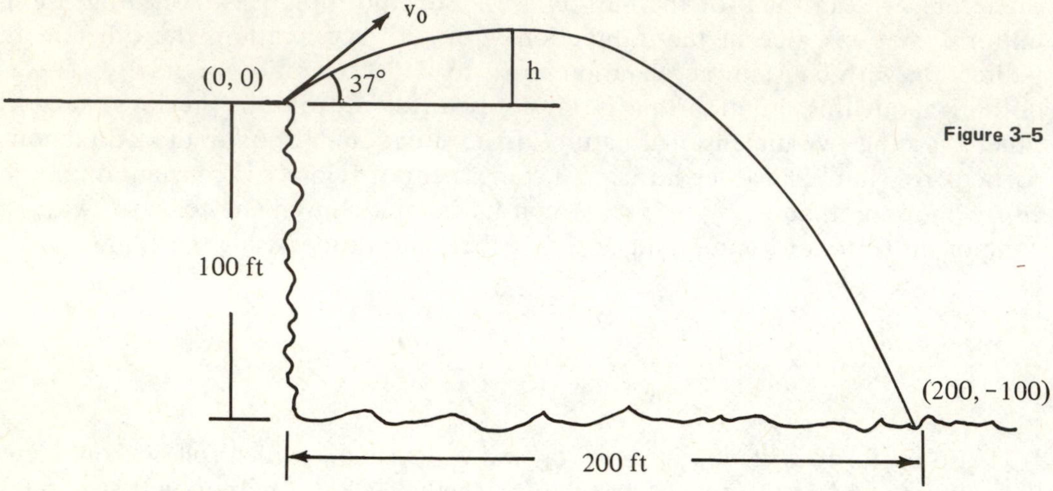

Figure 3-5

Solution

(a) We will take the origin of coordinates from the top of the cliff where the stone is thrown. We can use Equations (3.18) and (3.19) together to find the total time of flight t. The coordinates of the stone, when it hits the water, are (200, -100); therefore, substitution into these equations gives

$$200 = v_0 \cos(37)\, t$$

$$-100 = v_0 \sin(37) t - 16\, t^2$$

Substituting the value of v_0 from the first equation into the second gives

$$-100 = 200 \tan(37) - 16\, t^2$$

$$16\, t^2 = 150 + 100 = 250$$

$$t = 3.95 \text{ sec}$$

(b) Substituting this value of t into the first equation, and solving for v_0 gives

$$v_0 = \frac{200}{\cos(37)\, t} = \frac{200 \text{ ft}}{0.8\,(3.95) \text{ sec}} = 63.3 \text{ ft/sec}$$

(c) Since both θ_0 and v_0 are known, we can use Equation (3.21) to get h, the maximum altitude.

$$h = \frac{v_0{}^2 \sin^2 \theta_0}{2\, g} = \frac{(63.3)^2 \text{ ft}^2/\text{sec}^2\ (0.6)^2}{2\,(32) \text{ ft/sec}^2} = 22.5 \text{ ft}$$

3.5 PROGRAMMED EXERCISES

1.A

A particle moves along the x axis in such a way that its coordinate varies in time according to the expression

$$x = (3 - 5t + 6t^2)\ \text{m}$$

What is the coordinate of the particle at $t = 0$?

Since $x(t) = 3 - 5t + 6t^2$

$$x(0) = x_0 = 3\text{m} \qquad (1)$$

1.B

What is the initial velocity of the particle? Note that the particle moves with *constant* acceleration.

The expression for x is of the form

$$x = x_0 + v_{x0}\,t + \frac{1}{2}a_x t^2 \qquad (2)$$

Comparing this with (1), we see that $v_{x0} = -5$ m/sec. That is, the particle has a speed of 5 m/sec in the *negative* x direction at $t = 0$.

1.C

What is the acceleration of the particle?

Comparing (1) and (2), we see that

$$\frac{1}{2}a_x = 6$$

or

$$a_x = 12\ \text{m/sec}^2$$

---

†1.D

Determine the velocity and acceleration of the particle at any time $t \geqslant 0$.

From the definition of v and a, we have

$$v_x = \frac{dx}{dt} = \frac{d}{dt}(3 - 5t + 6t^2)$$

$$v_x = (-5 + 12t)\ \text{m/sec}$$

$$a_x = \frac{dv_x}{dt} = \frac{d}{dt}(-5 + 12t) = 12\ \text{m/sec}^2$$

1.E

Make plots of x *vs* t and v_x *vs* t. Note that the slope of x *vs* t gives v_x at any time t. Likewise, the slope of the curve v_x *vs* t gives the acceleration (which is constant in this example).

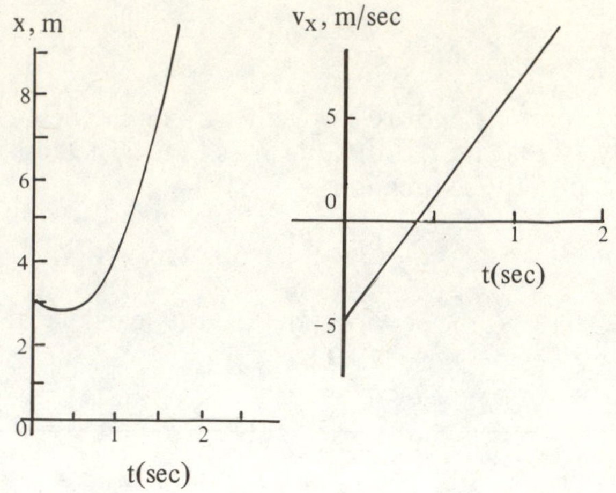

2.A

Ball A is thrown vertically upward with an initial speed of 32 ft/sec. One second later, ball B is thrown vertically upward with a speed of 64 ft/sec. What is the height of ball A at the time ball B is thrown?

$$y = v_{yo}\, t - \frac{1}{2} g t^2$$

Ball B is thrown at t = 1 sec; therefore,

$$y_A = 32\,(1) - \frac{1}{2}(32)\,(1)^2$$

$$y_A = 16 \text{ ft measured from the ground.}$$

2.B

Now let t = 0 be the time at which ball B is thrown. Find the time at which ball A reaches ball B. Note that the condition for this to occur is that the y coordinates of each ball *must be equal.*

$$y_B = v_{yo}\, t - \frac{1}{2} g t^2$$

$$y_B = 64\,t - 16\,t^2 \qquad (1)$$

Since A is already 16 ft above the ground at t = 0,

$$y_A = y_o + v_{yo}\, t - \frac{1}{2} g t^2$$

$$y_A = 16 + 32t - 16t^2 \qquad (2)$$

Now $y_A = y_B$ when (1) = (2).

$$64t - 16t^2 = 16 + 32t - 16t^2$$

t = 0.5 sec after B is thrown.

2.C

Find the speed of each ball when ball A reaches ball B. Explain your results.

We use $v_y = v_{yo} - gt$. (3)

For ball B, t = 0.5 sec, so

$$v_y = 64 - 32\,(0.5) = 48 \text{ ft/sec}$$

For ball A, t = 1.5 sec since it started its motion 1 sec before B.

$$v_y = 32 - 32\,(1.5) = -16 \text{ ft/sec.}$$

That is, ball **B** is moving *upward* and ball **A** is moving *downward* when they have the same y coordinate.

2.D

Find the total time of flight T for each ball. Note that when they are about to hit the ground, $v_y = -v_{yo}$.

We use (3), and set $v_y = -v_{yo}$.

$$T = \frac{2v_{yo}}{g}$$

For ball A, T = 2(32)/32 = 2 sec

For ball B, T = 4 sec

3.A

A projectile is fired with an initial speed v_0 at an angle θ_0 with the horizontal as shown. Sketch the horizontal and vertical components of velocity at the points indicated.

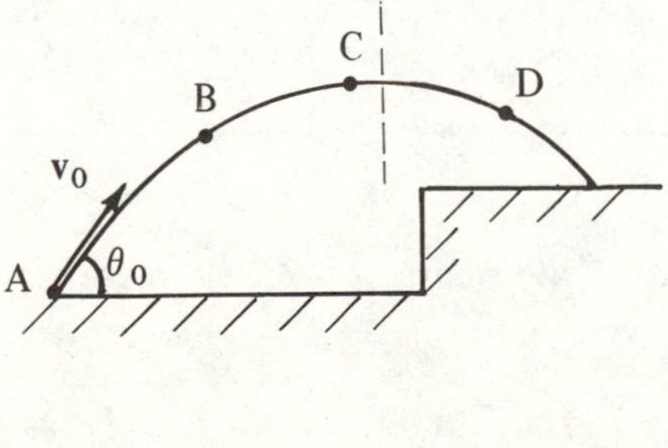

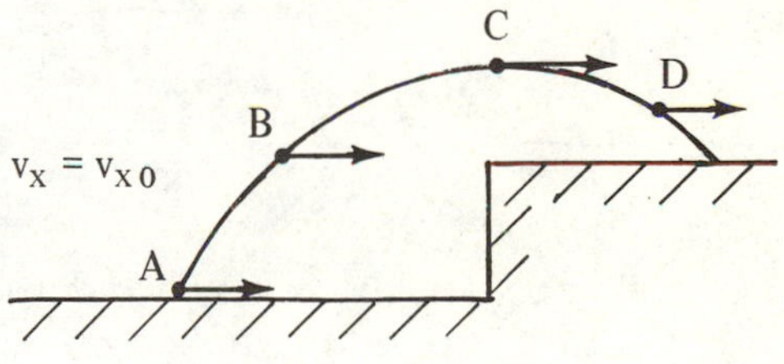

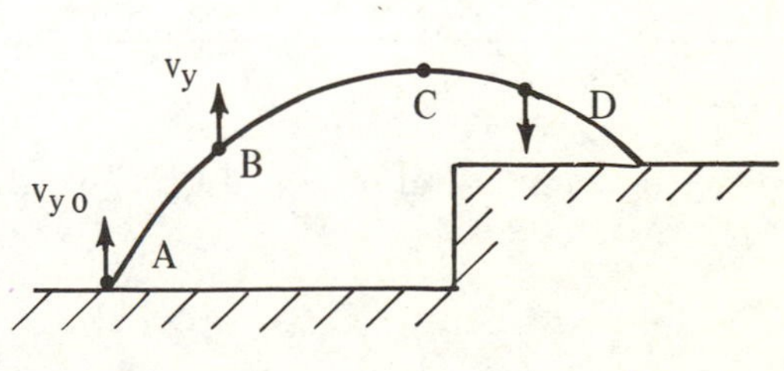

v_y is first positive, decreases to zero at the peak and becomes negative.

3.B

Sketch the acceleration vector at these points. What are the x and y components of acceleration when the projectile is in motion? Call y positive upwards.

$a_x = 0, a_y = -g$

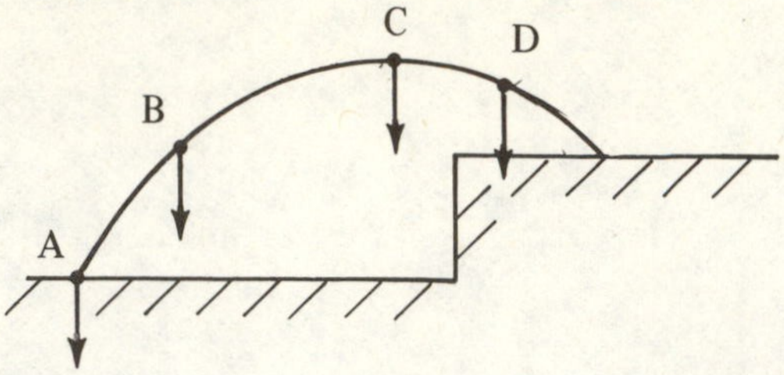

3.C

Why does the x component of velocity remain constant?

Since $a_x = 0$,

$$v_x = v_{x0} + a_x t = v_{x0} = v_0 \cos \theta_0$$

3.D

Why does the y component of velocity change in time?

v_y changes because $a_y = -g$. Therefore,

$$v_y = v_{y0} + a_y t = v_0 \sin \theta_0 - gt$$

3.E

What is the velocity of the projectile at B. Obtain both the magnitude of v_B and its orientation in terms of v_0, θ_0 and t.

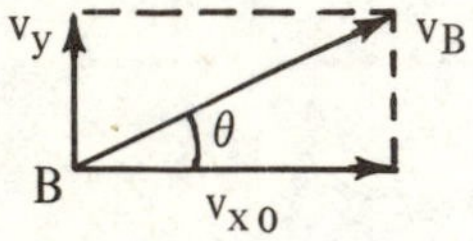

$$v_B = \sqrt{v_x^2 + v_y^2}$$

$$v_B = \sqrt{v_0^2 \cos^2 \theta_0 + (v_0 \sin \theta_0 - gt)^2}$$

$$v_B = \sqrt{v_0^2 + g^2 t^2 - (2v_0 \sin \theta_0)\, gt}$$

$$\theta = \arc\tan \frac{v_y}{v_{x0}} = \arc\tan \frac{(v_0 \sin \theta_0 - gt)}{v_0 \cos \theta_0}$$

3.F

Write expressions for the x and y coordinates at any time $t \geqslant 0$, taking the origin at A.

$$x = v_{x0} t = (v_0 \cos \theta_0) t$$

$$y = v_{y0} t - \frac{1}{2} g t^2 = (v_0 \sin \theta_0)\, t - \frac{1}{2} g t^2$$

3.G

Find the time t_1 it takes the projectile to reach C, the peak.

At the peak, $v_y = 0$; therefore,

$$v_y = 0 = v_{y0} - g t_1$$

$$t_1 = \frac{v_0 \sin \theta_0}{g}$$

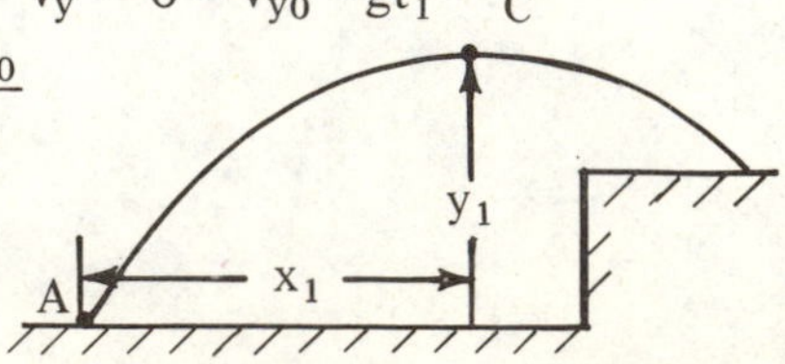

3.H

What is the maximum height y_1 reached by the projectile measured from A?

Using the results of 3.F and 3.G,

$$y_1 = (v_0 \sin\theta_0)\, t_1 - \frac{1}{2} g t_1^{\,2}$$

$$y_1 = \frac{v_0^{\,2} \sin^2\theta_0}{2\,g}$$

3.I

What is the x coordinate of the projectile at C relative to A?

Again, we apply the results of 3.F and 3.G.

$$x_1 = (v_0 \cos\theta_0) t_1 = \frac{v_0^{\,2} \sin\theta_0 \cos\theta_0}{g}$$

3.J

Let the distance from A to the foot of the cliff be d_1. At what time is the projectile directly above the edge of the cliff?

$$x = d_1 = (v_0 \cos\theta_0) t$$

$$t = \frac{d_1}{v_0 \cos\theta_0}$$

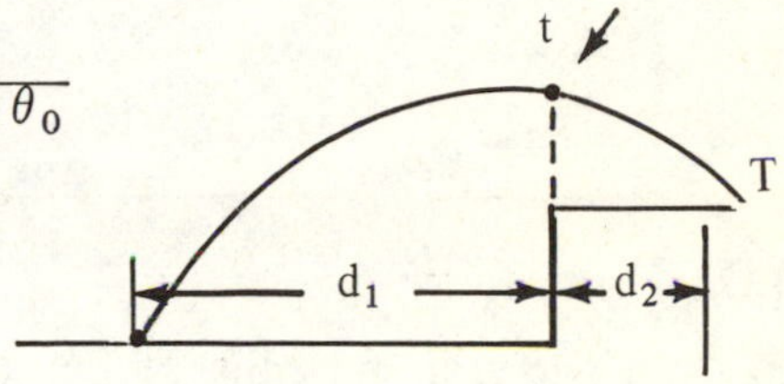

3.K

If the projectile hits the ground at a distance d_2 from the edge of the cliff, what is the total time of flight, T?

$$x = (d_1 + d_2) = (v_0 \cos\theta_0) T$$

$$T = \frac{d_1 + d_2}{v_0 \cos\theta_0}$$

Note: $d_1 + d_2$ is *not* the horizontal range R.

3.L

What is the y component of velocity just before it hits the ground?

$$v_y = v_0 \sin\theta_0 - gT \quad \text{at} \quad t = T$$

$$v_y = v_0 \sin\theta_0 - \frac{g(d_1 + d_2)}{v_0 \cos\theta_0}$$

3.M

What angle does **v** make with the horizontal just before the projectile hits the ground?

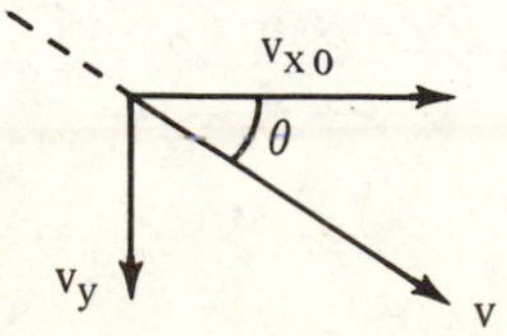

$$\theta = \arc\tan \frac{v_y}{v_{xo}}$$

where v_y is given in 3.L and $v_{xo} = v_0 \cos\theta_0$.

3.N

Suppose $v_0 = 30$ m/sec and $\theta_0 = 30°$. Determine the values of v_{xo} and v_{yo}, that is, the *initial* x and y components of velocity.

$v_{xo} = v_0 \cos\theta_0 = 30(0.87) = 26$ m/sec

$v_{yo} = v_0 \sin\theta_0 = 30(0.5) = 15$ m/sec

3.O

Now obtain expressions for the x and y coordinates, using (5) and (6).

$$x = (v_0 \cos\theta_0)t = 26t \text{ m} \qquad (9)$$

$$y = (v_0 \sin\theta_0)\,t - \frac{1}{2}gt^2$$

$$y = (15t - 4.9t^2) \text{ m} \qquad (10)$$

3.P

Use (1) and (2) to obtain the components of velocity at any time t.

$$v_x = v_{xo} = 26 \text{ m/sec} = \text{constant} \qquad (11)$$

$$v_y = v_0 \sin\theta_0 - gt$$

$$v_y = (15 - 9.8t) \text{ m/sec} \qquad (12)$$

3.Q

Explain the *full* meaning of (9), (10), (11), and (12).

(9) and (10) predict the x and y coordinates at *any* time t, where t is the time elapsed after leaving the origin (provided the projectile is still in flight). (11) and (12) predict the components of velocity at any time t. Note that v_x remains constant, while v_y changes in time.

3.R

Find the coordinates and velocity components for the projectile at t = 0.5 sec.

At t = 0.5 sec, we get

$$x = 26(0.5) = 13 \text{ m}$$

$$y = 15(0.5) - 4.9(0.5)^2 = 6.3 \text{ m}$$

$$v_x = 26 \text{ m/sec}$$

$$v_y = 15 - 9.8(0.5) \cong 10 \text{ m/sec}$$

3.S

Obtain the time it takes the projectile to reach its peak, and the maximum height of the projectile. Use the results of 3.G and 3.H.

$$t_1 = \frac{v_0 \sin\theta_0}{g} = \frac{15 \text{ m/sec}}{9.8 \text{ m/sec}^2} \cong 1.5 \text{ sec}$$

$$y_1 = \frac{v_0{}^2 \sin^2\theta_0}{g} = \frac{(15 \text{ m/sec})^2}{9.8 \text{ m/sec}^2} = 23 \text{ m}$$

4.A

Coin A is dropped from the edge of a table which is at a height h above the floor. At the *same instant,* a second coin B is projected horizontally from the edge of the table with an initial speed v_{x0}. Sketch the trajectories of the two coins.

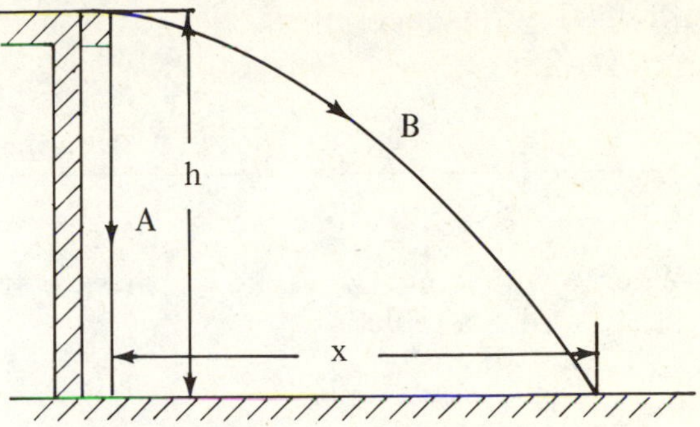

4.B

Sketch the x and y components of velocity for the two coins while they are in flight.

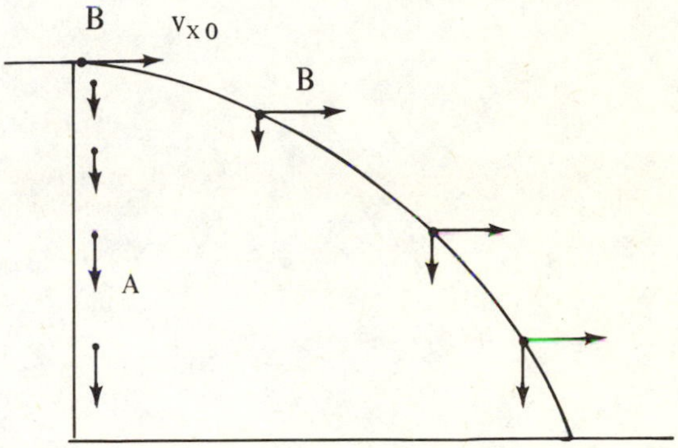

4.C

What is the acceleration of each coin while they are in motion?

They are both freely falling bodies; therefore, for *each* coin,

$$a_x = 0, \; a_y = -g$$

4.D

How long does it take coin A to reach the floor? Measure the coordinates from the top of the table. The table is at a height h above the floor.

$$y = v_{y0}t - \frac{1}{2}gt^2$$

But $v_{y0} = 0$. Therefore, the total time of flight, T, corresponds to $y = -h$;

$$\frac{1}{2}gT^2 = h$$

or,

$$T = \sqrt{\frac{2h}{g}}$$

4.E

How long does it take coin B to reach the floor? Try the experiment.

Again, $v_{yo} = 0$; therefore, following 4.D we see that $T = \sqrt{2h/g}$. That is, the coins are in flight for the same length of time, and will hit the floor *simultaneously*.

4.F

Where will coin A land relative to the bottom of the table?

$$x = v_{xo}\,T = v_{xo}\sqrt{\frac{2h}{g}}$$

3.6 SUMMARY

The kinematic equations for a particle moving along the x axis with a *constant* acceleration a_x are given by

$$v_x = v_{xo} + a_x t \tag{3.6}$$

$$x = x_0 + \frac{1}{2}(v_{xo} + v_x)\, t \tag{3.7}$$

$$x = x_0 + v_{xo} t + \frac{1}{2} a_x t^2 \tag{3.8}$$

$$v_x^2 = v_{xo}^2 + 2\, a_x (x - x_0) \tag{3.9}$$

For a *freely falling body* moving in the y direction in the Earth's gravitational field, replace x by y, v_x by v_y, and a_x by $-g$ in the expressions above.

Time of flight of a projectile:

$$T = \frac{2\, v_0 \sin \theta_0}{g}$$

Maximum height of a projectile:

$$h = \frac{v_0^2 \sin^2 \theta_0}{2\, g} \tag{3.21}$$

Range of a projectile:

$$R = \frac{v_0^2 \sin (2\theta_0)}{g} \tag{3.23}$$

3.7 PROBLEMS

1. A racing car increases its speed from 100 ft/sec to 200 ft/sec in a distance of 400 ft. Find (a) the magnitude of the acceleration, and (b) the time elapsed at this interval.

2. A particle starts from rest from the top of an inclined plane and slides down with constant acceleration. The inclined plane is 4m long and it takes 2 sec for the particle to reach the bottom. Find (a) the acceleration of the particle, (b) the speed of the particle at the bottom of the incline, (c) the time it takes the particle to reach the middle of the incline, and (d) the speed of the particle at the midpoint.

3. A bullet is fired through a 3 in thick board in such a way that its line of motion is perpendicular to the face of the board. If the initial speed of the bullet is 1000 ft/sec, and it emerges from the other side of the board with a speed of 500 ft/sec, find (a) the acceleration of the bullet while in contact with the board, and (b) the time it takes the bullet to pass through the board.

4. A boy kicks a football vertically upwards such that it just reaches the top of a 100 ft flag pole. (a) What was the initial velocity of the football? (b) How long is the football in flight?

5. In a drag race between cars A and B, car A is quicker to react at the start by 1 sec, and moves with a constant acceleration of 20 ft/sec^2. (a) If car B moves with a constant acceleration of 24 ft/sec^2, how far beyond the starting point will B overtake A? (b) What will be the speed of each vehicle at that instant?

6. In the opening of a football game, the ball is kicked at an angle of 37° above the horizontal. If the football travels 70 yards, (a) what was its initial speed? (b) How long was the ball in the air? (c) What was the maximum height reached by the ball?

7. An airplane flying parallel to the earth's surface has a constant speed of 300 mi/hr relative to the earth and is at an altitude of 2000 ft. A package is released from the airplane directly above an observer on the ground. (a) Compare the trajectories of the package as observed by the pilot and ground observer. (b) How long does it take the package to reach the earth? (c) Where does it land relative to the observer? Where does it land relative to the airplane? (d) What is the magnitude and direction of the package's velocity when it hits the earth?

8. A car is parked on a steep incline overlooking the ocean, where the angle of inclination is 37° from the horizontal. The negligent driver leaves the car in neutral and the emergency brakes are defective. The car rolls from rest down the incline with an acceleration of 4 m/sec^2 and travels a distance of 50 m to the edge of the cliff. The cliff is 30 m above the ocean level. (a) What is the speed of the car when it reaches the edge of the cliff, and how long does it take to get there? (b) What is the velocity of the car when it lands in the ocean? (c) What is the total time that the car is in motion? (d) Where does the car hit the ocean relative to the cliff?

9. An object is released from rest and falls freely in the earth's gravitational field. (a) Find the coordinate and velocity of the object 1 sec, 2 sec, and 4 sec after it is released. Use SI units and $g = 9.8$ m/sec^2. (b) Do the same calculations for an object dropped near the moon's surface where $g = 1.7$ m/sec^2. (c) Compare the results for parts (a) and (b) by making plots of y *vs* t and v_y *vs* t.

10. A rocket is launched at an angle of 53° from the horizontal with an initial velocity of 100 m/sec. It accelerates along its initial line of motion with an acceleration of 30 m/sec^2 for a period of 3 sec. At this time the rocket engines fail and the rocket proceeds to move as a free body. Find (a) the maximum altitude reached by the rocket, (b) the total time of flight for the rocket, and (c) the horizontal range of the rocket.

11. Suppose that a projectile is aimed directly at a target, whose initial polar coordinates are r, θ_0 relative to the projectile. Show that if the projectile is fired at the same instant that the target is dropped, a collision will always occur regardless of the initial velocity of the projectile.

†12. A particle moves in the x-y plane, and its vector position in meters varies in time according to the expression

$$\mathbf{r} = (2t - 3t^2 + t^3)\mathbf{i} + (4t^2 + 2t^3)\mathbf{j}$$

Find (a) expressions for the velocity and acceleration of the particle at some time t, and (b) values for **r**, **v** and **a** at t = 2 sec.

†13. (a) Prove that the horizontal range of a projectile is a maximum when $\theta_0 = \pi/4$. (b) Show that for an arbitrary initial velocity v_0, there are *two* values of θ_0 which would give the same range horizontal range R.

†14. The acceleration of a particle moving along the x axis varies in time according to the expression

$$a = -2t + 3t^2$$

If the speed of the particle is v_0 at $t_0 = 0$, and its coordinate is x_0 at $t_0 = 0$, find expressions for the speed and coordinate of the particle as a function of time.

15. In a recent Olympic long-jump, an athlete made a record leap of about 29 feet. Assuming the athlete leaves the ground at an angle of 25° with the horizontal, and assuming he is a point mass, find (a) the speed at which the athlete leaves the ground, and (b) the maximum height reached by the athlete.

4

CIRCULAR MOTION

4.1 CONCEPTS AND DEFINITIONS

When a particle is constrained to move in a *circle,* it is convenient to represent the position of the particle with the polar coordinates (r, θ) rather than the cartesian coordinates (x, y). The reason for this change in representation is that the problem will reduce to one in which only one coordinate, θ, changes in time, while r remains constant. [This is compared to (x, y) representation where both variables change in time.] If the particle starts its motion from $\theta = 0$, measured from the +x axis, as shown in Figure 4–1*a*, then the distance the particle moves along the circle is related to the angular displacement θ, through the relation

$$s = r\theta \tag{4.1}$$

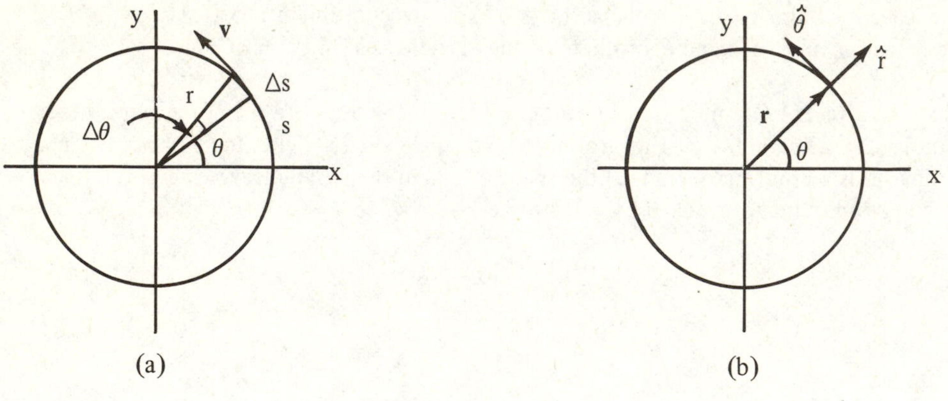

Figure 4–1 (*a*) Polar coordinates of a particle moving in a circle about a fixed axis z. (*b*) Unit vectors for a particle moving in a circle.

In circular motion, we commonly measure θ in *radians.* From Equation (4.1), we see that θ is the ratio of an arc length s to the radius r. Therefore, $\theta = 1$ radian corresponds to an arc length $s = r$. One complete revolution then corresponds to $s = 2\pi r$ (the circumference of the circle), and an angle $\theta = 2\pi$ radians. Consequently, $360° = 2\pi$ radians, and

$$\text{one radian} = \frac{360°}{2\pi} \cong 57.3°$$

To convert an angle in degrees to radians, we use the conversion

$$\theta \text{ (radians)} = \frac{\pi}{180}\theta \text{ (deg)}$$

For example, 60° equals $\pi/3$ radians and 90° equals $\pi/2$ radians. We will use the abbreviation rad for radians.

If the particle travels from (r, θ) to $(r, \theta + \Delta\theta)$ in a time Δt as shown in Figure 4–1(*a*), the distance it travels measured along the arc, Δs, is given by

$$\Delta s = r\Delta\theta$$

Dividing the left and right sides of this equation by Δt, and taking the limit as $\Delta t \to 0$, gives by definition the linear speed v, where **v** is always tangent to the circle.

$$v = \lim_{\Delta t \to 0} \frac{\Delta s}{\Delta t} = r \lim_{\Delta t \to 0} \frac{\Delta\theta}{\Delta t} = r\frac{d\theta}{dt} \tag{4.2}$$

The quantity $d\theta/dt$ is defined to be the *instantaneous angular* speed ω, so the relation between linear and angular speeds when *r is a constant* becomes

$$v = r\omega \tag{4.3}$$

where

$$\omega = d\theta/dt \tag{4.4}$$

Equation (4.3) is valid *only* when ω is expressed in rad/sec. If the particle moves in a circle in such a way that v is not constant in time, then obviously from Equation (4.3), ω is not constant in time. When this is the case, we say that the particle has an *instantaneous angular acceleration* α, defined by

$$\alpha = \frac{d\omega}{dt} = \frac{d^2\theta}{dt^2} \tag{4.5}$$

Since ω has units of rad/sec, we see from Equation (4.5) that α must have units of rad/sec^2. If we write Equation (4.5) in the form $d\omega = \alpha dt$, and let $\omega = \omega_0$ at $t = 0$, and α = constant, we can integrate the expression directly (as in linear motion) to give

$$\omega = \omega_0 + \alpha t \quad (\alpha = \text{constant}) \tag{4.6}$$

Likewise, substituting Equation (4.6) into Equation (4.4) and integrating once more (with $\theta = \theta_0$ at $t = 0$) gives

$$\theta = \theta_0 + \omega_0 t + \frac{1}{2}\alpha t^2 \tag{4.7}$$

If we eliminate t from Equation (4.6) and Equation (4.7), we get

$$\omega^2 = \omega_0{}^2 + 2\alpha(\theta - \theta_0) \tag{4.8}$$

Of course, when α = constant, we can also relate the average angular speed to the initial and final values with the expression

$$\omega_{av} = \frac{1}{2}(\omega_0 + \omega) \tag{4.9}$$

We see that these equations representing circular motion for constant angular acceleration are of the *same form* as those for linear motion with constant linear acceleration with the substitutions $x \to \theta$, $v \to \omega$ and $a \to \alpha$.

To proceed further, it is convenient to write a vector expression for the velocity of the particle. We can do this by defining the unit vectors $\hat{r}$ and $\hat{\theta}$, where $\hat{r}$ is a unit vector directed radially outward along the radius vector $\hat{r}$, and $\hat{\theta}$ is a unit vector tangent to the circular path. The direction of θ is in the direction of increasing $\hat{\theta}$, therefore it is counterclockwise as shown in Figure 4–1(*b*). It is important to note that both $\hat{r}$ and $\hat{\theta}$ vary in time relative to a stationary observer, since they are unit vectors which "move along with the particle." If the particle moves in the counterclockwise direction, we can write Equation (4.3) in vector form as

$$\mathbf{v} = r\omega\hat{\theta} \tag{4.10}$$

since $\mathbf{v}$ is always tangent to the circle, which is the direction of $\hat{\theta}$. Differentiating Equation (4.4) with respect to time gives the total acceleration of the particle.

$$\mathbf{a} = \frac{d\mathbf{v}}{dt} = \frac{d}{dt}(r\omega\hat{\theta}) = r\frac{d\omega}{dt}\hat{\theta} + r\omega\frac{d\hat{\theta}}{dt} \tag{4.11}$$

The first term on the right of Equation (4.10) is simply $r\alpha\hat{\theta}$. Since $\hat{\theta}$ is a unit vector in the direction tangent to the circle, we refer to this term as the *tangential acceleration* $\mathbf{a}_t$, which is related to the angular acceleration α through the relation

$$\mathbf{a}_t = r\alpha\hat{\theta} \tag{4.12}$$

Of course, $\mathbf{a}_t$ has units of linear acceleration, L/T^2, and is nonzero only if α is nonzero. The other component of $\mathbf{a}$ in Equation (4.11) is along the radial direction. This can be seen by noting the $d\hat{\theta}$ is along the direction $-\hat{r}$, therefore

$$\frac{d\hat{\theta}}{dt} = -\hat{r}\frac{d\theta}{dt}$$

so

$$r\omega\frac{d\hat{\theta}}{dt} = -r\omega\frac{d\theta}{dt}\hat{r} = -r\omega^2\hat{r} \tag{4.13}$$

But $\omega = v/r$ from Equation (4.3), therefore Equation (4.12) reduces to

$$\mathbf{a}_r = -\frac{v^2}{r}\hat{r} = -r\omega^2\hat{r} \tag{4.14}$$

Thus, we can write Equation (4.11) as

$$a = a_t + a_r = r\alpha\hat{\theta} - \frac{v^2}{r}\hat{r} \tag{4.15}$$

The component a_r is called the *radial* or *centripetal acceleration,* since it is *always* directed along the direction $-\hat{r}$. These components are shown in Figure 4–2.

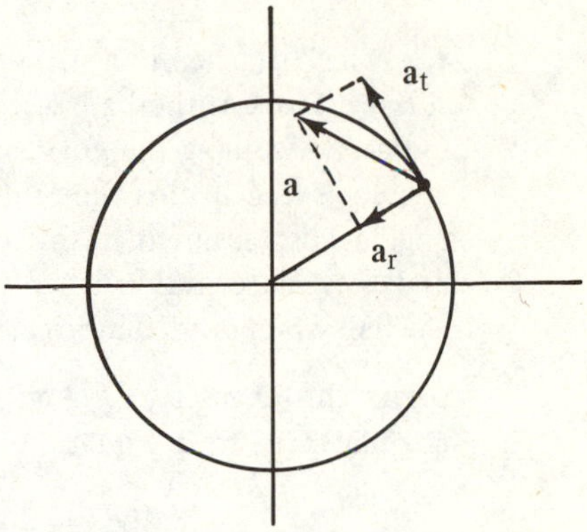

Figure 4–2 Tangential and radial components of acceleration for a particle moving in a circle when $\alpha \neq 0$. If $\alpha = 0$, $a_t = 0$, v = constant and $a = a_r$.

The magnitude of a is given by

$$a = \sqrt{a_t^2 + a_r^2} = \sqrt{r^2\alpha^2 + r^2\omega^4} = r\sqrt{\alpha^2 + \omega^4} \tag{4.16}$$

A special case occurs when $\alpha = 0$, or $a_t = 0$, v = constant. We see that a reduces to

$$a = -\frac{v^2}{r}\hat{r} \quad \text{(when v = constant)} \tag{4.17}$$

That is, the total acceleration of a particle moving in a circle with *constant speed* is directed towards the center of motion. This component arises from a *change* in the *direction* of v (where Δv is always inwards towards the center of motion).

Example 4.1

A wheel 18 inches in diameter starts from rest and rotates with uniform acceleration to 100 rev/sec in a time of 4 sec. Find the angular acceleration of the wheel and the speed of a point on its rim [$\omega = 2\pi/T$, where T is the time for *one* revolution].

Since the wheel starts from rest, $\omega_0 = 0$. The frequency of rotation, f, after 4 sec is given (that is, the number of revolutions per second). The inverse of f is the time it takes for one revolution. Therefore, the angular velocity after 4 sec is

$$\omega = 2\pi f = 2\pi(100) \text{ rad/sec} = 200\pi \text{ rad/sec}$$

Since α = constant, we can apply Equation (4.6) with $\omega_0 = 0$.

$$\omega = \alpha t$$

$$\alpha = \frac{200\pi \text{ rad/sec}}{4 \text{ sec}} = 50\pi \text{ rad/sec}^2$$

To find the speed of a point on the rim, we can make use of Equation (4.3).

$$v = r\omega = \frac{9\text{ in}}{12\text{ in/ft}} \quad 200\pi\text{ rad/sec} = 150\pi\text{ ft/sec}^2$$

Note: The radian is not a fundamental unit. It only indicates an *angular displacement.*

Example 4.2

A ball attached to a string 1 m in length whirls in a horizontal circle on a frictionless surface. It takes the ball 2 sec to go from an angular speed of 4 rad/sec to a speed of 20 rad/sec. Assuming the *angular* acceleration is constant, find the following: (a) the average angular speed in this time interval, (b) the angular acceleration of the ball, (c) the total angular displacement in this time interval, (d) the components of the acceleration vector and total acceleration when its angular speed is 4 rad/sec, and (e) the linear velocity of the ball when $\omega = 20$ rad/sec.

(a) In this example, r = 1 m, $\omega_0 = 4$ rad/sec, $\omega = 20$ rad/sec, and t = 2 sec. The average angular speed from Equation (4.9) is

$$\omega_{av} = \frac{1}{2}(\omega_0 + \omega) = \frac{1}{2}(24)\text{ rad/sec} = 12\text{ rad/sec}$$

(b) We can use Equation (4.6) to get α.

$$\alpha = \frac{\omega - \omega_0}{t} = \frac{20\text{ rad/sec} - 4\text{ rad/sec}}{2\text{ sec}} = 8\text{ rad/sec}^2$$

(c) The total angular displacement from Equation (4.7) is

$$\Delta\theta = \theta - \theta_0 = \omega_0 t + \frac{1}{2}\alpha t^2 = \frac{4\text{ rad}}{\text{sec}}(2\text{ sec}) + \frac{1}{2}\frac{8\text{ rad}}{\text{sec}^2}(2\text{ sec})^2$$

$$\Delta\theta = 8\text{ rad} + 16\text{ rad} = 24\text{ rad}$$

Since 2π rad = 1 rev, this corresponds to $\frac{24}{2\pi} \cong 3.8$ rev. We could also obtain this result by using the fact that $\Delta\theta = \omega_{av}t$.

(d) From Equation (4.15), when $\omega_0 = 4$ rad/sec, we have

$$a_t = r\alpha = 1\text{ m }(8\text{ rad/sec}^2) = 8\text{ m/sec}^2$$

$$a_r = v^2/r = r\omega_0{}^2 = 1\text{ m}\left(\frac{4\text{ rad}}{\text{sec}}\right)^2 = 16\text{ m/sec}^2$$

Hence,

$$\mathbf{a} = (8\hat{\theta} - 16\hat{r})\text{ m/sec}^2$$

$$a = \sqrt{a_t{}^2 + a_r{}^2} = \sqrt{(8\text{ m/sec}^2)^2 + (16\text{ m/sec}^2)^2} = 17.9\text{ m/sec}^2$$

(e) When $\omega = 20$ rad/sec,

$$v = r\omega = 1\text{ m }(20\text{ rad/sec}) = 20\text{ m/sec}$$

At this time, $a_r = \frac{v^2}{r} = 400\ \text{m/sec}^2$, so the total acceleration has increased to

$$\mathbf{a} = (8\hat{\theta} - 400\,\hat{\mathbf{r}})\ \text{m/sec}^2$$

That is, although α = constant, **a** *is not* constant. The tangential component of acceleration remains constant in time, but the radial component changes since v changes in time.

†Example 4.3

A rotating disc has an angular displacement described by the expression

$$\theta = \pi/3 + 2\,t + 3\,t^2 - t^3$$

where θ is measured in radians and t in seconds. (a) Find expressions for the angular speed and angular velocity as a function of time. (b) Determine the values of ω and α at t = 2 sec. (c) Find the angular displacement and the number of revolutions made at the end of 2 sec.

(a) Equations (4.4) and (4.5) define ω and α.

$$\omega = \frac{d\theta}{dt} = \frac{d}{dt}\left(\frac{\pi}{3} + 2\,t + 3\,t^2 - t^3\right) = 2 + 6\,t - 3\,t^2$$

$$\alpha = \frac{d\omega}{dt} = \frac{d}{dt}(2 + 6t - 3t^2) = 6 - 6t$$

(b) Using the results of (a) gives

$$\omega(t = 2) = 2 + 6\,(2) - 3\,(2)^2 = 2\ \text{rad/sec}$$

$$\alpha(t = 2) = 6 - 6\,(2) = -6\ \text{rad/sec}^2$$

(c) At t = 0, $\theta_0 = \pi/3$ rad. When t = 2 sec, $\theta = \pi/3 + 2\,(2) + 3\,(2)^2 - (2)^3 = (\pi/3 + 8)$ rad. Therefore, $\Delta\theta = \theta - \theta_0 = 8$ rad.

$$8\ \text{rad corresponds to}\ \frac{8\ \text{rad}}{2\,\pi\ \text{rad/rev}} = 1.27\ \text{rev}$$

4.2 PROGRAMMED EXERCISES

1.A

A record 12 inches in diameter rotates at 45 rev/min. What is the frequency of rotation?

$$f = 45\text{ rev/min}\left(\frac{1}{60\text{ sec/min}}\right)$$

$$f = 0.75\text{ rev/sec}$$

1.B

What is the period of the record, that is, the time for one complete revolution?

$$T = \frac{1}{f} = \frac{1}{0.75\text{ rev/sec}} = 1.33\text{ sec}$$

1.C

What is the angular speed of the record?

$$\omega = 2\pi f = \frac{2\pi}{T} = 1.5\pi\text{ rad/sec}$$

1.D

What is the angular acceleration of the record?

Since ω = constant,

$$\alpha = \frac{d\omega}{dt} = 0$$

1.E

Calculate the speed of a point on the rim of the record.

Using the results of 1.C, we have

$$v = r\omega = \left(\frac{6\text{ in}}{12\text{ in/ft}}\right) 1.5\pi\ \frac{\text{rad}}{\text{sec}}$$

$$v = 0.75\pi\text{ ft/sec}$$

1.F

What is the angular displacement of the record in two seconds?

Since $\alpha = 0$, Equation (4.7) gives

$$\Delta\theta = \omega t = (1.5\pi)\frac{\text{rad}}{\text{sec}}(2\text{ sec})$$

$$\Delta\theta = 3\pi\text{ rad (or } \Delta\theta = 540°)$$

1.G

How many revolutions does the record make in two seconds?

From 1.F, $\Delta\theta = 3\pi$ rad, therefore the number of revolutions in two seconds is

$$\frac{3\pi\ \text{rad}}{2\pi\ \text{rad/rev}} = 1.5\text{ rev}$$

1.H

Find the radial and tangential components of acceleration of a point on the rim of the record.

Since ω = constant, $\alpha = 0$, so $a_t = r\alpha = 0$.

$$a_r = \frac{v^2}{r} = \frac{(0.75\ \pi)^2\ ft^2/sec^2}{1/2\ ft} = 1.12\ \pi^2 \frac{ft}{sec^2}$$

---

†2.A

From the definition of α, derive Equation (4.6) for circular motion with constant acceleration.

Let $\omega = \omega_0$ at $t = 0$.

$$d\omega = \alpha dt$$

$$\int_{\omega_0}^{\omega} d\omega = \int_0^t \alpha dt = \alpha \int_0^t dt$$

$$\omega = \omega_0 + \alpha t \qquad (1)$$

2.B

Using the definition of ω, and Equation (4.6), derive Equation (4.7) letting $\theta = \theta_0$ at $t = 0$.

$$d\theta = \omega dt$$

$$\int_{\theta_0}^{\theta} d\theta = \int_0^t \omega dt = \int_0^t (\omega_0 + \alpha t)\ dt$$

$$\theta = \theta_0 + \omega_0 T + \frac{1}{2}\alpha t^2 \qquad (2)$$

2.C

Use the results of 2.A and 2.B to show that $\omega^2 = {\omega_0}^2 + 2\alpha(\theta - \theta_0)$, which is Equation (4.8).

Solve for t in (1). $t = \dfrac{\omega - \omega_0}{\alpha}$

Substitute this in (2).

$$\theta = \theta_0 + \omega_0 \left(\frac{\omega - \omega_0}{\alpha}\right) + \frac{1}{2}\alpha \left(\frac{\omega - \omega_0}{\alpha}\right)^2$$

This simplifies to the result. Try it.

2.D

Write the basic equations of linear motion with constant acceleration, and the analogous equations for circular motion.

$$v = v_0 + at \quad \sim \quad \omega = \omega_0 + \alpha t$$

$$x = x_0 + v_0 t + \frac{1}{2}at^2 \quad \sim \quad \theta = \theta_0 + \omega_0 t + \frac{1}{2}\alpha t^2$$

$$v^2 = {v_0}^2 + 2a(x - x_0) \quad \sim \quad \omega^2 = {\omega_0}^2 + 2\alpha(\theta - \theta_0)$$

$$v_{av} = \frac{1}{2}(v_0 + v) \quad \sim \quad \omega_{av} = \frac{1}{2}(\omega_0 + \omega)$$

2.E

Suppose that the angular acceleration of a particle varies in time according to the expression $\alpha = At^2$, where A is a constant. Determine the angular speed ω, letting $\omega = \omega_0$ at t = 0.

$$d\omega = \alpha dt$$

$$\int_{\omega_0}^{\omega} d\omega = \int_0^t \alpha dt = \int_0^t At^2\, dt = \frac{At^3}{3}$$

$$\omega = \omega_0 + \frac{At^3}{3}$$

2.F

Using the definition of ω, calculate the angular displacement for the particle described in 2.E. Let $\theta = \theta_0$ at t = 0.

$$d\theta = \omega dt = \left(\omega_0 + \frac{At^3}{3}\right) dt$$

$$\int_{\theta_0}^{\theta} d\theta = \int_0^t \omega_0\, dt + \frac{A}{3}\int_0^t t^3\, dt$$

$$\theta = \theta_0 + \omega_0 t + \frac{A}{12} t^4$$

3.A

The relations between cartesian and polar coordinates in two dimensions are given by

$$x = r\cos\theta \qquad y = r\sin\theta$$

as defined in Figure 4-1.A. Rewrite these expressions for the case of a particle moving in a circle with *constant angular* speed ω, and find v_x, v_y. Note that $\theta = \omega t$ under these conditions.

When ω = constant, $\theta = \omega t$,

$$x = r\sin\omega t$$

$$y = r\cos\omega t$$

$$v_x = \frac{dx}{dt} = r\frac{d}{dt}(\sin\omega t) = r\omega\cos\omega t$$

$$v_y = \frac{dy}{dt} = r\frac{d}{dt}(\cos\omega t) = -r\omega\sin\omega t$$

For circular motion, r = constant

---

†3.B

Find the x and y components of acceleration from the expressions for v_x and v_y.

$$a_x = \frac{dv_x}{dt} = r\omega\frac{d}{dt}(\cos\omega t) = -r\omega^2\sin\omega t$$

$$a_y = \frac{dv_y}{dt} = -r\omega\frac{d}{dt}(\sin\omega t) = -r\omega^2\cos\omega t$$

3.C

Using the results to 3.A, find the total speed of the particle in terms of r and ω. Is the result familiar?

$$v = \sqrt{v_x^2 + v_y^2} = \sqrt{r^2 \omega^2 \cos^2 \omega t + r^2 \omega^2 \sin^2 \omega t}$$

$$v = r\omega \sqrt{\cos^2 \omega t + \sin^2 \omega t} = r\omega$$

Since $\cos^2 \theta + \sin^2 \theta = 1$. This expression for v is equivalent to Equation (4.3),

$$v = r\omega$$

3.E

In this problem, the important point is that although ω = constant, the particle still has an acceleration $\mathbf{a}_r$ directed towards the center of motion. Sketch the acceleration vector and velocity vector as a function of time at different points on the trajectory.

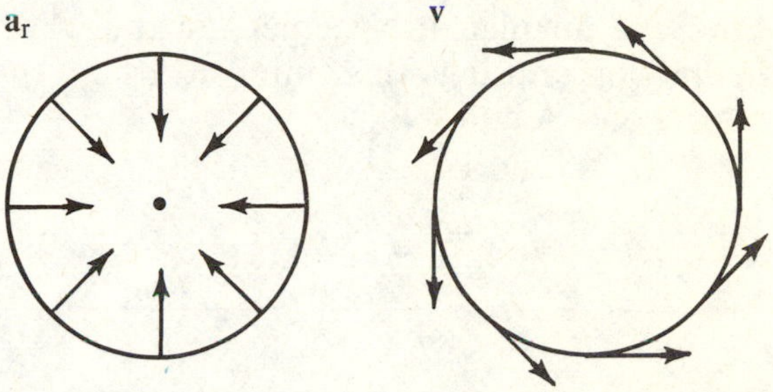

Note: $|\mathbf{v}|$ = constant

3.D

Using the results to 3.B, find the total acceleration of the particle in terms of r and ω. Is this result familiar?

$$a = \sqrt{a_x^2 + a_y^2} = \sqrt{r^2 \omega^4 \sin^2 \omega t + r^2 \omega^4 \cos^2 \omega t}$$

$$a = r\omega^2 \sqrt{\sin^2 \omega t + \cos^2 \omega t} = r\omega^2$$

But $v = r\omega$, so $a = v^2/r = a_r$. This is equivalent to Equation (4.14), where the particle has a *centripetal acceleration only,* since ω = constant.

4.A

The rotation of a wheel decreases uniformly in a counterclockwise manner from 300 rev/min to 60 rev/min in a time of 2 sec. Compute the initial and final angular speeds in rad/sec.

$$1 \text{ rev} = 2\pi \text{ rad}$$

$$\omega_0 = 300 \frac{\text{rev}}{\text{min}} \; 2\pi \frac{\text{rad}}{\text{rev}}$$

$$= 600\pi \frac{\text{rad}}{\text{min}} \frac{1}{60 \text{ sec/min}}$$

$$\omega_0 = 10\pi \text{ rad/sec}$$

Likewise, $\omega = 2\pi$ rad/sec.

4.B

Determine the angular acceleration of the wheel using the results to 4.A, and Equation (4.6). What does the sign of α imply?

$$\omega = \omega_0 + \alpha t$$

$$2\pi = 10\pi + \alpha\,(2)$$

$$\alpha = -4\pi \ \ \text{rad/sec}^2$$

The negative sign implies a deceleration.

4.C

Find the angular displacement of the wheel in radians during this interval using the results of 4.A and 4.B.

$$\Delta\theta = \omega_0 t + \frac{1}{2}\alpha t^2$$

$$\Delta\theta = 10\pi(2) + \frac{1}{2}(-4\pi)\,(2)^2$$

$$\Delta\theta = 12\pi \ \text{rad}$$

4.D

If the wheel has a radius of 0.5 m, calculate the velocity of a point on its rim when $\omega = 2\pi$ rad/sec.

$$\mathbf{v} = r\omega\hat{\theta}$$

$$\mathbf{v} = (0.5\ \text{m})\left(2\pi\frac{\text{rad}}{\text{sec}}\right)\hat{\theta}$$

$$\mathbf{v} = \pi\hat{\theta}\ \text{m/sec}$$

4.E

Calculate the centripetal acceleration of the wheel when $\omega = 2\pi$ rad/sec.

$$\mathbf{a}_r = -r\omega^2\hat{r} = -(0.5)\ \text{m}\left(2\pi\frac{\text{rad}}{\text{sec}}\right)^2\hat{r}$$

$$\mathbf{a}_r = -2\pi^2\ \hat{r}\ \text{m/sec}^2$$

4.F

Calculate the tangential accelration of a point on the rim when $\omega = 2\pi$ rad/sec.

$$\mathbf{a}_t = r\alpha\hat{\theta} = (0.5\text{m})\left(-4\pi\frac{\text{rad}}{\text{sec}^2}\right)\hat{\theta}$$

$$\mathbf{a}_t = -\ 2\pi\hat{\theta}\ \text{m/sec}^2$$

4.G

Write an expression for the total acceleration vector when $\omega = 2\pi$ rad/sec, and find its magnitude.

Using the results to 4.E and 4.F gives

$$\mathbf{a} = \mathbf{a}_t + \mathbf{a}_r = (-2\pi\hat{\theta} - 2\pi^2\ \hat{r})\ \text{m/sec}^2$$

$$a = \sqrt{a_t^{\,2} + a_r^{\,2}} = 2\pi\sqrt{1+\pi^2}\ \frac{\text{m}}{\text{sec}^2}$$

4.3 SUMMARY

Angular velocity: $$\omega = \frac{d\theta}{dt} \qquad (4.4)$$

Angular acceleration: $$\alpha = \frac{d\omega}{dt} \qquad (4.5)$$

where θ is the angular displacement in *radians.*

The kinematic equations for a particle moving in a circle with *constant* angular acceleration are given by

$$\omega = \omega_0 + \alpha t \qquad (4.6)$$

$$\theta = \theta_0 + \omega_0 t + \frac{1}{2}\alpha t^2 \qquad (4.7)$$

$$\omega^2 = \omega_0{}^2 + 2\,\alpha\,(\theta - \theta_0) \qquad (4.8)$$

Relation between linear and angular speeds: $$v = r\omega \qquad (4.10)$$

Total acceleration of a particle rotating in a circle: $$a = r\alpha\hat{\theta} - \frac{v^2}{r}\,\hat{r} \qquad (4.15)$$

Tangential acceleration: $$a_t = r\alpha\hat{\theta} \qquad (4.12)$$

Radial (or centripetal) acceleration: $$a_r = -\frac{v^2}{r}\,\hat{r} \qquad (4.14)$$

4.4 PROBLEMS

1. A racing car moves on a circular track having a radius of 1200 ft. The car moves with a constant speed of 180 miles per hour. (a) What is the angular speed of the car? (b) Find the magnitude and direction of the car's acceleration.

2. The racing car in problem 1 starts from rest and accelerates uniformly to a speed of 180 miles per hours in a time of 30 sec. (a) What is the average angular speed of the car in this time interval? (b) What is the angular acceleration of the car? (c) What is the magnitude of the car's acceleration at t = 10 sec? (d) Find the total distance traveled by the car in the first 30 seconds.

3. A wheel having a radius of 6 inches is rotating at a constant rate of 1200 rev/min about an axis through its center. (a) What is the angular speed of the wheel? (b) Determine the linear speed of the points 2 inches, 4 inches and 6 inches from the center. Make a plot of v *vs* r. (c) What is the radial acceleration of a point on the rim? (d) Find the total distance that a point on the rim moves in a time interval of 10 sec.

4. A wheel 4 ft in diameter rotates *counterclockwise* about an axle with a constant angular acceleration of 4 rad/sec. The wheel starts at rest at t = 0, and the radius vector to a point A on the rim makes an angle of 57.3° with the horizontal at this time. (a) Find the angular speed of the wheel at t = 1 and 2 sec. (b) Determine the velocity and acceleration of the point A at t = 1 and 2 sec. (c) Determine the position of point A at t = 1 and 2 sec.

5. The moon revolves around the earth about once every 28 days, and its average distance from the earth is 3.84×10^8 m. Assuming that the moon moves in a circular orbit about the earth, calculate its (a) angular velocity, (b) linear velocity and (c) centripetal acceleration.

6. The time it takes the earth to make one revolution about the sun is about 365 days. Assuming the earth travels in a circular orbit around the sun, with a radius of 1.49×10^{11} m, calculate (a) the angular speed of the earth, (b) the orbital speed and (c) the centripetal acceleration of the earth due to its rotation about the sun.

†7. A wheel rotates in such a way that its angular displacement is given by $\theta = at + bt^3$, where t is in seconds, a and b are constants, and θ is in radians. At t = 2 sec, determine (a) the angular displacement, (b) the angular speed and (c) the angular acceleration of the wheel.

5

THE LAWS OF MOTION-PARTICLE DYNAMICS

5.1 BASIC DEFINITIONS AND CONCEPTS

In previous chapters, we described the motion of particles without giving any specific description of the forces governing the motion. We simply made use of such quantities as displacement, velocity and acceleration and the relationships derived from these quantities. However, it is possible to obtain the same relationships using a formulation first described by Isaac Newton (1642-1727). These laws of motion are based on three principles which are, for the most part, the result of common observations. We must introduce the concepts of *force* and *mass* in order to deal with these laws. The concept of *force* is related to the effect of the environment on the motion of the particle. In fact, the acceleration experienced by the particle depends on the environment, or on the applied forces. For example, an object placed on a horizontal table will move only if some external influence acts on it, such as the force of our hand. The acceleration of the object will increase as the applied force increases. If we release the object, it will eventually come to rest due to the friction between the object and the table. We say that there exists a *frictional force* between the object and the surface which produces a decrease in velocity or a deceleration. If there was no friction present, the object would continue to move indefinitely once we released it. In practice, there is always some friction present, but there are some laboratory demonstrations which approximate frictionless surfaces, such as a flat puck riding on an air cushion.

The concept of *mass,* which is a *scalar* quantity, is a measure of the resistance of a body to undergo acceleration when an external force acts on it. We sometimes refer to this property as inertia. For example, we are aware of the inertia of a body even in free space where gravitational forces are absent. If an astronaut kicks an object in a spacecraft while on a journey to the moon, he will sense the presence of the object by the pain in his foot. The more massive the object, the greater is its inertia, and the louder the "ouch"!

The *first law of motion* as stated by Newton is written as follows: "Every body persists in its state of rest or of uniform motion in a straight line unless it is compelled to change that state by forces impressed on it." In simpler terms, we can state the first law in an alternate manner: *when the external force acting on a body is zero, its acceleration is zero.* This law is sometimes referred to as the *law of inertia,* since it applies to bodies which are in an inertial frame of reference. *Inertial frames of reference* are either fixed with respect to the distant stars or move with a uniform velocity with respect to the distant stars. Therefore, inertial frames cannot rotate since rotations give rise to accelerations (due to changes in the direction of **v**). The earth is usually approximated to be an inertial frame of reference. One important point about the first law is the fact that a body at rest *or* a body moving with constant velocity are *identical* as far as the net force acting is concerned. In both cases, the net force must be zero since the body has zero acceleration with respect to an observer in an inertial frame of reference. Therefore, we can conclude that a free particle (a particle subject to no forces or a net force of zero) always moves with constant velocity (that is, zero acceleration), or is at rest relative to an observer in any inertial frame.

Newton's second law of motion states that the *acceleration of a particle is equal to the resultant of all external forces acting on the particle divided by the mass of the particle.* The second law can be written as the vector equation

$$\Sigma \mathbf{F} = m\mathbf{a} \tag{5.1}$$

where it is important to note the left hand side of Equation (5.1) is a *vector sum over all external forces.* In component form, we can write Equation (5.1) as three expressions:

$$\Sigma F_x = ma_x, \quad \Sigma F_y = ma_y, \quad \Sigma F_z = ma_z \tag{5.2}$$

Equation (5.1) is *not* the most general form of Newton's second law since it is only valid if the mass remains constant during the motion. In fact, this is a very good approximation for particle speeds that are small compared to the speed of light, $c \cong 3 \times 10^8$ m/sec. Unless stated otherwise, we will use this assumption for discussing the motion of such objects as laboratory carts, baseballs, satellites and planets. The general form of Newton's second law, which includes the possibility of *variations* of the mass with velocity, can be written in the form

$$\Sigma \mathbf{F} = \frac{d}{dt}(m\mathbf{v}) \tag{5.3}$$

In either case, the acceleration is measured by an observer in an inertial frame; therefore, we can conclude that *an inertial frame is one in which Newton's laws are valid.*

Newton's third law of motion, sometimes referred to as the *law of action and reaction,* states that "To every action there is always opposed an equal reaction; or, the mutual actions of two bodies upon each other are always equal, and directed to contrary parts." That is, *when two particles A and B interact, the force of A on B is equal and opposite to the force of B on A.* We can conclude that *isolated forces do not exist* in nature. For example, if a man attempts to push a stalled automobile to get it started, he exerts a force $\mathbf{F}_1$ on the car (the action force). Since the car has mass (or inertia), the man is quite aware of its presence, and the car "pushes back" at the man with a force $\mathbf{F}_2$ (the reaction force). According to Newton's third law,

$\mathbf{F}_1 = -\mathbf{F}_2$. Another example of Newton's third law involves gravitational forces. When we say that we "weigh" 150 lb, we mean that the earth exerts an equivalent force of 150 lb on our body. What is the reaction to this force? In this case, the reaction is that our body exerts a force of 150 lb on the earth! We will encounter many other examples of Newton's third law.

The three systems of units which are commonly used are SI(mks), cgs and Engineering units. The units of force in these three systems are the Newton (N), dyne and pound (lb), as described in Table 5-1. For example, a force of one N is that force which will accelerate a one kg mass at one m/sec^2.

TABLE 5-1 UNITS OF FORCE, BASED ON $\mathbf{F} = m\mathbf{a}$

System of Units	Mass	Acceleration	Force
SI (mks)	kg	m/sec^2	N = kg m/sec^2
cgs	g	cm/sec^2	dyne = g cm/sec^2
Engineering	slug	ft/sec^2	lb = slug ft/sec^2

Many of the examples we will treat involve a force of *friction* between the object and surface. If the object is at rest, and an external force acts which has a component parallel to the surface, the force of friction acts opposite to the applied force. We call this the force of *static friction,* $\mathbf{f}_s$. Experiments show that this force has a *maximum* value whose magnitude is proportional to N, the normal force (which is the force of the surface on the object). Therefore, we can write

$$f_s \leqslant \mu_s N \tag{5.4}$$

where μ_s is a dimensionless constant called the *coefficient of static friction.* The *maximum* value $\mu_s N$ acts when the object is on the *verge* of slipping. Once the object begins to move, the force of friction *decreases* but is still proportional to N. The force of friction which acts when the body is in motion is called the *force of kinetic friction,* $\mathbf{f}_k$, where $\mathbf{f}_k$ *opposes* the motion of the object. This force is given by the expression

$$f_k = \mu_k N \tag{5.5}$$

where μ_k is called the *coefficient of kinetic friction.* In general, it is found that $\mu_k \leqslant \mu_s$.

5.2 SOME APPLICATIONS OF NEWTON'S LAWS

The problems we will deal with in this chapter will involve Newton's second and third laws. We will reserve the application of Newton's first law to a later chapter dealing with the equilibrium of bodies. The first law of motion is, in fact, a special case of the second law. That is, when the resultant force acting on a particle is zero, its acceleration is zero, which implies that the particle is either at rest or is moving at a uniform velocity relative to any inertial frame. Thus, the two laws are not independent.

Example 5.1

A man pulls a 192 lb crate horizontally with a light rope across a rough surface as in Figure 5-1. The man exerts a force of 150 lb on the rope, and the coefficient of sliding friction between the crate and surface is 0.5. (a) What is the reaction force to the man's force on the rope? (b) What force pulls the crate? (c) Draw a free body diagram for the crate and identify all the forces involved. (d) What is the resultant force on the crate? (e) What is the acceleration of the crate?

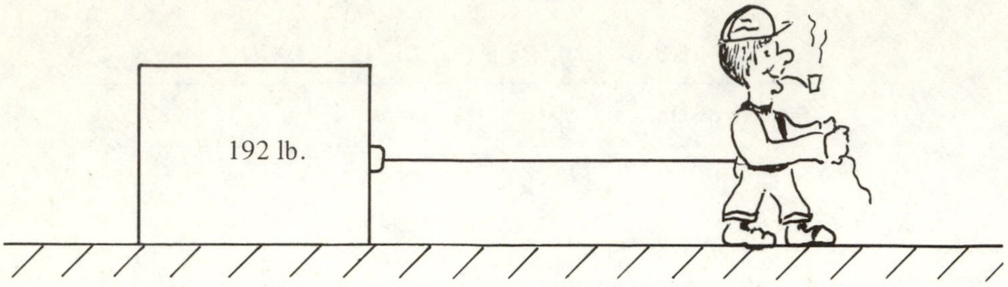

Figure 5-1

(a) The reaction force is the force of the rope on the man, or the force of *tension* is the rope. This force is to the left and equals 150 lb.

(b) The force of tension in the rope pulls the crate. Ropes or strings *always* pull on objects. (Try pushing a crate with a rope!) This force is to the right and equals 150 lb.

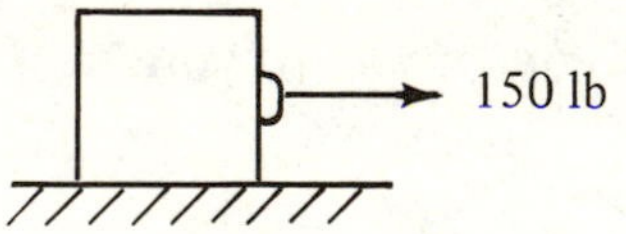

(c) The free body diagram is shown below. The 150 lb force is the force of the rope on the crate; the force **W** is the force of the earth on the crate, or the weight of the crate; the force **N** is the force of the floor on the crate; **f** is the force of friction acting on the crate, which is opposite to the direction of motion. Note that **N** *is not* the reaction force to **W**. The reaction force to **W** is the force of the crate on the earth (upward). The reaction of **N** is the force of the crate on the floor (downward). These reaction forces are not indicated in the free body diagram. Since we are interested in the motion of the crate, we only need to concern ourselves with the action forces, or the external forces acting on the crate.

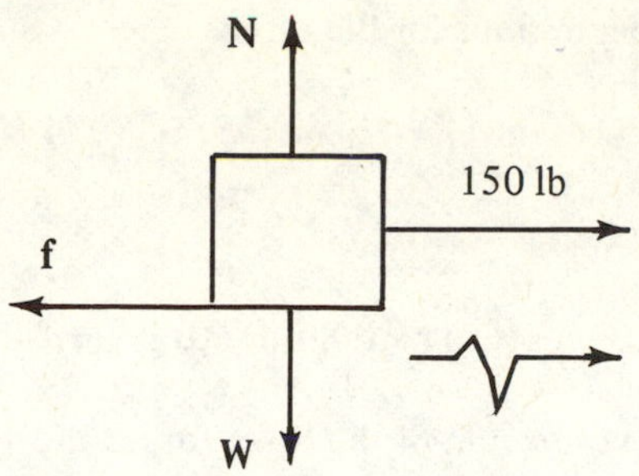

(d) Since there is no motion in the vertical direction, we see from the free body diagram that N = W = 192 lb. The *maximum* frictional force is given by

$$f = \mu_k N = \mu_k W = 0.5\ (192)\ \text{lb} = 96\ \text{lb}$$

Since this force is to the left, and the force of the rope on the crate is to the right, the resultant force on the crate is to the right.

$$\Sigma F \text{ on the crate} = 150\ \text{lb} - 96\ \text{lb} = 54\ \text{lb}$$

(e) From Newton's second law applied to the crate, we have

$$\text{Resultant force on crate} = \text{mass of crate} \times \text{acceleration of crate}$$

$$\Sigma F = ma$$

$$54\ \text{lb} = \left(\frac{192}{32}\ \text{slugs}\right) a$$

$$a = \frac{54\ \text{lb}}{6\ \text{slugs}} = 9\ \text{ft/sec}^2$$

where **a** is to the right.

Example 5.2

A flat plate weighing 4 lb is located on a frictionless horizontal surface and has three external forces applied to it as shown in Figure 5-2. Determine the acceleration of the plate assuming all forces act through the center of the plate.

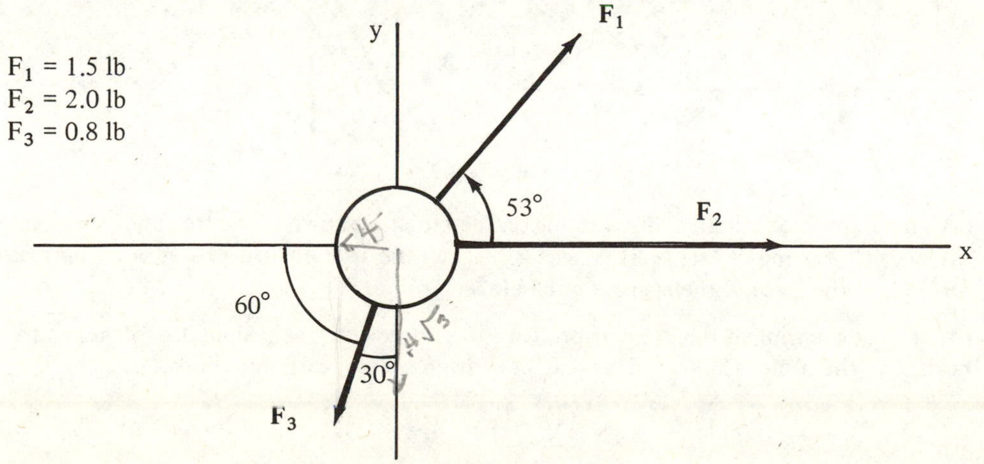

Figure 5-2

First, let us write vector expressions for the three forces.

$$\mathbf{F}_1 = 1.5 \cos(53)\,\mathbf{i} + 1.5 \sin(53)\,\mathbf{j} = (0.9\mathbf{i} + 1.2\mathbf{j}) \text{ lb}$$

$$\mathbf{F}_2 = 2.0\mathbf{i} \text{ lb}$$

$$\mathbf{F}_3 = -0.8 \cos(60)\,\mathbf{i} - 0.8 \sin(60)\,\mathbf{j} = (-0.4\mathbf{i} - 0.7\mathbf{j}) \text{ lb}$$

The resultant force is the *vector* sum of the three forces above:

$$\Sigma\mathbf{F} = \mathbf{F}_1 + \mathbf{F}_2 + \mathbf{F}_3 = (2.5\mathbf{i} + 0.5\mathbf{j}) \text{ lb}$$

Applying the second law, and noting that $m = W/g = \dfrac{4 \text{ lb}}{32 \text{ ft/sec}^2} = \dfrac{1}{8}$ slugs, we have

$$\Sigma\mathbf{F} = m\mathbf{a} = (2.5\mathbf{i} + 0.5\mathbf{j}) \text{ lb}$$

$$\mathbf{a} = (20\mathbf{i} + 4\mathbf{j}) \text{ ft/sec}^2$$

Example 5.3

A 3 kg block slides down a rough plane inclined at an angle of 37° with the horizontal as in Figure 5-3. The block accelerates from rest at the top of the incline to the bottom, a total distance of 2 m, and the coefficient of sliding friction is 0.2. (a) Draw a free body diagram for the block, identifying all the forces involved. (b) Write the equations of motion for the block and calculate the frictional force. (c) Determine the acceleration of the block. (d) Calculate the speed of the block when it reaches the bottom of the incline.

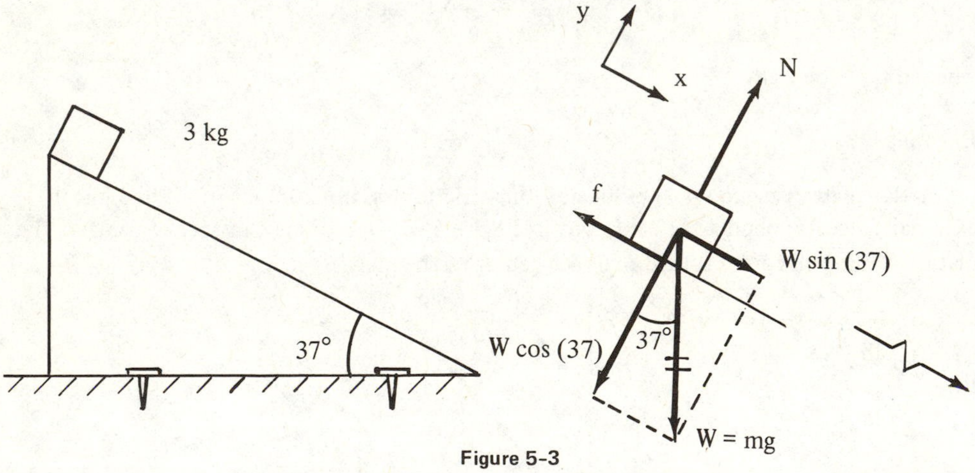

Figure 5-3

(a) There are three external forces acting on the block as shown in the free body diagram: the weight $W = mg = 3 \text{ kg } (9.8) \text{ m/sec}^2 = 29.4$.N, the frictional force f and the normal force N, or the force of the incline on the block.

(b) It is convenient in this type of problem to choose the x axis along the plane, and y normal to the plane. The second law can be written as two scalar equations,

$$(1)\quad \Sigma F_x = ma_x$$

$$(2)\quad \Sigma F_y = ma_y$$

In this example, we resolve the weight vector into components along x and y. Since $a_y = 0$, (1) and (2) become

$$(3) \quad W \sin(37) - f = ma_x$$

$$(4) \quad N - W \cos(37) = 0$$

Equations of Motion for the Block

Since the maximum frictional force $f = \mu_k N$, we have from (4),

$$(5) \quad f = \mu_k N = \mu_k \, W \cos(37) = (0.2)(29.4)(0.8)\,N = 4.7\,N$$

(c) We can solve for the acceleration using (3) and (5).

$$a_x = \frac{1}{m}[W \sin(37) - f] = \frac{1}{3\text{ kg}}[29.4\,(0.6)\,N - 4.7\,N]$$

$$a_x = \frac{12.9\text{ kg m/sec}^2}{3\text{ kg}} = 4.3\text{ m/sec}^2$$

(d) Since the acceleration is constant, we can obtain the speed of the block from the following expression for linear motion:

$$v^2 = v_0{}^2 + 2a_x\,(x - x_0)$$

In this case, $v_0 = 0$ at the top, and $x - x_0 = 2$ m; therefore, the speed at the bottom can be obtained directly.

$$v^2 = 2\,(4.3\text{ m/sec})\,2\text{ m} = 17.2\,\frac{\text{m}^2}{\text{sec}^2}$$

$$v = 4.1\text{ m/sec}$$

Example 5.4

Two blocks having masses of 2 kg and 3 kg are in contact on a fixed *smooth* inclined plane as in Figure 5-4. (a) Treating the two blocks as a composite system, calculate the force F that will accelerate the blocks up the incline with an acceleration of 2 m/sec²; (b) Draw free body diagrams for each block, and identify all the forces. (c) Calculate the contact force between the blocks, and show that the acceleration of *each* block is consistent with the resultant force acting on them.

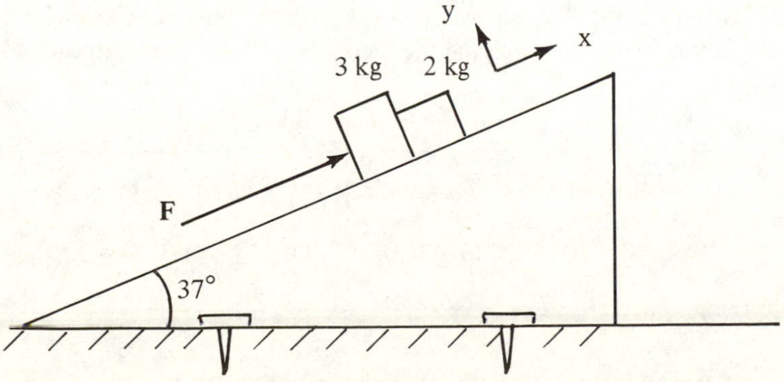

Figure 5-4

(a) We can replace the two blocks by an equivalent 5 kg block. Letting the x axis be along the incline, the resultant force *on the system* (the two blocks) in the x direction gives

$\Sigma F_x = F - W \sin(37) = ma_x$

$F - 5\,g\,(0.6) = 5\,(2)$ N

$F = 39.4$ N

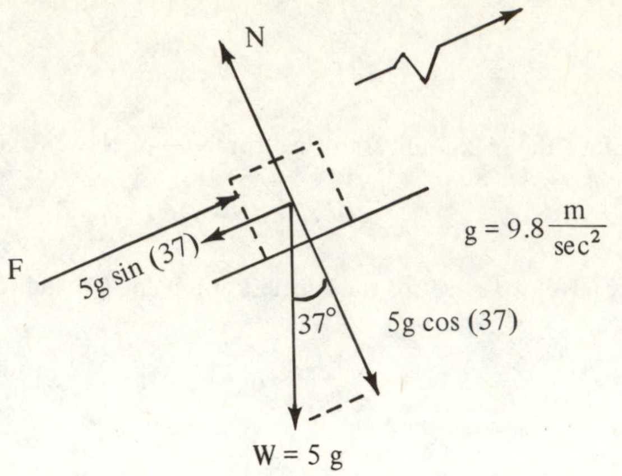

(b)

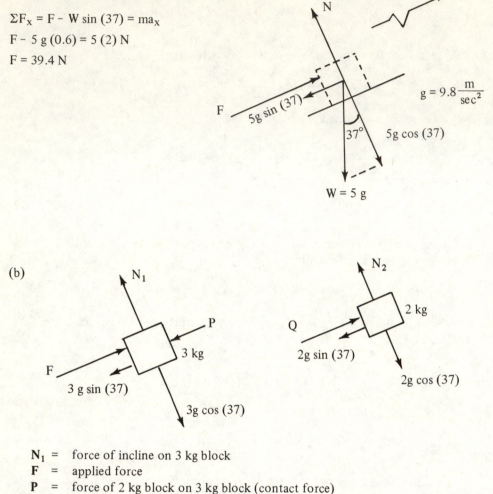

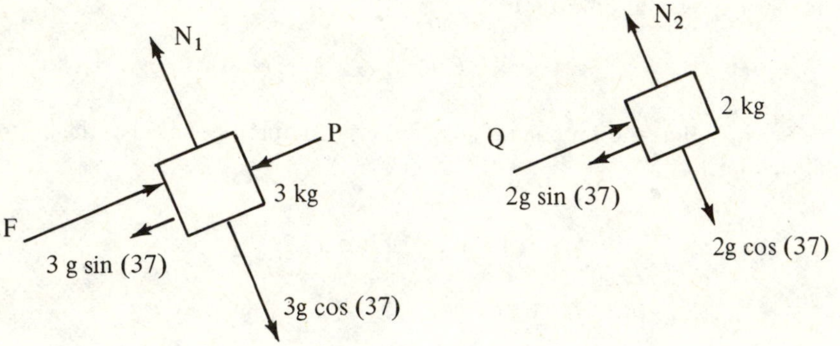

N_1 = force of incline on 3 kg block
F = applied force
P = force of 2 kg block on 3 kg block (contact force)
N_2 = force of incline on 2 kg block
Q = force of 3 kg block on 2 kg block
The remaining forces are the components of weight along the x and y axes.

(c) From Newton's third law, we would expect that P = Q. Let us check this. We can calculate the contact force by applying the second law, $\Sigma F_x = ma_x$, to each block:

3 kg	2 kg
$F - P - 3g \sin(37) = m_3 a_x$	$Q - 2\,g \sin(37) = m_2 a_x$
$39.4 - P - 3(9.8)(0.6) = 3(2)$	$Q - 2\,(9.8)(0.6) = 2\,(2)$
$P = (39.4 - 6 - 17.6)$ N	$Q = 15.8$ N
$P = 15.8$ N	or, $Q = P$

Therefore, the two calculations check. That is, the acceleration of *each* block considered separately is consistent with the resultant force acting on them. Note that it is the *combination of F and the counteracting contact force* which accelerates the 3 kg block, *while only the contact force accelerates* the 2 kg block.

Example 5.5

Two blocks are connected by a light string over a frictionless pulley as in Figure 5-5. The coefficient of sliding friction between m_1 and the surface is μ. Find the acceleration of the two blocks and the tension in the string.

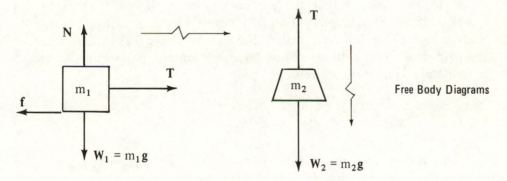

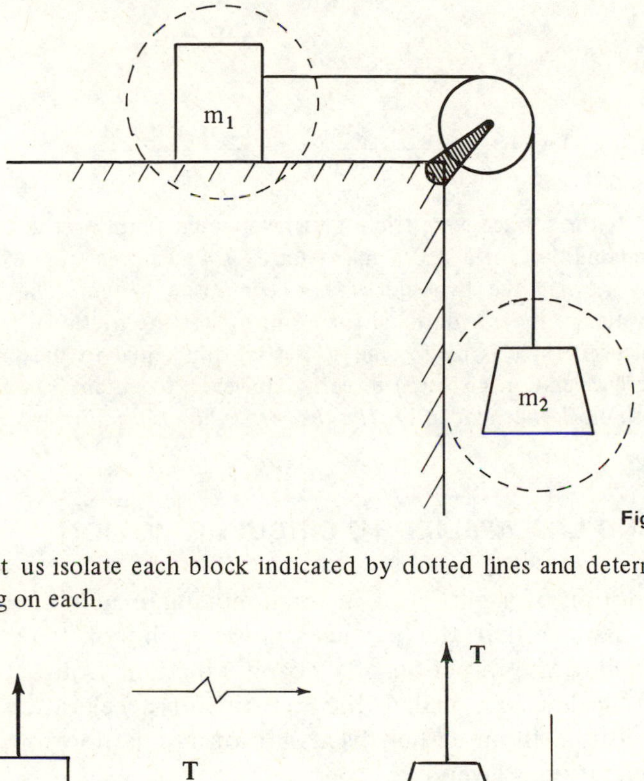

Figure 5-5

First, let us isolate each block indicated by dotted lines and determine the external forces acting on each.

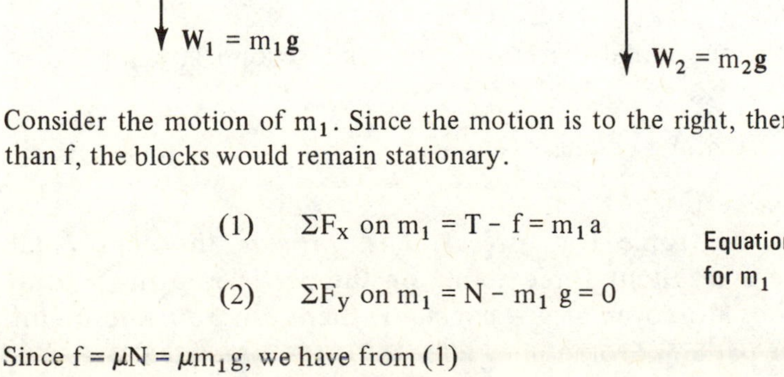

Free Body Diagrams

Consider the motion of m_1. Since the motion is to the right, then $T > f$. If T were less than f, the blocks would remain stationary.

(1) ΣF_x on $m_1 = T - f = m_1 a$

(2) ΣF_y on $m_1 = N - m_1 g = 0$

Equations of Motion for m_1

Since $f = \mu N = \mu m_1 g$, we have from (1)

(3) $T = m_1 (a + \mu g)$

For m_2, the motion is downward, therefore $m_2 g > T$. Note that the force T is uniform through the rope. That is, the force which accelerates m_1 to the right is also the force which keeps m_2 from falling freely. The equation of motion for m_2 is:

$$(4) \qquad \Sigma F_y \text{ on } m_2 = T - m_2 g = - m_2 a; \quad T = m_2(g - a)$$

where we have called the upward direction positive. Subtracting (4) from (3) gives

$$m_1 (a + \mu g) - m_2 (g - a) = 0$$

$$(5) \qquad a = \left(\frac{m_2 - \mu m_1}{m_1 + m_2}\right) g$$

Substituting (5) into (4) gives T.

$$(6) \qquad T = m_2 \left(1 - \frac{m_2 - \mu m_1}{m_1 + m_2}\right) g = \frac{m_1 m_2 (1 + \mu) g}{m_1 + m_2}$$

Comments: If the surface were frictionless, we would simply set $\mu = 0$ in (5) and (6). Of course, the condition for $a > 0$ is $m_2 > \mu m_1$, as we can see from (5). We could also obtain (5) by viewing the two blocks as a composite system. The "net force" accelerating the system is the gravitational force on m_2 less the frictional force on m_1. That is, the unbalanced force is $m_2 g - \mu m_1 g$. Setting this equal to the "mass of the system" times the acceleration gives (5) directly. However, to obtain T, we would still have to analyze each block independently. *The first procedure is usually preferred.*

5.3 NEWTON'S SECOND LAW APPLIED TO CIRCULAR MOTION

Now consider the motion of a particle of mass m moving in a *circular orbit.* For simplicity, let us first assume that the particle moves with constant speed. Its velocity is *not* constant since the direction of v changes in time. Consequently, the particle undergoes an acceleration, called the centripetal acceleration, directed towards the center of motion. In an earlier discussion of circular motion, we found that the centripetal acceleration is given by

$$\mathbf{a}_r = - \frac{v^2}{r} \hat{r}$$

where $\hat{r}$ is a *unit vector* along the radial direction. Newton's second law applied to this motion gives

$$\Sigma F \text{ along } \hat{r} = m a_r = - \frac{mv^2}{r} \hat{r} \qquad (5.6)$$

In other words, some external force (or forces) must *provide* the centripetal acceleration. There *must* be a resultant force acting on the particle having a component in the radial direction. Moreover, if v = constant, there can be *no* resultant force tangent to the circular path. We sometimes refer to forces directed along $-\hat{r}$ as *central forces.* Such forces can be provided by gravity, by strings attached to a whirling mass, or by electrostatic forces that exist between charged particles. Let us consider these cases separately. In Examples (6) and (7) we will deal with circular motion of particles having a constant speed v. In such cases, the angular acceleration $\alpha = 0$; therefore, the tangential acceleration $\mathbf{a}_t = 0$, and the particles only have a centripetal acceleration $\mathbf{a}_r$.

Example 5.6

A satellite of mass m moves in a circular orbit around the earth at an altitude h above the earth's surface, with constant speed v. The force of *attraction* between any two gravitational masses is given by $F = Gm_1m_2/r^2$, where G is the universal gravitational constant and r is the separation of m_1 and m_2. (a) Draw a free body diagram for the satellite. (b) Write Newton's law of motion for the satellite and, from this, determine v in terms of G, h, the radius of the earth R and the mass of the earth M.

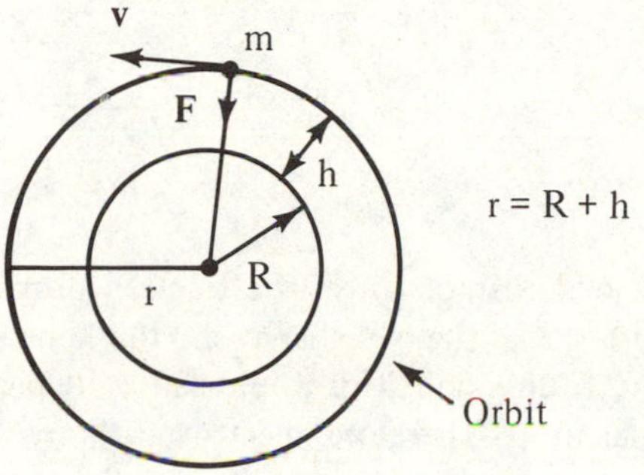

(a) Since gravitational forces are *always* attractive,

$$\mathbf{F} = -\frac{GmM}{(R+h)^2}\hat{r}$$

(b) Applying Equation (5.6), we have

$$\Sigma\mathbf{F} \text{ in the radial direction} = -\frac{GmM}{(R+h)^2}\hat{r} = -\frac{mv^2}{(R+h)}\hat{r}$$

Solving for v gives

$$v = \sqrt{\frac{GM}{R+h}}$$

Note: The gravitational force **F** provides the centripetal acceleration.

Example 5.7

The hydrogen atom consists of a light negatively charged particle, the electron, moving approximately in a circular orbit about a heavy positively charged particle, the proton. Charged particles are known to exert an electrostatic force between each other given by $\mathbf{F} = (kq_1q_2/r^2)\hat{r}$, where k is a constant, q_1 and q_2 are the charges on each particle and r is their separation. The mass of the electron is taken to be m_e, and its speed is a constant equal to v. The charges of the electron and proton are equal in magnitude, opposite in sign. Hence, we take $q_1 = -e$, $q_2 = +e$. (a) Find the speed of the electron in terms of k, r, m_e and e. (b) What force provides the centripetal acceleration of the electron?

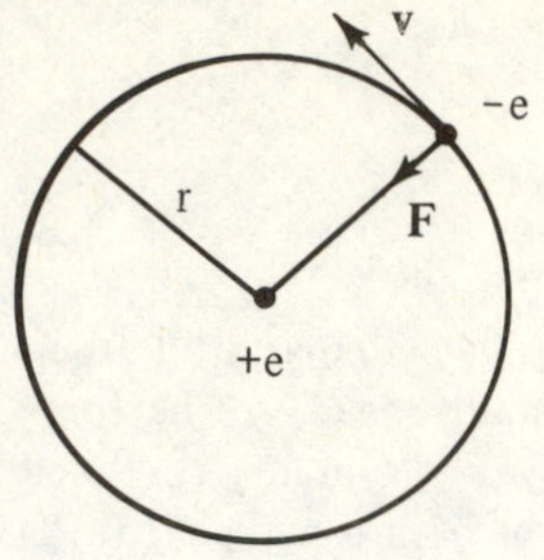

(a) The force on the electron is

$$\mathbf{F} = -\frac{ke^2}{r^2}\hat{r} = -\frac{m_e v^2}{r}\hat{r}$$

$$\therefore \qquad v = \sqrt{\frac{ke^2}{m_e r}}$$

(b) In this case, the electrostatic force of attraction provides the centripetal acceleration of the electron. Note that if the two charges are the same sign, **F** would be *repulsive;* that is, **F** would be in the $+\hat{r}$ direction. This would cause the particles to fly apart; hence, there would be no circular orbit. Therefore, electrostatic forces can be attractive *or* repulsive depending on the signs of the charges, while gravitational forces are *always* attractive.

Now let us consider a particle moving in a circular orbit in such a way that the speed v is *not* constant in time. This can only occur if there is a component of force along the direction tangent to the path of the particle, or a so-called *tangential force.* Such a force provides a tangential acceleration $\mathbf{a}_t$ and an angular acceleration α.

Example 5.8

A ball of mass m attached to the end of a string is whirled in a *vertical* circular path about a fixed point as in Figure 5-6. The radius of the circle is R, and gravity acts on the ball. (a) Draw a free body diagram for the ball. (b) Write Newton's equations of motion for the radial direction and for the tangential direction. (c) Calculate the tension in the string when the ball makes an angle θ with the vertical and has a speed v. At what point along the path is T a maximum? Where is T a minimum? (d) What force (or forces) provide the centripetal acceleration?

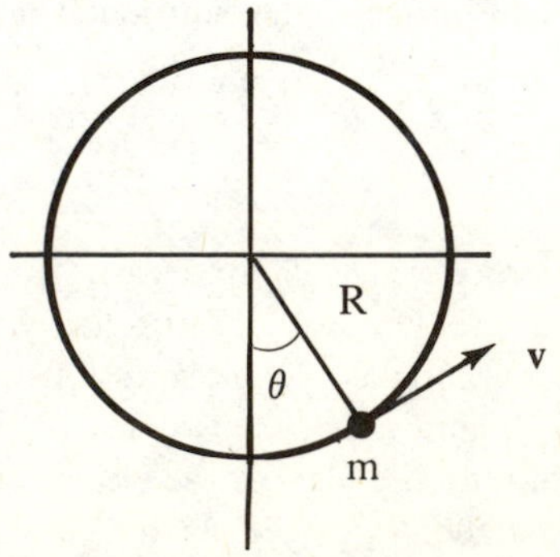

Figure 5-6

(a) $\Sigma\mathbf{F} = \mathbf{T} + \mathbf{W} = m\mathbf{a}$

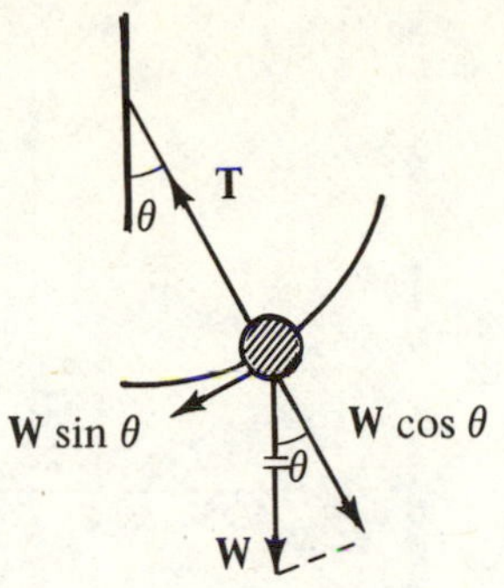

(b) $\Sigma\mathbf{F}$ along $\hat{r} = \mathbf{T} + \mathbf{W}\cos\theta = m\mathbf{a}_r$. But $\mathbf{T}$ is along $-\hat{r}$, and $\mathbf{W}\cos\theta$ is along $+\hat{r}$; therefore,

$$(W\cos\theta - T)\,\hat{r} = -\frac{mv^2}{R}\,\hat{r}$$

$$T = W\cos\theta + \frac{mv^2}{R} \qquad (1)$$

$$\Sigma\mathbf{F} \text{ along } \hat{\theta} = \mathbf{W}\sin\theta = m\mathbf{a}_t$$

But W sin θ is along $-\hat{\theta}$ ($+\hat{\theta}$ is in clockwise direction). Since W = mg, we have

$$a_t = -g\sin\theta \qquad (2)$$

The negative sign signifies that the gravitational force is tending to bring the mass back to its position of stable equilibrium, $\theta = 0$. At this point, $a_t = 0$. Note, that $a_t = -g$ at $\theta = \pi/2$ and at $\theta = 3\pi/2$, where the ball is in a horizontal position.

(c) The tension T is given by (1). T is a maximum when cos θ is a maximum. This occurs when $\theta = 0$, where $\cos\theta = 1$. That is, T is a maximum when the ball is at its lowest position. The speed v is also a maximum at $\theta = 0$. T is a minimum when cos θ is a minimum, that is, when $\cos\theta = -1$. This corresponds to $\theta = \pi$, or when the ball is at the top of the circle. The speed v is a minimum at this angle.

(d) From (1), we see that the centripetal acceleration v^2/R is provided by the force of tension **T** *and* the component of weight along $\hat{r}$.

5.4 PROGRAMMED EXERCISES

1.A

A force **F** is applied to a block on a *rough* surface as shown below. Draw a free body diagram for the block.

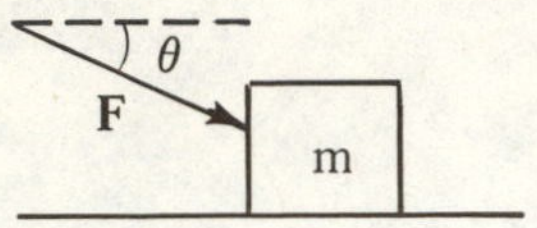

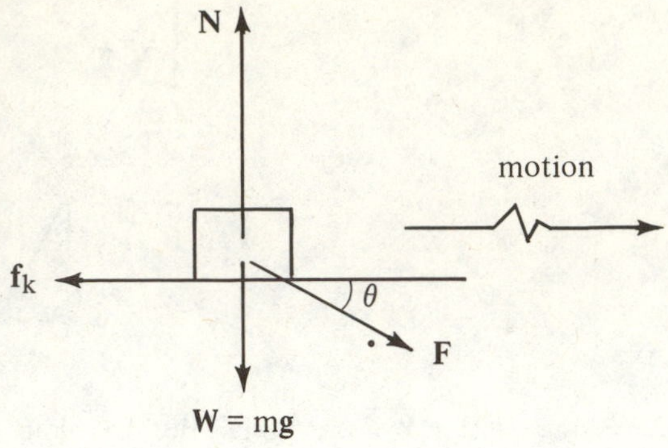

1.B

What are the horizontal and vertical components of F?

$F_x = F\cos\theta$ (to right)

$F_y = -F\sin\theta$ (down)

1.C

Apply Newton's second law for motion in the x direction.

ΣF_x on m $= F\cos\theta - f_k = ma_x$ (1)

1.D

What information does ΣF_y give in this case? Compare your answer to the case where **F** is horizontal.

ΣF_y on m $= N - mg - F\sin\theta = 0$

$$N = mg + F\sin\theta$$

N is *larger* than the weight mg in this case, since **F** has a component in the negative y direction.

1.E

Using the result of 1.D, determine the force of kinetic friction and compare your result with the case when **F** is horizontal.

$f_k = \mu N = \mu(mg + F\sin\theta)$ (2)

So f_k is *greater* than μmg by an amount $\mu F\sin\theta$. It is *important* to note that forces of friction can be greater or less than μmg if there is another force component normal to the surface other than mg.

1.F

Determine the acceleration of the block using (1) and (2).

Substituting (2) into (1) gives

$$a_x = \frac{1}{m}[F\cos\theta - \mu(mg + F\sin\theta)]$$

$$a_x = \frac{1}{m}[F(\cos\theta - \mu\sin\theta) - \mu mg]$$

2.A

Two weights are connected by a light string over a frictionless pulley as shown, where $W_1 > W_2$. Draw a free body diagram for each weight.

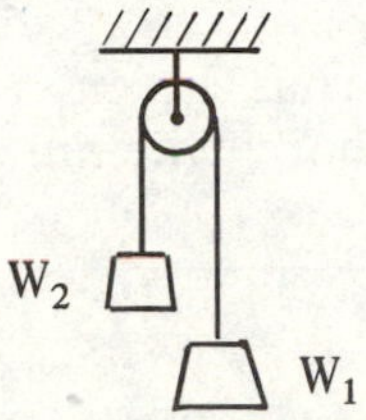

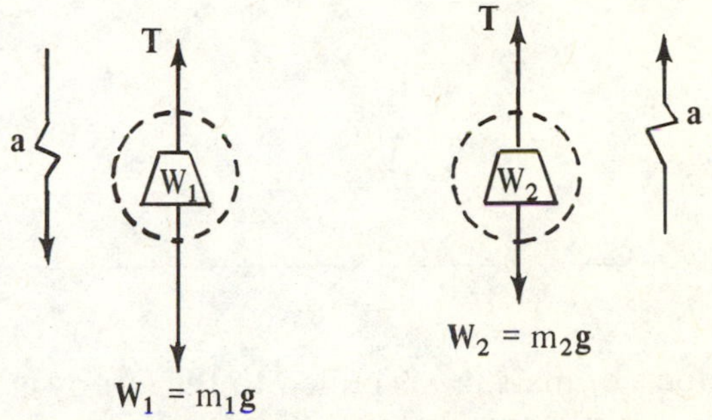

2.B

Knowing that $W_1 > W_2$, determine the direction of **a** for each weight.

a is *down* for W_1 (or $W_1 > T$)

a is *up* for W_2 (or $T > W_2$)

2.C

Write vector expressions for Newton's second law applied to each weight.

$\Sigma\mathbf{F}$ on $W_1 = \mathbf{T} + \mathbf{W}_1 = m_1\mathbf{a}$

$\Sigma\mathbf{F}$ on $W_2 = \mathbf{T} + \mathbf{W}_2 = m_2\mathbf{a}$

2.D

Write the equations of motion for W_1 and W_2 in component form, making use of the free body diagrams, and the result of 2.B.

ΣF_y on $W_1 = T - W_1 = -m_1 a$ (1)

ΣF_y on $W_2 = T - W_2 = m_2 a$ (2)

$\Sigma F_x = 0$ for both weights

2.E

Calculate the acceleration of the weights using (1) and (2) and the fact that

$$W_1 = m_1 g, \quad W_2 = m_2 g$$

Subtracting (1) from (2) gives

$$W_1 - W_2 = (m_2 + m_1)a$$

$$a = \left(\frac{m_1 - m_2}{m_1 + m_2}\right) g \qquad (3)$$

2.F

Determine the tension in the string using (3) and (1), [or (3) and (2)].

Substituting (3) into (1) gives

$$T = W_1 - m_1 a = m_1 g - m_1 \left(\frac{m_1 - m_2}{m_1 + m_2} \right) g$$

$$T = \frac{2\, m_1 m_2}{m_1 + m_2}\, g \qquad (4)$$

2.G

If $m_1 \gg m_2$, what value does a approach? What physical meaning does this result have?

From (3), we can neglect m_2, so

$$a \rightarrow \left(\frac{m_1 - 0}{m_1 + 0} \right) g = g$$

That is, m_1 behaves like a freely falling body when $m_1 \gg m_2$.

3.A

A block of mass m_1 is pulled to the *left* by a force **F** as shown below. The block is on a *rough* surface and is connected to a second block of mass m_2, over a pulley. Draw free body diagrams for m_1 and m_2.

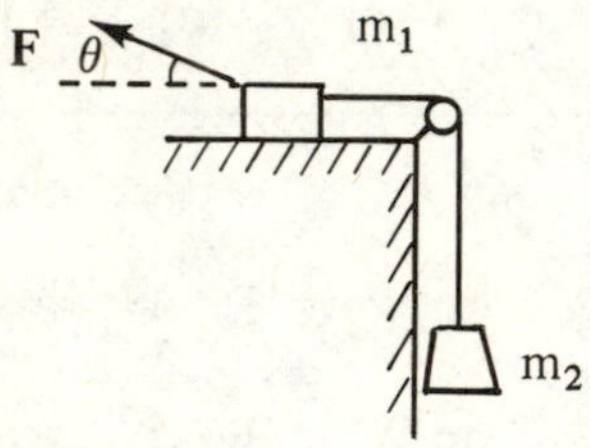

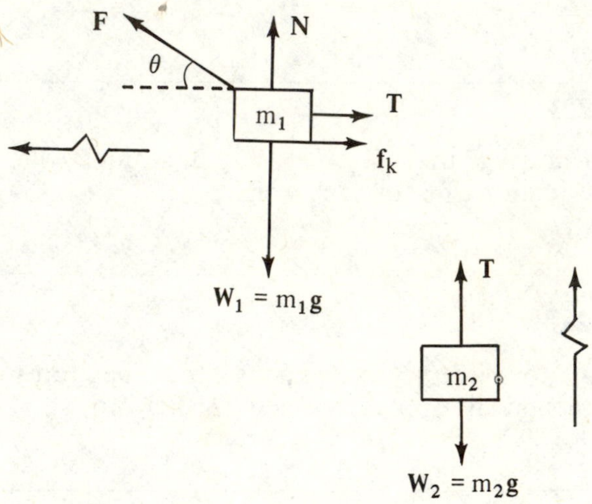

3.B

What are the horizontal and vertical components of F.

$F_x = -\, F \cos\theta$

$F_y = F \sin\theta$

3.C

Write an equation of motion for m_1 in the x direction, and for m_2 in the y direction.

$$\Sigma F_x \text{ on } m_1 = T + f_k - F\cos\theta = -m_1 a \qquad (1)$$

$$\Sigma F_y \text{ on } m_2 = T - m_2 g = m_2 a \qquad (2)$$

Note: **a** is to the left for m_1, **a** is up for m_2.

3.D

Write an equation for ΣF_y for m_1, and find f_k.

ΣF_y on $m_1 = N + F \sin\theta - m_1 g = 0$

$f_k = \mu N = \mu(m_1 g - F \sin\theta)$ (3)

3.E

Find the acceleration of the two blocks using the results of 3.C and 3.D.

Subtract (2) from (1)

$$f_k - F\cos\theta + m_2 g = -(m_1 + m_2)a \quad (4)$$

Now use (3) in (4), solve for a. The rest is algebra.

3.F

Find the tension in the string.

Substitute the value of a form (4) into (2), and solve for T.

4.A

A system of weights is assembled as shown. The top blocks accelerate to the right with an acceleration a. All surfaces are rough. Draw free body diagrams for m_3 and for m_1 and m_2 considered as a unit.

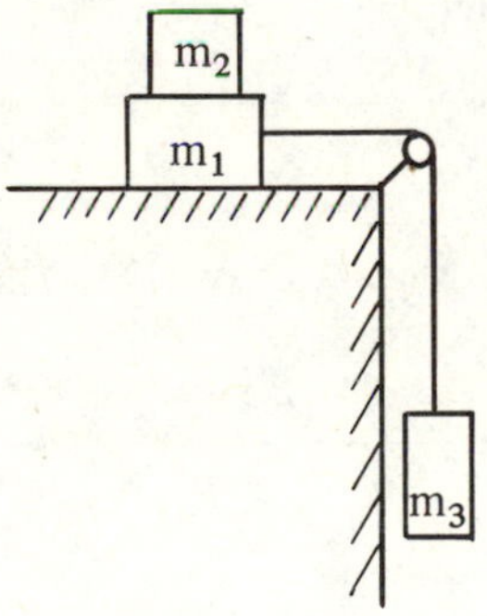

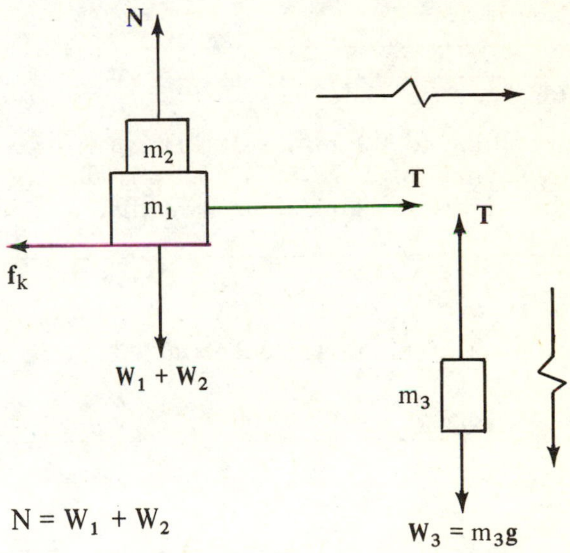

$N = W_1 + W_2$

$f_k = \mu(W_1 + W_2)$

4.B

Write equations of motion for m_1 and m_2 as a system and for m_3.

ΣF_x on $(m_1 + m_2) = T - f_k = (m_1 + m_2)a$ (1)

ΣF_y on $m_3 = T - m_3 g = -m_3 a$ (2)

(m_3 accelerates down, so **a** is negative)

4.C

Using the fact that $f_k = \mu(W_1 + W_2)$, find a and T. That is, solve (1) and (2) simultaneously.

Subtracting (2) from (1) gives a,

$$a = \frac{m_3 - \mu(m_1 + m_2)}{m_1 + m_2 + m_3} g \qquad (3)$$

From (2),

$$T = \frac{m_3 (m_1 + m_2)(1 + \mu)}{m_1 + m_2 + m_3} g \qquad (4)$$

4.D

Draw free body diagrams for m_1 and m_2 separately. Explain any "new" forces involved.

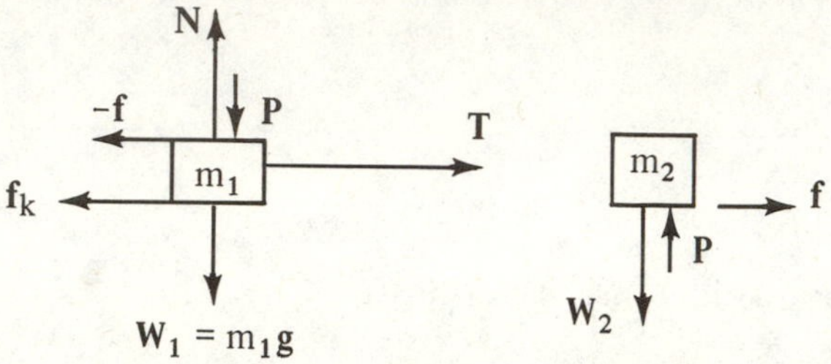

P = contact force between m_1 and m_2

f = force of static friction between m_1 and m_2

4.E

According to the results of 4.D, what force accelerates m_2? Is it T? This is not an obvious result. Think about it carefully.

No. The force that accelerates m_2 is **f**, so from the second law, ΣF_x on $m_2 = f = m_2 a$ where a is given by (3).

4.F

What is the contact force P equal to?

From 4.D, we see that $P = m_2 g$ (from either diagram).

5.A

A coin rests on a record revolving at a constant angular speed ω. The coefficient of static friction is μ, and the coin has a mass m. Draw a free body diagram for the coin.

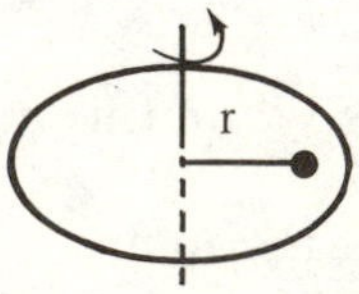

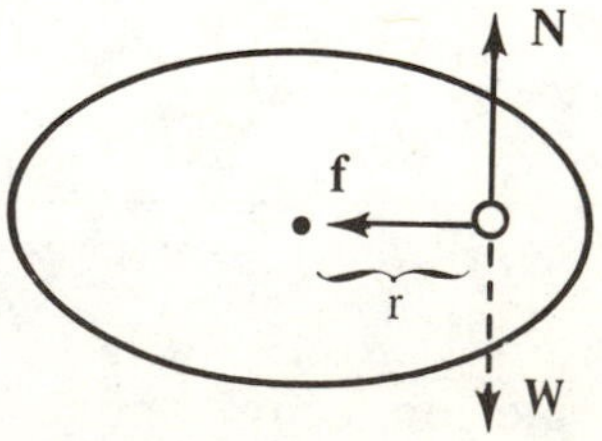

f = force of static friction

N = W = mg

5.B

If the coin is at a distance r from the center, what is the linear speed of the coin?

$v = r\omega$

5.C

What is the centripetal acceleration $\mathbf{a}_r$ and the tangential acceleration $\mathbf{a}_t$ of the coin?

$$\mathbf{a}_r = -\frac{v^2}{r}\hat{r}$$

$\mathbf{a}_t = 0$ since ω = constant

5.D

Write Newton's second law describing the motion of the coin. What force provides $\mathbf{a}_r$?

$\Sigma\mathbf{F}$ along $\hat{r} = \mathbf{f} = m\mathbf{a}_r$

$$-f\hat{r} = -\frac{mv^2}{r}\hat{r}$$

$$f = \frac{mv^2}{r} = mr\omega^2 \qquad (1)$$

f provides the centripetal acceleration

5.E

Suppose ω could be varied. Explain how you could experimentally find a value for μ for a fixed r. Your result for μ should be *independent* of m.

When the coin is on the *verge of slipping* $f = \mu N = \mu mg$; therefore this can be used in (1) to give $\mu = r\omega^2/g$. So we can slowly increase ω until the coin slips. The value of ω at which the coin first sips gives μ.

5.F

What other technique can be used to get μ?

You can keep ω constant, and increase r until the coin is on the verge of slipping. This value of r, together with a known ω can be used in the expression $\mu = r\omega^2/g$ to get μ.

5.5 SUMMARY

The *first law of motion* states that a body at rest or in uniform motion (constant velocity) will maintain that state unless an external force acts on the body.

The *second law of motion* states that the acceleration of a body of constant mass m equals the *resultant* force acting on it divided by its mass. That is,

$$\Sigma F = ma \tag{5.1}$$

The *third law of motion,* or the law of action-reaction, states that if body A exerts a force **F** on body B, then body B exerts an equal and opposite force −**F** on body A.

The force of kinetic friction between an object and another surface it moves on is given in magnitude by the expression

$$f_k = \mu_k N \tag{5.5}$$

where μ_k is a dimensionless constant called the coefficient of kinetic friction and N is the normal force. The force of *static friction* has a *maximum* value given by $f_s = \mu_s N$ which acts when the object is about to move. The constant μ_s is called the *coefficient of static friction,* and generally $\mu_s > \mu_k$. Forces of friction *always* oppose the motion of the object.

5.6 PROBLEMS

1. A block starts from rest from the top of a frictionless, inclined plane whose length is 2 m. The plane makes an angle of 10° with the horizontal. Find (a) the acceleration of the block after it is released, (b) the time it takes to reach the bottom, and (c) the speed of the block at the bottom.

2. A horizontal force of 15 N acts on a block situated on a rough horizontal surface. If the block has a mass of 2 kg, and the coefficient of kinetic friction is 0.4, determine the acceleration of the block.

3. Two masses m_1 and m_2 lying on a rough horizontal surface, are connected by a light string as in Figure 5-7. The coefficients of kinetic friction between each block and the surface is μ. If a force F acts to the right to cause an acceleration of the system, determine (a) the acceleration of each block, and (b) the tension in the string.

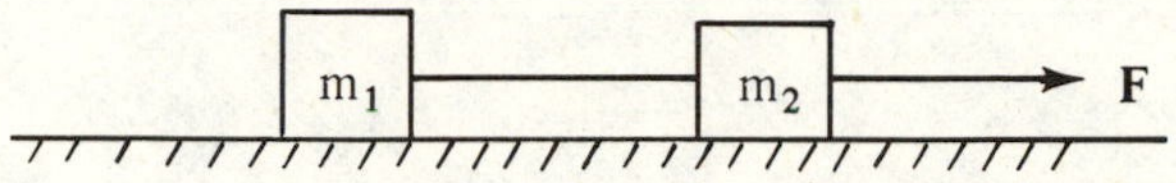

Figure 5-7

4. A block weighing 20 lb is placed on an inclined plane and is connected to a 10 lb block by a light string over a frictionless pulley as in Figure 5-8. When the angle of inclination is 53°, the 20 lb block moves *down* the plane with an acceleration of 4 ft/sec^2. (a) Draw free body diagrams for the two blocks. (b) Find the tension in the string. (c) Determine the coefficient of kinetic friction between the 20 lb block and the incline.

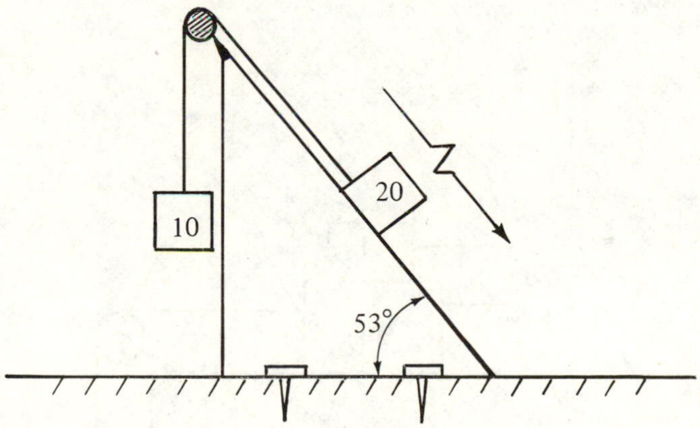

Figure 5-8

5. Two blocks of masses m_1 and m_2 are connected by a light string over a frictionless, light pulley as in Figure 5-9. The coefficient of kinetic friction between the blocks and the surface is μ and $m_1 > m_2$. (a) Draw free body diagrams for each mass. (b) Find the acceleration of the blocks and the tension in the string.

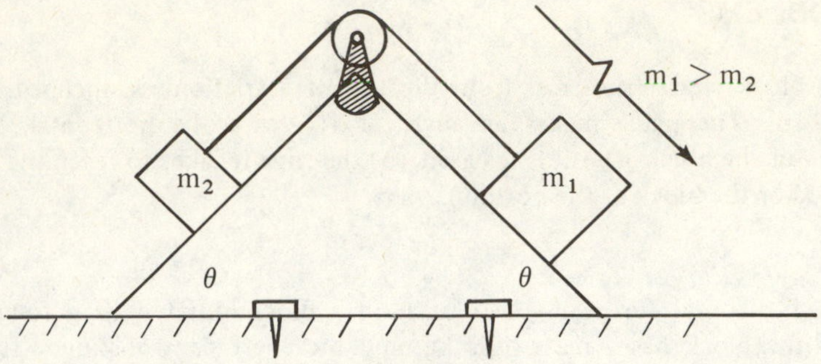

Figure 5-9

6. A 2 kg block is placed on top of a 5 kg block as in Figure 5-10. The coefficient of kinetic friction between the 5 kg block and the surface is 0.05. (a) Draw free body diagrams for both blocks if a force **F** is applied to the 5 kg block as shown. (b) Calculate the force necessary to pull the blocks to the right with an acceleration of 3 m/sec^2. (c) What force accelerates the 2 kg block? (d) Find the minimum coefficient of friction between the two blocks such that the 2 kg block does not slip for this value of the acceleration.

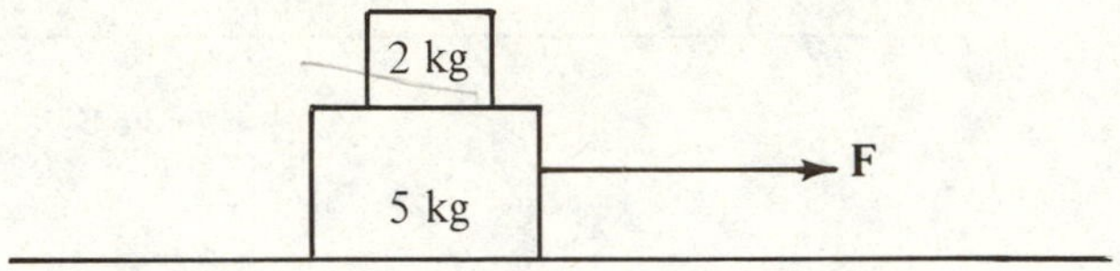

Figure 5-10

7. A 2 lb block is placed on a 3 lb block, and a string is attached between the two as in Figure 5-11. A 10 lb block is attached to the 3 lb block by a string over a frictionless pulley. The coefficient of kinetic friction between all surfaces is 0.30. (a) Draw free body diagrams for the three blocks. (b) Determine the acceleration of all the blocks. (c) Find the tensions in the strings.

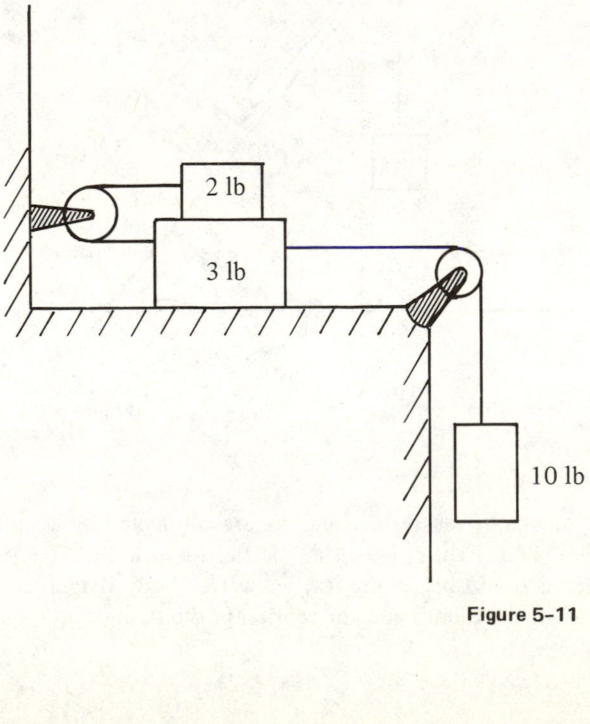

Figure 5-11

***8.** A daring young boy named Clyde wants to reach an apple in a tree the hard way. He sits in a chair which is connected to a rope over a frictionless pulley as in Figure 5-12. He then pulls on the loose end of the rope in such a way that the spring scale reads 60 lb. Clyde's true weight is 64 lb., and the chair weighs 32 lb. (a) Draw a free body diagram for Clyde and the chair as separate systems. (b) Write equations of motion for Clyde and the chair. (c) Show that the acceleration is *upward,* and find its value. (d) Find the force that Clyde exerts on the chair. (e) Show that the same acceleration is obtained if one considers Clyde and the chair as *one* system.

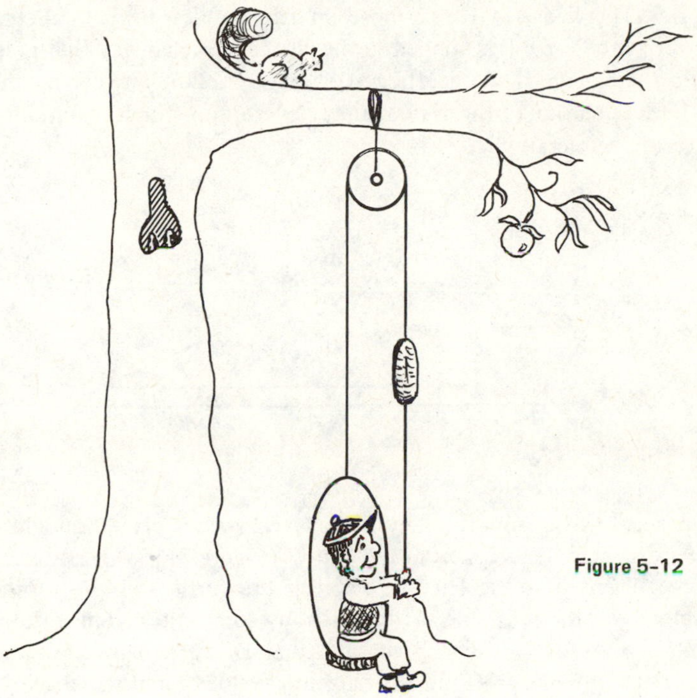

Figure 5-12

9. A bowling ball attached to a spring scale is suspended from the ceiling of an elevator. The scale reads 16 lb when the elevator is at rest, which corresponds to the "true" weight of the blowling ball. (a) If the elevator accelerates *upwards* at the rate of 8 ft/sec^2, what will the scale read? (b) If the elevator accelerates *downwards* at the rate of 8 ft/sec^2, what will the scale read? (c) If the supporting rope can withstand a maximum tension of 25 lb, what is the maximum acceleration the elevator can have before the rope breaks? (d) If the spring scale weighs 5 lb which rope breaks first? Why?

10. An entrance ramp of a highway (Figure 5-13) is designed for a speed of 30 miles per hour. The radius of curvature of the entrance ramp is R = 200 ft. (a) Show that the proper banking angle θ is given by $\tan \theta = v^2/Rg$. (b) Determine θ for these conditions.

Figure 5-13

11. A car moves with an initial velocity v_0 down an inclined road, where the angle of inclination to the horizontal is θ. The driver slams on his breaks at some instant. The coefficient of friction between the tires and road is μ. Assume the tires never skid, and the frictional force is a maximum. (a) Find the deceleration of the car. (b) Determine the distance the car will move before coming to rest after the brakes are applied. (c) Repeat parts (a) and (b) when θ is a *very* small angle. (d) If v_0 is 60 mph, $\theta = 10^\circ$ and $\mu = 0.6$, find numerical values for the deceleration and the distance traveled.

***12.** A mass m rests on a rough, inclined face of a wedge of mass M, and angle of inclination θ. The wedge is free to move on a frictionless horizontal surface as in Figure 5-14. A horizontal force F is applied to the wedge in such a way that m is on the *verge* of slipping *up* the incline. If the coefficient of static friction between m and the wedge is μ, (a) find the acceleration of the system and (b) determine the horizontal force F necessary to produce this acceleration.

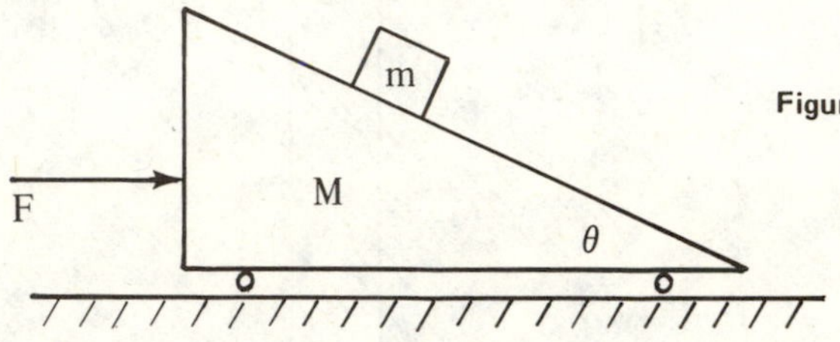

Figure 5-14

13. A small turtle, appropriately named "Dizzy", is placed on a horizontal rotating turntable a distance of 20 cm from the center (Figure 5-15). Dizzy has a mass of 50 g, and the coefficient of static friction between the turtle and the turntable is 0.3. The angular velocity of the turntable can be varied by some external means. (a) Determine the *maximum* value of the angular speed such that Dizzy remains stationary at this point relative to the turntable. (b) Find the tangential velocity and radial acceleration of Dizzy when he is on the verge of slipping.

Figure 5-15

14. A 4 kg mass is placed on a horizontal turntable, and is connected to a string of length 0.5 m, which, in turn, is fastened to the center of the table. The string can support a mass of 10 kg before breaking, and the coefficient of static friction is 0.6. If the *maximum* friction force acts on the mass when the turntable rotates, what is the angular speed of the turntable when the string is on the verge of breaking?

6

WORK AND THE ENERGY PRINCIPLE

In this chapter, we will treat a number of problems using the concepts of work and energy. As we shall see, the energy of a mechanical system is a very useful quantity for describing the motion of a system. In fact, the energy approach alludes to the application of Newton's second law, and is especially useful when the force or forces acting on the system are not constant. The energy method is also useful in the treatment of problems in electricity and magnetism, atomic and nuclear physics. It is one of the most powerful tools that the physicist has at his disposal in solving relatively complex problems. The usefulness for its utility lies in the fact that energy can take on various forms, and there is a conservation theorem that links these various forms together.

6.1 WORK DONE BY A CONSTANT FORCE

The concept of *work* as defined by the physicist is perhaps best understood by first defining the work done by a constant force **F** on a body. Later, we shall see that work and energy are related to each other by an important theorem. If a body undergoes a displacement **s** under the action of a *constant* force **F**, the work done by the *constant* force **F** is defined to be

$$W = \mathbf{F} \cdot \mathbf{s} \qquad (\text{constant } \mathbf{F}) \tag{6.1}$$

Since W is defined by a *scalar* product, it is itself a *scalar* quantity. Its units are N-m, dyne-cm and ft-lb in SI, cgs and engineering units, respectively. These systems of units, and the definitions of the joule (J) and erg are given in Table 6–1. It is a simple exercise to show that 1 J = 10^7 ergs.

TABLE 6–1 ENERGY UNITS IN THE THREE COMMON SYSTEMS

System	Unit of Work	Name of Combined Unit
SI	Newton-meter (N-m)	Joule (J)
cgs	dyne-centimeter (dyn-cm)	erg
Engineering	pound-foot (lb-ft)	foot-pound (ft-lb)

Equation (6.1) states that the work done by a constant force **F** is nonzero if the force has a component along the direction of the displacement vector **s**. Therefore, when **F** is perpendicular to **s**, the work done by **F** is *zero*. Of course, if the body does not move under the action of **F**, then W is also zero since s = 0. For computational purposes, it is useful to write Equation (6.1) in the form

$$W = Fs \cos \theta \tag{6.2}$$

where θ is the angle between **F** and **s**.

Example 6.1

A force of 10 N is applied to a crate at an angle of 37° with the horizontal (Figure 6-1). The crate moves a total distance of 5 m. (a) What is the work done by the 10 N force?

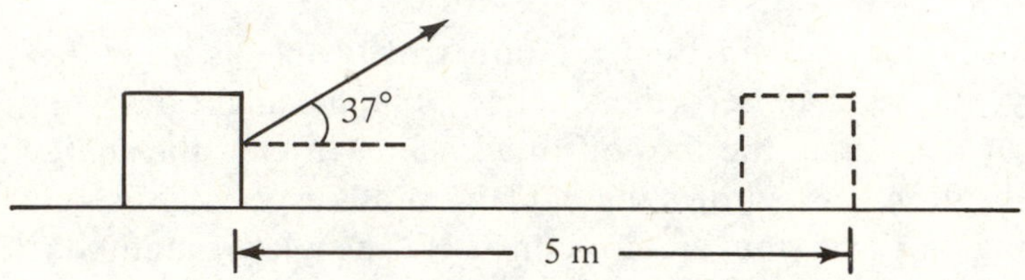

Figure 6-1

Solution

$$W = Fs \cos \theta = 10\text{ N }(5\text{m}) \cos (37)$$

$$W = 40\text{ Nm} = 40\text{ J}$$

Alternate Solution

We could also write **F** = 10 cos (37) **i** + 10 sin (37) **j** and **s** = 5**i**, and apply Equation (6.1). The results are the same. Try it!

Comments: We see that the calculation of the work done by a constant force **F** requires only a knowledge of **F** and the displacement **s**. However, in this problem there are other forces acting on the crate. If the surface is frictionless, there are two other forces, namely the weight **W** and the normal force **N**. However, the work done by these forces is *zero* since they are both ***perpendicular*** to **s**. In other words, for **N**, $\theta = \pi/2$, and since $\cos (\pi/2) = 0$, W = 0. Similarly for **W**, $\theta = -\pi/2$ and W = 0.

(b) Suppose that in this example a frictional force of 5 N is acting on the crate in the negative x direction. What is the work done by this force?

Solution

In this case, the frictional force is in the opposite direction to **s** as it always is, since friction always opposes the motion of an object. Using Equation (6.2), gives

$$W = fs \cos (\pi) = -fs = -5\text{N }(5\text{m})$$

$$W = -25\text{ Nm} = -25\text{ J}$$

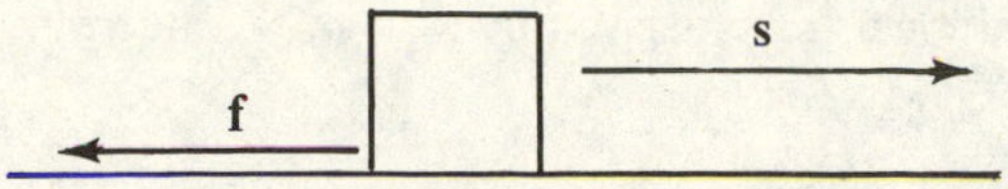

Alternate Solution

f = −5**i**, **s** = 5**i**, W = **f**·**s** = −5**i**·5**i** = −25J. That is, the frictional force does *negative* work on the crate. This is equivalent to saying the crate does work on the environment, since the surface heats up during the motion, and so forth. Whenever $\theta < \pi/2$, W is positive or **F** has a component along **s**, whereas when $\theta > \pi/2$, W is negative, meaning **F** has a component in the direction −**s**.

Example 6.2

A 3 kg block is pushed up an inclined plane a distance of 2 m by a horizontal force of 40 N (Figure 6-2). The coefficient of sliding friction is 0.1 and the angle of inclination is 37°. (a) What is the work done by the 40 N force? (b) What is the work done by gravity? (c) What is the work done by friction?

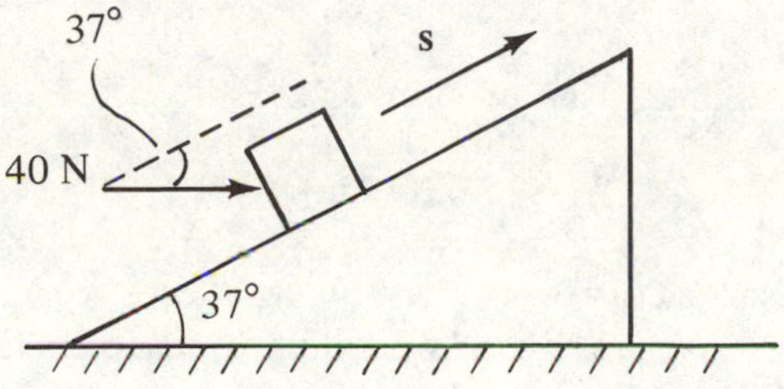

Figure 6-2

Solution

(a) The displacement **s** is up the plane and equals 2 m. The horizontal 40 N force has a component 40 cos (37) along the incline, in the direction of **s**, therefore

$$W_F = Fs \cos(37) = 40\text{ N }(2\text{ m})(0.8) = 64\text{ J}$$

(b) The weight vector **W** = m**g** can be resolved into components perpendicular and parallel to the incline. The component perpendicular to the plane does *no* work, since it is also perpendicular to **s**, so W = 0. The component 3 g sin (37) points down the incline and opposes **s**; hence $\theta = \pi$, and the work done by gravity is

$$W_g = Fs \cos\theta = 3\text{ g} \sin(37)(2)\cos(\pi)$$

$$W_g = -3\,(9.8)\text{ N }(0.6)(2)\text{ m}$$

$$W_g = -35\text{ J}$$

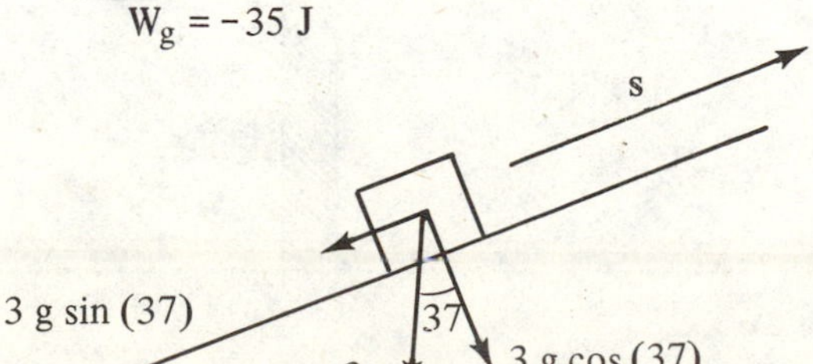

(c) The maximum frictional force is given by $f_k = \mu_k N$. But from the free body diagram we see that

$$N = 3\ g \cos(37) + 40 \sin(37) = 56\ N$$

Therefore, f_k becomes

$$f_k = \mu_k N = 0.1\ (56) = 5.6\ N$$

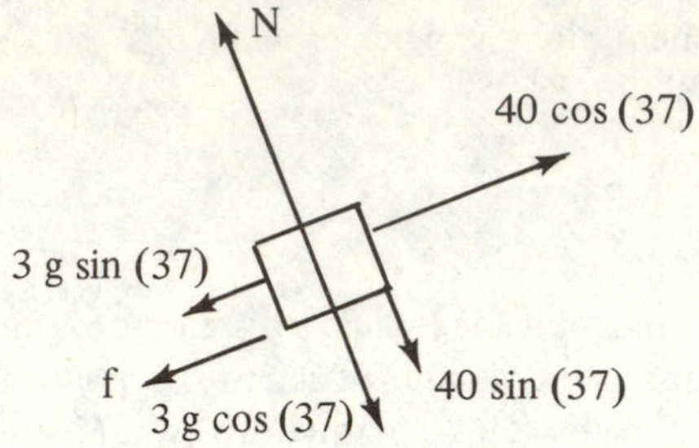

Since f_k is opposite to s, $\theta = \pi$, and the work done by f_k is

$$W_f = f_k s \cos(\pi) = -f_k s$$

$$W_f = -5.6\ N\ (2\ m) = -11\ J$$

Note that we have calculated the work done by each force separately, even though all forces act on the block at once. The *total* work done by *all* forces is the sum of parts (a), (b) and (c), or $W_{tot} = 18$ J.

6.2 GENERAL DEFINITION OF WORK

Consider a body in motion, where the forces acting on it change as the particle moves from one point to another. When this is the case, one cannot use Equations (6.1) and (6.2) to calculate the work done. Suppose that the displacement of the body is along the x axis, and the force acting on the body is also in the x direction, where F_x varies with the position x according to Figure 6-3.

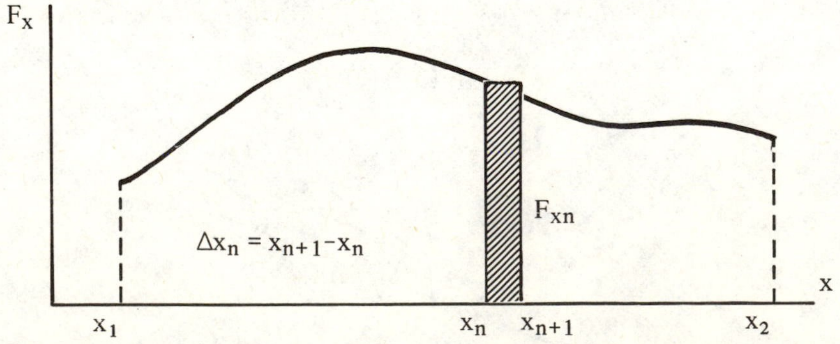

Figure 6-3 The work done by the variable force F_x is the area under the curve F_x *vs* x from $x = x_1$ to $x = x_2$.

First, we see that the work done by F_x for a small displacement Δx is $F_x \Delta x$. Therefore, if we divide the curve of F_x *vs* x into small thin strips of equal width Δx_n as shown, the work done for each displacement depends on the value of F_x at that point. Therefore, the work done by F_{xn} in displacing the body an amount Δx_n is

$$\Delta W_n = F_{xn} \Delta x_n$$

Now if we add up all the areas of all these intervals from x_1 to x_2, we get an approximate expression for the total work done by F for the displacement x_1 to x_2.

$$W = \Sigma \Delta W_n = \sum_n F_{xn} \Delta x_n \tag{6.3}$$

As the intervals Δx_n are made smaller and smaller, the work calculated according to Equation (6.3) involves a larger number of terms, but the result *approaches* the *true* area under the curve bounded by F_x and the x axis. Taking the limit of Equation (6.3) as $\Delta x_n \to 0$ gives, by definition, an integral quantity for W which *is* the true area under the curve F_x *vs* x from x_1 to x_2.

$$W = \lim_{\Delta x_n \to 0} \sum_n F_{xn} \Delta x_n = \int_{x_1}^{x_2} F_x \, dx \tag{6.4}$$

Note that Equation (6.1) is a *special case* of Equation (6.4). That is, if F_x is constant, it can be taken outside of the integral expression, and the value of W becomes

$$W = F_x \int_{x_1}^{x_2} dx = F_x (x_2 - x_1) = F_x s$$

where s is the displacement along the x axis, and $F_x = F \cos \theta$. Therefore, Equations (6.1) and (6.4) are equivalent when F is constant.

If a situation involves an arbitrary force **F** and corresponding vector displacements $\Delta \mathbf{r}_n$, we can generalize Equation (6.4) to include two or three dimensional problems. In those situations, we calculate W using the expression

$$W = \int_a^b \mathbf{F} \cdot d\mathbf{r} \tag{6.5}$$

where the limits a to b represent the initial and final coordinates of the particle. The integral given by Equation (6.5) is sometimes referred to as a line integral; however, it can only be evaluated if the variation of **F** with position is known. It is a simple exercise to show that Equation (6.5) reduces to

$$W = \int_{x_1}^{x_2} F_x \, dx + \int_{y_1}^{y_2} F_y \, dy + \int_{z_1}^{z_2} F_z \, dz \tag{6.6}$$

where x_1, y_1, z_1 are the initial coordinates and x_2, y_2, z_2 are the final coordinates of the object. Comparing Equation (6.6) with Equation (6.4), we see that the general situation simply includes the work done by **F** for displacements along x, y, and z.

Example 6.3

A mass m lies on a horizontal frictionless surface and is connected to a spring whose force constant is k, as shown in Figure 6-4(*a*). The equilibrium position of the mass is taken to be at $x = 0$. The force required to move the mass a distance x from its equilibrium position is given by $F = kx$. Calculate the work done by the external force F in displacing m from $x = 0$ to the position x_0.

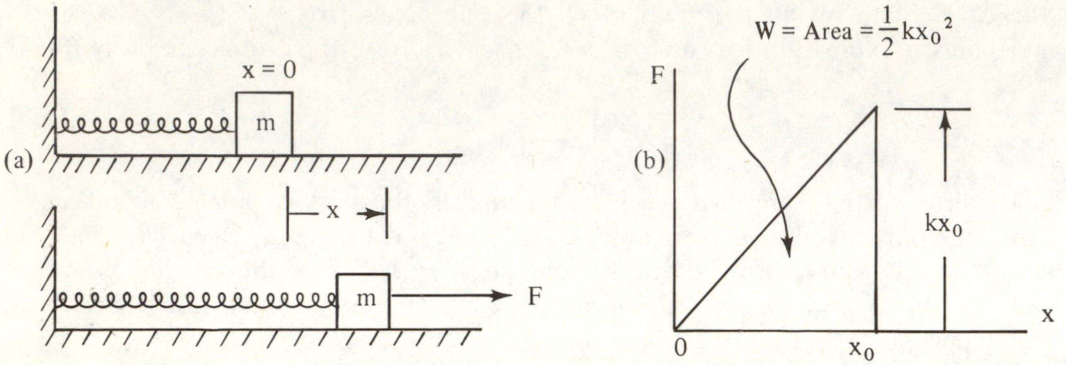

Figure 6-4 (*a*) An example of a system where the force depends on the particle displacement. In this example, the spring force obeys Hooke's law, that is, $F_s = -kx$. The external force F must be at least equal and opposite to F_s to displace m from equilibrium. (*b*) The work done in stretching the spring is the area under the F *vs* x curve.

We can apply Equation (6.4), using the limits $x_1 = 0$ to $x_2 = x_0$. This gives

$$W = \int_0^{x_0} F\,dx = \int_0^{x_0} kx\,dx = k\int_0^{x_0} x\,dx = \frac{1}{2}kx_0^2$$

Note that this result is simply the area under the curve F *vs* x, in this case a triangle whose height is $F = kx_0$ and base equal to x_0 [Figure 6-4(*b*)]. The quantity $\frac{1}{2}kx_0^2$ is sometimes referred to as the *elastic potential energy* of the spring. In this calculation, it is important to note that the initial and final velocities of the mass are assumed to be *zero*. We will generalize the situation later to include the possibility of a change in velocity of m when an external force is applied. When this is the case, the work done by F is not simply equal to the elastic potential energy.

6.3 RELATION BETWEEN WORK AND KINETIC ENERGY

The *kinetic energy* K of a particle whose mass is m and whose velocity is v (where we assume $v \ll c$ in the non-relativistic limit) is defined to be

$$K = \frac{1}{2}m\mathbf{v}\cdot\mathbf{v} = \frac{1}{2}mv^2 \qquad (6.7)$$

This definition implies that kinetic energy is a form of energy associated with the motion of a body.

Kinetic energy, like work, is a scalar quantity and has the same units as work. When the resultant force acting on a body is nonzero, obviously the body accelerates (or decelerates), its velocity changes; therefore, its kinetic energy changes. It is left as an exercise to prove that the *work done by the resultant force on a body equals the change in kinetic energy of the body.* This statement, known as the work-energy theorem, is a very useful concept in mechanics. We can write the theorem in the following way:

$$W = \int \mathbf{F} \cdot d\mathbf{r} = K - K_0 \tag{6.8}$$

where **F** is the resultant force acting on the object, K and K_0 are the final and initial kinetic energies.

$$K = \frac{1}{2}mv^2; \quad K_0 = \frac{1}{2}mv_0{}^2 \tag{6.9}$$

Sometimes, we will write Equation (6.8) as

$$W = \Delta K \tag{6.10}$$

It follows from the work-energy theorem that the kinetic energy of a body *increases* when W is *positive*, while the kinetic energy *decreases* when W is *negative.* This is not surprising, since a positive W corresponds to a resultant force in the direction of the displacement, and an increase in velocity. When W is negative, the resultant force is opposite to the displacement, the body slows down and loses kinetic energy. Of course, when W = 0, corresponding to a resultant force of zero, the change in kinetic energy is zero.

Example 6.4

In this example, we will show that Equation (6.8) is satisfied for a particle undergoing a *constant acceleration* as the result of a constant resultant force. The work done by a constant force F_x in displacing a particle at distance x is given by Equation (6.1).

$$W = F_x x = max$$

The following relations apply for linear motion with constant acceleration:

$$x = \frac{1}{2}(v + v_0)t$$

$$a = \frac{v - v_0}{t}$$

Substituting these into the expression for W gives

$$W = m\left(\frac{v - v_0}{t}\right)\frac{1}{2}(v + v_0)\,t = \frac{1}{2}mv^2 - \frac{1}{2}mv_0{}^2$$

Therefore, Equation (6.8) is satisfied. This example represents a proof of the work-energy theorem when the resultant force is constant; however, it also applies when **F** varies both in magnitude and direction (see Problem 25).

Example 6.5

A *constant* force of 5 N acts on a 4 kg mass. The mass starts from rest at the origin at t = 0, and moves a distance of 2 m. (a) Find the final velocity of the 3 kg mass using the work-energy theorem. (b) Verify your results to part (a) by applying the expressions for linear motion with constant acceleration.

(a) $W = F_x x = 5\text{ N} \times 2\text{ m} = 10\text{ J}$

$$\frac{1}{2}mv^2 = 10\text{J} \qquad (v_0 = 0)$$

$$v = \sqrt{\frac{2 \times 10\text{ J}}{4\text{ kg}}} = \sqrt{5\frac{\text{Nm}}{\text{kg}}} = \sqrt{5\frac{\text{kgm}^2/\text{sec}^2}{\text{kg}}} = \sqrt{5}\,\frac{\text{m}}{\text{sec}}$$

(b) $v^2 = \not{v_0^2} + 2\,ax = 2\,\frac{F_x}{m}x$

$$v^2 = 2\frac{(5\text{ N})}{4\text{ kg}}2\text{m} = 5\text{m}^2/\text{sec}^2$$

$$v = \sqrt{5}\text{ m/sec}$$

6.4 ELASTIC POTENTIAL ENERGY

Forces in nature can be divided into two categories, *conservative* and *non-conservative.* We say that a force is *conservative* if the work done by that force on a body for any closed path is *zero.* In addition, the work done by a *conservative force* in taking the body between two points a and b is *independent of the path* between a and b. Only the end points are important. Examples of conservative forces are the restoring force in a spring, gravitational forces and electrostatic forces between charged particles. *Nonconservative forces* are those forces for which the work done for any closed path is *nonzero.* Therefore, the calculation of the work done by a nonconservative force acting on a body that moves from a to b *depends on the path.* A common example of a nonconservative force is the force of friction.

First, let us consider the restoring force in a spring. The procedure for showing that this force is conservative is straightforward. Following Example 6.4, we can say that the work done by the external force $F = kx$ (equal and opposite to the restoring force) in moving m from x_0 to x is

$$W = \int_{x_0}^{x} F dx = \int_{x_0}^{x} kx dx = \frac{1}{2}kx^2 - \frac{1}{2}kx_0{}^2$$

Since the restoring force of the spring $F_x = -kx = -F$, the work of the restoring force during any process where the spring extends (or compresses) from x_0 to x is equal to $-W$.

$$W_s = -\left(\frac{1}{2}kx^2 - \frac{1}{2}kx_0{}^2\right) \tag{6.11}$$

Now for any closed path process, $x = x_0$; therefore, $W_s = 0$. Hence, we conclude that the restoring force of a spring is conservative. The quantity $\frac{1}{2}kx^2$ is referred to as the *elastic potential energy* of the spring, denoted by the symbol U(x).

$$U(x) = \frac{1}{2}kx^2 \tag{6.12}$$

Note that the elastic potential energy of a spring is *zero* when $x = 0$; that is, when the spring is in equilibrium. U(x) is a *maximum* when the spring has reached a maximum extension or maximum compression. In addition, U(x) is *always* positive since it is proportional to x^2, and k is a positive force constant. It is a scalar quantity, like work and kinetic energy, and has units of energy.

Now we can apply the work-energy theorem to the mass on the end of a spring. If we call W the work done by external forces, then the total work done on m is $W + W_s$. Applying Equation (6.10) gives

$$W + W_s = \Delta K$$

or

$$W = \left(\frac{1}{2}mv^2 - \frac{1}{2}mv_0{}^2\right) + \left(\frac{1}{2}kx^2 - \frac{1}{2}kx_0{}^2\right) \tag{6.13}$$

We can think of the last two terms in Equation (6.13) as the *change* in the elastic potential energy ΔU.

$$\Delta U = \frac{1}{2}kx^2 - \frac{1}{2}kx_0{}^2$$

If there are *no external forces* acting on the system, $W = 0$, Equation (6.13) becomes

$$\left(\frac{1}{2}mv^2 + \frac{1}{2}kx^2\right) - \left(\frac{1}{2}mv_0{}^2 + \frac{1}{2}kx_0{}^2\right) = 0 \tag{6.14}$$

This is a statement of the *conservation of mechanical energy* for the mass-spring system. The *total* mechanical energy E is the sum of the kinetic and potential energies.

$$E = \frac{1}{2}mv^2 + \frac{1}{2}kx^2 \tag{6.15}$$

Equation (6.13) can be written as

$$W = E - E_0$$

where E_0 and E are the initial and final *total* energies, respectively. Therefore, if W is positive, the mechanical energy increases; if W is negative, the mechanical energy decreases (for example, when friction is present); if $W = 0$, the mechanical energy remains constant or is conserved.

Example 6.6

A mass of 0.2 kg lying on a horizontal surface is attached to a spring whose force constant is 20 N/m, as in Figure 6-4(*a*). The mass is stretched from equilibrium by an external force to a distance of 20 cm and is then released from rest. (a) Assuming the surface is frictionless, calculate the speed of the mass when it reaches the equilibrium position.

As soon as the mass is released, there are *no* external forces acting on the mass-spring system. The force that accelerates m is the restoring force of the spring which is *internal* to the system. Therefore, we can apply Equation (6.14) using the conditions $x_0 = 20$ cm $= 0.2$ m, $x = 0$ and $v_0 = 0$.

$$\frac{1}{2}mv^2 = \frac{1}{2}kx_0{}^2$$

$$v^2 = \frac{k}{m}x_0{}^2 = \frac{20\ \text{N/m}}{0.2\ \text{kg}}(0.2\ \text{m})^2$$

$$v = 2\ \text{m/sec}$$

From an energy viewpoint, we can say that the total energy of the system is initially equal to the elastic potential energy stored in the spring; (the initial kinetic energy is zero since $v_0 = 0$). When the mass reaches the equilibrium position, $x = 0$, the elastic potential energy stored in the spring is zero, and the total energy of the system is the kinetic energy of the mass. Therefore, the speed we calculated above is the *maximum* speed of the mass. The total energy of the system remains *constant* during the motion. As the motion commences, energy is continually being transferred between elastic potential energy and kinetic energy. However, the energy of the system has to be supplied by the external force, which, of course, initiates the motion. For any arbitrary value of $|x| < 20$ cm, the total energy is a combination of kinetic and potential energy according to Equation (6.15).

(b) What is the speed of the 0.2 kg mass at $x = 0$ if a frictional force acts on it? Assume the coefficient of friction between the mass and surface is 0.5.

In this case, the total mechanical energy is *not* constant since the nonconservative force of friction acts on the system. Let us apply Equation (6.13) to this situation. The only *external* force acting on the mass-spring system is the force of friction, **f**, which opposes the motion *and* the displacement. Therefore, negative work is done by **f**, as in Example 6.1(b).

$$f = \mu_k N = \mu_k mg = 0.5\ (0.2\ \text{kg})\ (9.8\ \text{m/sec}^2) = 0.98\ \text{N}$$

Therefore, the work done by **f** in the displacement $x = 0.2$ m, to $x = 0$ is given by

$$W = -fx_0 = -0.98\ \text{N}\ (0.2\ \text{m}) \cong -0.20\ \text{J}$$

Substituting this into Equation (6.13) gives

$$-0.20\ \text{J} = \frac{1}{2}mv^2 - \frac{1}{2}kx_0{}^2$$

$$\frac{1}{2}mv^2 = \frac{1}{2}kx_0{}^2 - 0.20 = 0.40 - 0.20 = 0.20\ \text{J}$$

$$v^2 = \frac{2\ (0.20)\ \text{J}}{0.2\ \text{kg}} = 2\ \text{m}^2/\text{sec}^2$$

$$v = \sqrt{2}\ \text{m/sec}$$

This is *less* than the value obtained in part (a). The result should not be surprising, since one would expect the force of friction to retard the motion of the system.

6.5 GRAVITATIONAL POTENTIAL ENERGY

Let us consider the work done by the gravitational force. Suppose a particle of mass m is initially located at the vector position $\mathbf{r}_0$, and is then moved to position $\mathbf{r}$ under the action of several forces (Figure 6–5).

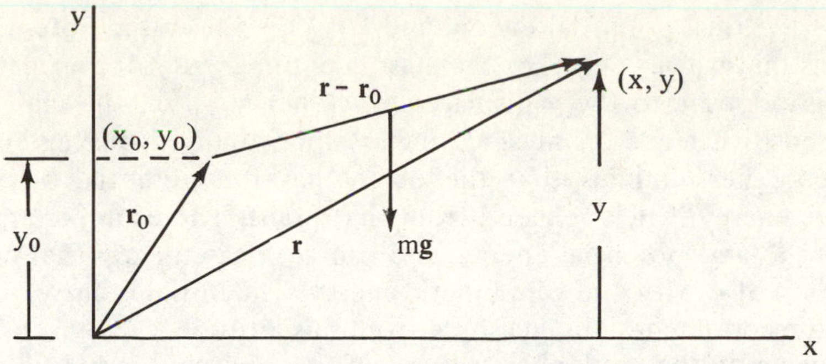

Figure 6-5 Displacement of a mass m from r_0 to r under the action of gravity.

For the time being, we will neglect variations in the acceleration of gravity with altitude. This is a good approximation if the displacement of the particle in the y direction is small compared to the earth's radius. Then the force of gravity is *constant* and is given by

$$\mathbf{F} = m\mathbf{g} = -mg\mathbf{j}$$

where we are calling y positive upwards. The work done by gravity when the particle moves from $\mathbf{r}_0$ to $\mathbf{r}$ along some arbitrary path is

$$W_g = \int_{\mathbf{r}_0}^{\mathbf{r}} \mathbf{F}\cdot d\mathbf{r} = \mathbf{F}\cdot\int_{\mathbf{r}_0}^{\mathbf{r}} d\mathbf{r} = \mathbf{F}\cdot(\mathbf{r} - \mathbf{r}_0)$$

Note that this result is *independent* of how the particle moves from $\mathbf{r}_0$ to $\mathbf{r}$. It only depends on $\mathbf{F}$ (which is constant) and the displacement vector $\mathbf{r} - \mathbf{r}_0$. Only the beginning and end points are important for calculating W in this case. We can write $\mathbf{r}_0$ and $\mathbf{r}$ as

$$\mathbf{r}_0 = x_0\mathbf{i} + y_0\mathbf{j}$$

$$\mathbf{r} = x\mathbf{i} + y\mathbf{j}$$

Therefore the work done by gravity is

$$W_g = \mathbf{F}\cdot\mathbf{r} - \mathbf{F}\cdot\mathbf{r}_0 = -mg\mathbf{j}\cdot(x\mathbf{i} + y\mathbf{j}) + mg\mathbf{j}\cdot(x_0\mathbf{i} + y_0\mathbf{j})$$

or,

$$W_g = mgy_0 - mgy \tag{6.16}$$

Let us define the *gravitational potential energy* as

$$U(y) = mgy \tag{6.17}$$

In this manner, Equation (6.16) can be written as

$$W_g = -\Delta U = -(mgy - mgy_0) \tag{6.18}$$

where U(y) is the final potential energy and $U(y_0)$ is the initial potential energy of m. This means that when $y > y_0$, or the mass goes up, gravity does negative work on m since mg is opposite to the y displacement; when $y < y_0$, the mass goes down, gravity does positive work on m since mg is along the y displacement. The term potential energy here implies that the object has the potential or capability of gaining kinetic energy if it is released from that point. Therefore, a large value of y corresponds to a large potential energy, and a mass released from that point will fall under the action of gravity and gain kinetic energy. The initial point y_0 can be taken to be zero if the coordinates are measured from this altitude.

Now let us apply the work-energy theorem to a body in the presence of gravity. We can write Equation (6.10) as

$$W + W_g = \Delta K \tag{6.19}$$

where W represents the work done on the body by all forces *other* than gravity. Substituting Equation (6.18) into Equation (6.19), we have

$$W = \Delta K + \Delta U \tag{6.20}$$

where $\Delta K = K - K_0$ and $\Delta U = U - U_0$. The *total* energy of the body E can be taken to be the sum of the kinetic and potential energies. That is,

$$E = K + U$$

Therefore, Equation (6.20) is sometimes written as

$$W = E - E_0 \tag{6.21}$$

Again, E and E_0 are the final and initial total energies of the body, respectively. If $W = 0$, the total energy of the body remains constant, that is, $E = E_0$. This is a statement of the *law of conservation of energy.*

Example 6.7

A particle of mass m falls from rest from a height h above the ground as in Figure 6-6. Determine the speed of the particle when it is at a height y above the gound.

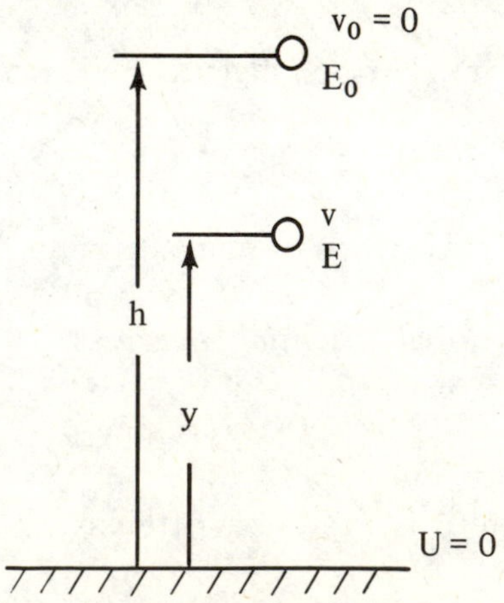

Figure 6-6

Solution

When the particle is at a height h above the ground, its kinetic energy $K_0 = 0$ and its potential energy $U_0 = mgh$; therefore, its total energy $E_0 = mgh$. When the particle is a distance y above the ground, its kinetic energy $K = \frac{1}{2}mv^2$ and its potential energy $U = mgy$. Therefore, its total energy $E = \frac{1}{2}mv^2 + mgy$. Since there are *no* forces other than gravity acting on m, $W = 0$ in Equation (6.21), and we have

$$E - E_0 = 0$$

$$\frac{1}{2}mv^2 + mgy - mgh = 0$$

$$v^2 = 2\,g(h - y)$$

Note that this result is consistent with one of the equations developed for linear motion with constant acceleration, with $v_0 = 0$. If the mass had an initial velocity v_0 at the initial altitude h, the analysis would be the same, but the initial energy E_0 would include kinetic $\frac{1}{2}mv_0{}^2$. In that case, we have

$$E - E_0 = \left(\frac{1}{2}mv^2 + mgy\right) - \left(\frac{1}{2}mv_0{}^2 + mgh\right) = 0$$

or

$$v^2 = v_0{}^2 + 2\,g(h - y)$$

The student should show that the same analysis can be used to obtain the speed of a *projectile* at any height h, when thrown from the ground with a speed v_0.

Example 6.8

A 2 kg block slides down a rough inclined plane 1 m in length as in Figure 6-7. The block starts from the top at rest and experiences a constant force of friction of 3 N. If the angle of inclination is 37°, determine the speed of the block when it reaches the bottom of the incline. Check your results using Newton's law of motion.

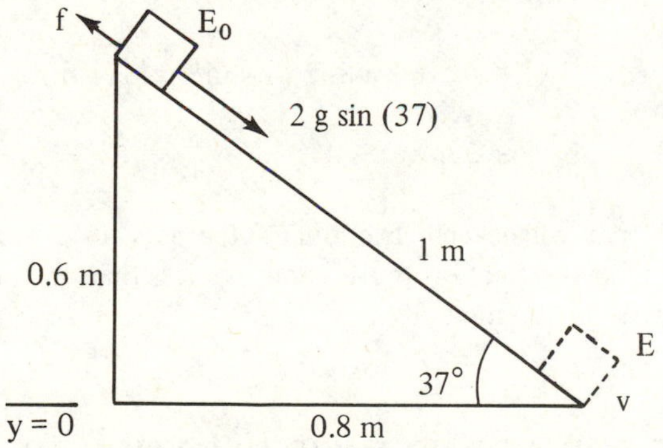

Figure 6-7

Solution

Since $v_0 = 0$, the initial kinetic energy is zero. Calling $y = 0$ the bottom of the incline, we have $y_0 = 0.6$ m. So the *total* energy at the top is potential energy and equals

$$E_0 = U_0 = mgy_0 = 2\text{ kg}\left(9.8\,\frac{\text{m}}{\text{sec}^2}\right)0.6\text{ m} = 11.8\text{ J}$$

When the block reaches the bottom, it has kinetic energy $\frac{1}{2}$ mv², but its potential energy is *zero* since we have called y = 0 the bottom of the incline. Therefore, $E = \frac{1}{2} mv^2$ at the bottom. But now we must be careful. $E \neq E_0$, since there is another force acting which does work on the block, namely the force of friction. [The forces normal to the incline do *no* work since they are perpendicular to the displacement.] Since f = 3 N, and **f** is *opposite* to the displacement vector s we have

$$W = -fs = -3 \text{ N} (1 \text{ m}) = -3 \text{ J}$$

That is, some mechanical energy is lost due to the presence of friction. Applying the work-energy theorem gives

$$W = E - E_0$$

$$-3 = \frac{1}{2} mv^2 - 11.8$$

$$\frac{1}{2} mv^2 = 8.8 \text{ J}$$

$$v^2 = \frac{17.8 \text{ J}}{2 \text{ kg}} = \frac{8.9 \cancel{\text{kg}} \text{ m}^2/\text{sec}^2}{\cancel{\text{kg}}}$$

or,

$$v \cong 3 \text{ m/sec}$$

Let us check this result using Newton's second law to get the acceleration.

$$\Sigma F \text{ along the plane} = mg \sin(37) - f = ma$$

$$a = g \sin(37) - \frac{f}{m} = \left[9.8(0.6) - \frac{3}{2}\right] \text{m/sec}^2$$

$$a = 4.4 \text{ m/sec}^2$$

Since a = constant, we have

$$v^2 = \cancel{v_0^2} + 2 as = 2(4.4 \text{ m/sec}^2)(1 \text{ m})$$

$$v \cong 3 \text{ m/sec}$$

We can conclude that whenever a frictional force acts on a body, the work done by friction is *negative*, so the final energy E is always less than the initial energy E_0 if there are no other forces acting on the body.

6.6 CONSERVATIVE FORCES AND THEIR RELATION TO POTENTIAL ENERGY

In the previous sections we have seen that the concept of potential energy is related to the configuration or *position* of the particle. The elastic potential energy of a stretched or compressed spring is proportional to the square of the elongation of the spring, while gravitational potential energy is proportional to the y coordinate of the body. However, both cases demonstrate that potential energy may be regarded as the energy associated with the *position* of a body. The potential energy of a mechanical system refers to that form of energy which can be converted into kinetic energy. However, we have seen that *both kinetic* and *potential energy* are *measures of a system's ability to perform work.* A potential energy function can be associated *only* with a conservative force. One *cannot* associate a potential energy function with nonconservative forces such as friction.

Suppose we consider a conservative force F acting in the x direction. From the definition of work, Equation (6.4), and its relation to the change in potential energy, Equation (6.18), we can write the work energy theorem as

$$W = \int_{x_0}^{x} F(x)dx = -\Delta U \tag{6.22}$$

This expression enables us to calculate the change in potential energy of a body undergoing a displacement from x_0 to x under the action of a *conservative* force F. Thus, if we know how F varies with position, we can calculate U. For any value of x, the *total mechanical energy* in one dimension is defined to be

$$E = \frac{1}{2}mv^2 + U(x) \tag{6.23}$$

If we are dealing with a three dimensional problem, then both v and U are functions of x, y and z.

Differentiating Equation (6.22), we find that

$$F(x) = -\frac{dU}{dx} \tag{6.24}$$

This expression says that the conservative force F equals the negative derivative of the potential energy function. This expression is valid *only* for *conservative* forces. We can check this expression immediately for the two examples already discussed. The potential energy of the stretched or compressed spring is

$$U(x) = \frac{1}{2}kx^2$$

Therefore,

$$F(x) = -\frac{dU}{dx} = -kx$$

This is the restoring force of the spring. The gravitational potential energy $U(y) = mgy$, therefore $F = -mg$, the gravitational force downwards.

Since $\Delta U = U(x) - U(x_0)$ in one dimension, we can rewrite Equation (6.22) as

$$U(x) = U(x_0) - \int_{x_0}^{x} F(x)dx \tag{6.25}$$

This expression says that the function U(x) can be defined only to within an *arbitrary* constant $U(x_0)$. In other words, the choice of origin for our potential energy function is arbitrary, therefore $U(x_0)$ is arbitrary. The only quantity that has physical meaning is the potential energy difference $U(x) - U(x_0)$. Sometimes we choose the origin of coordinates in such a way that $U(x_0) = 0$.

Finally, if the *only* force that acts on a body is the conservative force F(x) as it moves from x_0 to x, we have the important law of conservation of mechanical energy,

$$E - E_0 = 0 \tag{6.26}$$

or,

$$E = K + U(x) = \text{constant} \tag{6.27}$$

If there is more than one conservative force acting, we can still write an equation for conservation of mechanical energy but we would have to include a potential energy function for *each* conservative force.

6.7 THE INVERSE SQUARE LAW OF FORCE AND ASSOCIATED POTENTIAL ENERGY

Consider a particle of mass m at a distance r from a second particle whose mass is M. Newton's *law of universal gravitation* states that the force between these two masses is *attractive,* is inversely proportional to the square of their separation, and acts along r. We will deal with this law in more detail in a subsequent chapter. We can state the inverse square law of force as

$$\mathbf{F} = -\frac{GMm}{r^2}\hat{r}$$

where **F** is the force on m due to M, and is directed radially inward, as in Figure 6–8. Let us calculate the work done by **F** in moving m from $\mathbf{r}_0$ to **r**. Since dr points radially outward in the direction of $\hat{r}$, we can write equation (6.5) as

$$W = \int_{r_0}^{r} \mathbf{F} \cdot d\mathbf{r} = -GMm \int_{r_0}^{r} \frac{dr}{r^2} = +\frac{GMm}{r}\Bigg|_{r_0}^{r}$$

$$W = GMm\left[\frac{1}{r} - \frac{1}{r_0}\right] \tag{6.28}$$

Note that this result can also be obtained by taking m along the paths ab and bc shown in Figure 6–8. The work done by **F** along path bc is *zero*, since **F** is perpendicular to the displacement for this path. We could also choose to go from $\mathbf{r}_0$ to **r** along the path adc. In this case, the work done by **F** along ad is zero. The work done by **F** along dc equals the work done along ab, where the value of W is given above. Therefore, W is independent of the path taken between r_0 and r, and **F** is a conservative force.

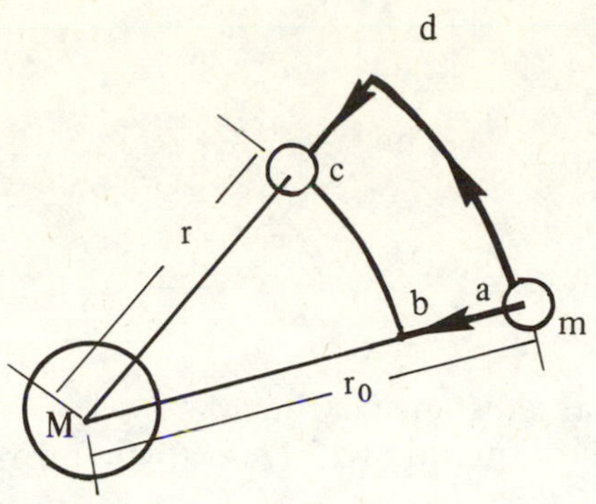

Figure 6-8 Initial and final positions of a mass m in the presence of a second mass M.

Now let us apply Equation (6.22) to this situation.

$$W = -\Delta U = -\ [U(r) - U(r_0)]$$

Solving for U(r), the potential energy at r, we have

$$U(r) = U(r_0) - GMm \left[\frac{1}{r} - \frac{1}{r_0}\right] \qquad (6.29)$$

It is customary to *choose* $U(\infty) = 0$, corresponding to $r_0 = \infty$. In that case, the potential energy of m at r becomes

$$U(r) = -\frac{GMm}{r} \qquad (6.30)$$

The function U(r), then, represents the work done by the gravitational force **F** as the mass m moves from ∞ to r.

Suppose M represents the earth, and m any particle above the earth's surface. The negative sign in Equation (6.30) says that the potential energy starts off at zero when $r_0 = \infty$ and becomes more and more negative for finite values of r. This is consistent with the fact that **F** is an attractive force and as U(r) becomes more negative, the kinetic energy of m *increases.* Since we are dealing with a conservative force, the total energy of the system remains constant. In other words, if we neglect the motion of M ($M \gg m$), the conservation of energy equation for the two particle system can be written as

$$E = \frac{1}{2} mv^2 + U(r) = \text{constant}$$

If m starts off at rest at $r_0 = \infty$, then $K_0 = 0$, and since $U(\infty) = 0$, $E_0 = 0$. So the equation above becomes

$$\frac{1}{2} mv^2 + U(r) = 0$$

or,

$$\frac{1}{2}mv^2 = \frac{GMm}{r}$$

$$v = \sqrt{\frac{2\,GM}{r}} \qquad (6.31)$$

where v is the speed of m at a distance r from M.

Finally, let us show that Equation (6.24) is satisfied for the inverse square law of force.

$$F = -\frac{dU}{dr} = -\frac{d}{dr}\left(-\frac{GMm}{r}\right) = -\frac{GMm}{r^2}$$

Again, this relationship *must* be satisfied if F is a conservative force. The gravitational potential energy function given by Equation (6.30) is a property of the *system,* meaning the combination of m and M. However, when $m \ll M$, as in the case of a light particle in the vicinity of the earth, we sometimes refer to U(r) as the potential energy of m, the light mass, since it acquires most of the kinetic energy when a change in separation occurs.

Example 6.9

What is the *minimum* speed which a particle must have at the earth's surface in order that it escape from the earth? This is known as *escape velocity*. The mass of the earth is $M = 5.98 \times 10^{24}$ kg, the earth's radius $R \cong 6.4 \times 10^6$ m, and the value of G is 6.67×10^{-11} N m^2/kg^2.

This can be answered by energy considerations. In order for m to reach $r = \infty$ with a final speed of zero, it must have enough initial kinetic energy such that its final kinetic energy at ∞ is *zero.* Since $U(\infty) = 0$, then the total energy must be zero. If the particle has a total energy greater than zero, then it will have some residual kinetic energy at $r = \infty$. Its total energy at the earth's surface is the sum of kinetic and potential energy, which must then add up to zero, since energy is conserved.

$$E = \frac{1}{2}mv_{esc}^2 - \frac{GMm}{R} = 0$$

$$v_{esc} = \sqrt{\frac{2\,GM}{R}}$$

$$v_{esc} = \left[\frac{2 \times (6.67 \times 10^{-11}\ \text{N m}^2/\text{kg}^2) \times (5.98 \times 10^{24}\ \text{kg})}{6.4 \times 10^6\ \text{m}}\right]^{1/2} \cong 1.1 \times 10^4\ \text{m/sec}$$

This can also be calculated by noting that at the surface of the earth, the gravitational force equals the weight of the particle, mg.

$$\frac{GMm}{R^2} = mg$$

or,

$$g = \frac{GM}{R^2} = 9.8 \text{ m/sec}^2$$

Therefore,

$$v_{esc} = \sqrt{2\,gR} = [2 \times 9.8 \text{ m/sec}^2 \times 6.4 \times 10^6 \text{ m}]^{1/2} = 1.1 \times 10^4 \frac{\text{m}}{\text{sec}}$$

6.8 POWER

Power is defined as the time rate at which energy is transferred from one system to another. If we denote power by the symbol P, we have

$$P = \frac{dE}{dt} \tag{6.32}$$

If there is a force **F** acting on the particle, then work is done in displacing the particle by an amount Δs. If the time interval for this displacement is Δt, then the *average power* supplied to the particle is

$$\langle P \rangle = \frac{W}{\Delta t} = \frac{\mathbf{F} \cdot \Delta \mathbf{s}}{\Delta t} \tag{6.33}$$

We also can define the instantaneous power P as the limit of the average power as $\Delta t \to 0$. Using the definition of instantaneous velocity, and Equation (6.33), we see that

$$P = \lim_{\Delta t \to 0} \mathbf{F} \cdot \frac{\Delta \mathbf{s}}{\Delta t} = \mathbf{F} \cdot \mathbf{v} \tag{6.34}$$

In SI units, power is expressed in watts (W), where

$$1 \text{ W} = 1 \text{ J/sec} = 1 \text{ kg m}^2/\text{sec}^3$$

Another common unit of power is the horsepower (hp), where

$$1 \text{ hp} = 550 \text{ ft lb/sec} = 746 \text{ W}$$

Example 6.10

An elevator has a mass of 1000 kg, and can carry a maximum load of 800 kg. A constant frictional force of 4000 N retards its motion upwards. (a) What minimum horsepower motor is needed to lift the elevator at a constant speed of 3 m/sec?

Solution

The motor must supply the force T which pulls the elevator upward. Since $v = 3$ m/sec = constant, $a = 0$, so

$$T = Mg + f$$

where M is the *total* mass 1800 kg. Therefore,

$$T = 1.8 \times 10^3 \text{ kg } (9.8 \text{ m/sec}^2) + 4 \times 10^3 \text{ N} = 2.16 \times 10^4 \text{ N}$$

and the power that must be supplied is

$$P = Tv = (2.16 \times 10^4 \text{ N})\ 3 \text{ m/sec} = 6.5 \times 10^4 \text{ W}$$

or,

$$P = 65 \text{ kW} = 87 \text{ hp}$$

(b) Suppose the elevator accelerates upwards at a constant rate of 1 m/sec². What *average* horsepower is required to bring it from rest to a speed of 6 m/sec?

Solution

The displacement of the elevator, Δy, can be determined from the expression

$$v^2 = \cancel{v_0^2} + 2a\Delta y$$

$$\Delta y = \frac{(6 \text{ m/sec})^2}{2\ (1 \text{ m/sec}^2)} = 18 \text{ m}$$

The time it takes the elevator to obtain a speed of 6 m/sec can be obtained from the expression

$$v = \cancel{v_0} + at$$

or

$$t = \frac{v}{a} = \frac{6 \text{ m/sec}}{1 \text{ m/sec}^2} = 6 \text{ sec}$$

The tension in the cable, T, is *greater* than that obtained in part (a), since the elevator accelerates upwards.

$$\Sigma F_y = T - Mg - f = Ma$$

$$T = M(a + g) + f$$

$$T = 1.8 \times 10^3 \text{ kg } (2 + 9.8) \text{ m/sec}^2 + 4000 \text{ N} = 2.52 \times 10^4 \text{ N}$$

The average power required is

$$\langle P \rangle = \frac{W}{\Delta t} = \frac{T\Delta s}{\Delta t} = \frac{(2.52 \times 10^4 \text{ N})\ 18 \text{ m}}{6 \text{ sec}}$$

$$\langle P \rangle = 7.52 \times 10^4 \text{ W} \approx 100 \text{ hp}$$

(c) What horsepower must the motor deliver at the instant the speed of the elevator is 6 m/sec, and a = 1 m/sec² upwards?

Using the value of T obtained in part (b), we have

$$P = Tv = (2.52 \times 10^4 \text{ N})\ 6 \text{ m/sec} = 15.1 \times 10^4 \text{ W} \cong 200 \text{ hp}$$

6.9 PROGRAMMED EXERCISES

1.A

Work. A constant force **F** acts on a particle which undergoes a displacement **s**. What is the work done by **F**?

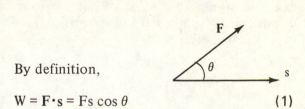

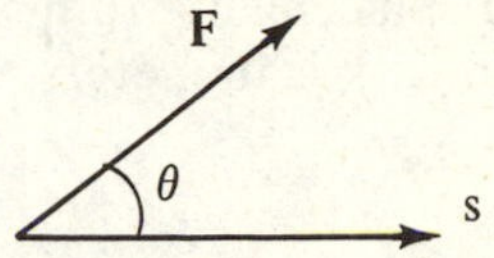

By definition,

$$W = \mathbf{F} \cdot \mathbf{s} = Fs \cos \theta \quad (1)$$

1.B

Can the work done by **F** ever be negative? Explain.

Yes. If **F** has a component in the direction $-\mathbf{s}$, W is negative. This occurs for $3\pi/2 > \theta > \pi/2$.

1.C

What are the units of work in SI, cgs and Engineering units?

SI $W \rightarrow$ Nm = Joule

cgs $W \rightarrow$ dyne-cm = erg

Eng. $W \rightarrow$ ft lb

1.D

For a given force **F**, for what values of θ is W a maximum? When is W zero? [In both cases, assume a finite displacement **s**, and $\mathbf{F} \neq 0$.]

W is a maximum for $\theta = 0$, since $\cos(0) = 1$, $W = Fs$

W = zero for $\theta = \pi/2$ or $3\pi/2$, since

$$\cos(\pi/2) = \cos(3\pi/2) = 0$$

1.E

Suppose two constant forces, $\mathbf{F}_1$ and $\mathbf{F}_2$, act on a particle which undergoes a displacement **s**. What is the total work done?

$$W = \mathbf{F}_1 \cdot \mathbf{s} + \mathbf{F}_2 \cdot \mathbf{s} = (\mathbf{F}_1 + \mathbf{F}_2) \cdot \mathbf{s}$$

or,

$$W = \mathbf{F} \cdot \mathbf{s}$$

where **F** is the *resultant* force.

1.F

Is work *always* defined to be equal to $\mathbf{F} \cdot \mathbf{s}$? Explain.

No. If **F** is *not* constant (either in magnitude, direction or both), we *cannot* use $\mathbf{F} \cdot \mathbf{s}$, to get W. This *only* applies if **F** is constant.

†1.G

How do you calculate the work done by **F** if it is known to vary with position? For example, suppose $F_x = 2x^3$. What is W for a displacement from x = 0 to x = 1, where F is in N, x is in meters.

Use the integral expression

$$W = \int_a^b \mathbf{F} \cdot d\mathbf{r}$$

In one dimension,

$$W = \int F_x\, dx = \int_0^1 2x^3\, dx = \left.\frac{2x^4}{4}\right|_0^1$$

W = 0.5 J

2.A

Kinetic Energy. A particle of mass m moves with a speed v. What is its kinetic energy?

By definition,

$$K = \frac{1}{2}mv^2$$

2.B

Is kinetic energy *always* given by $\frac{1}{2}mv^2$? Explain.

No. This relation is valid *only* if $v \ll c$, where c is the speed of light.

2.C

Suppose the particle has velocity vector components v_x, v_y and v_z. What is its kinetic energy?

$$K = \frac{1}{2}m\,(v_x{}^2 + v_y{}^2 + v_z{}^2), \text{ since}$$

$$v^2 = v_x{}^2 + v_y{}^2 + v_z{}^2$$

2.D

Show that the units of kinetic energy are the same as work.

$[mv^2] = ML^2/T^2$

$[\text{Work}] = [Fs] = \frac{ML}{T^2} \times L = ML^2/T^2$

2.E

Is kinetic energy a scalar or vector quantity?

A scalar. It can be written as

$$\frac{1}{2}m\mathbf{v}\cdot\mathbf{v} = \frac{1}{2}mv^2$$

2.F

Can kinetic energy ever be negative? Can it be zero? Explain.

Kinetic energy can *never* be negative since it goes as v^2, which is always positive. Kinetic energy is zero *only* when v = 0.

2.G

What is the kinetic energy of a 4 kg mass moving with a speed of 20 m/sec?

$$K = \frac{1}{2} mv^2 = \frac{1}{2}(4)\ \text{kg}\ (20\ \text{m/sec})^2$$

$$K = 800\ \text{J}$$

2.H

What is the kinetic energy of a 10 g mass moving with a velocity of $\mathbf{v} = (8\mathbf{i} + 3\mathbf{j})$ cm/sec?

$$v^2 = \mathbf{v} \cdot \mathbf{v} = (8\mathbf{i} + 3\mathbf{j}) \cdot (8\mathbf{i} + 3\mathbf{j})\ \text{cm}^2/\text{sec}^2$$

$$v^2 = 64 + 9 = 73\ \text{cm}^2/\text{sec}^2$$

$$K = \frac{1}{2} mv^2 = \frac{1}{2}(10\ \text{g})\ 73\ \text{cm}^2/\text{sec}^2$$

$$K = 365\ \text{ergs}$$

3.A

Elastic Potential Energy. A mass m is attached to a spring and moves on a horizontal smooth surface. If the mass is pulled a distance x_0 from its equilibrium position, what is the elastic potential energy of the system?

$$U = \frac{1}{2} k{x_0}^2$$

where k is the force constant of the spring.

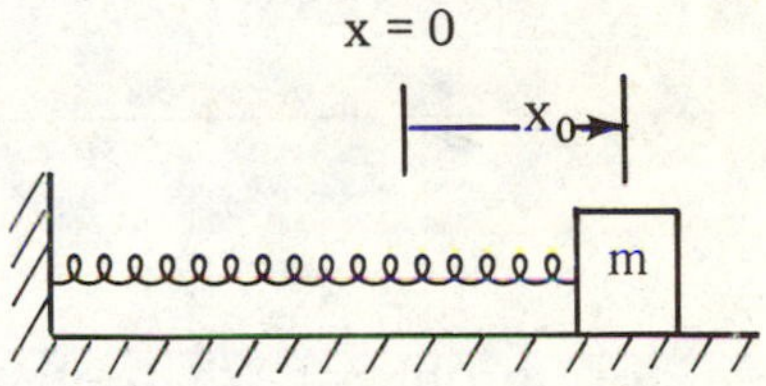

3.B

What is the point of zero elastic potential energy for this system?

When $x = 0$, the elastic potential energy $U = 0$. This corresponds to the unstretched spring.

3.C

What is the elastic potential energy when the spring is compressed a distance x from equilibrium?

$$U = \frac{1}{2} kx^2$$

3.D

When the spring is compressed, x is *negative.* Does this correspond to a negative potential energy?

No. Elastic potential energy is always greater than or equal to zero since it goes as x^2.

3.E

If m is released at $x = x_0$, and it moves to $x = 0$, the elastic potential energy becomes zero. What happens to this energy?

The energy is transformed into the kinetic energy of m. At $x = 0$, $K = \frac{1}{2} m{v_m}^2$, where v_m is the maximum speed of m.

3.F

What is the total energy of the system when $x = \pm x_0$, when $x = 0$, and for an arbitrary value of x?

At $x = \pm x_0$, $E = U = \frac{1}{2}kx_0{}^2$

At $x = 0$, $E = K = \frac{1}{2}mv_m{}^2$

At x, $E = U + K = \frac{1}{2}kx^2 + \frac{1}{2}mv^2$

3.G

Find an expression for the maximum velocity of m in terms of k and x_0.

Since the total energy is conserved, we have from 3.F,

$$\frac{1}{2}mv_m{}^2 = \frac{1}{2}kx_0{}^2$$

$$v_m = \pm\sqrt{k/m}\,x_0$$

3.H

What do the positive and negative signs signify in the solution for v_m?

The positive and negative signs correspond to the m moving to the right and left, respectively, as it passes through $x = 0$.

3.I

We have assumed that there are no frictional forces in this problem. What other basic assumptions have we made?

We have assumed that the mass of the spring is negligible. If this were not so, the kinetic energy of the spring would have to be included in the calculations.

4 A block of mass m is dragged up a rough incline by a constant force F parallel to the incline. It starts from the bottom with a speed v_0 and reaches the top with a speed v. The height of the incline is h, and the angle of inclination is θ.

4.A

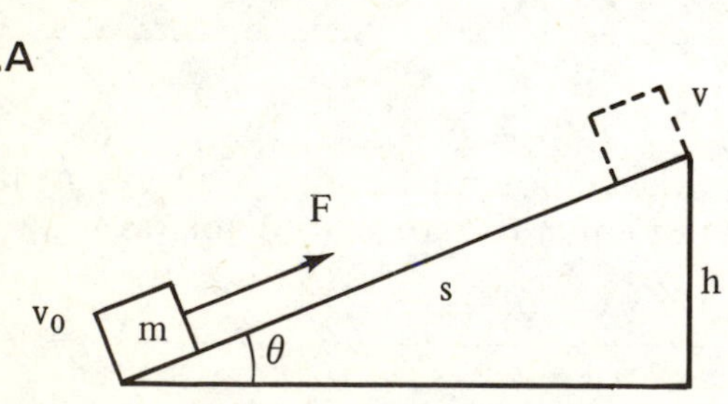

What is the work done by F in moving m from the bottom to the top?

The displacement can be determined from the right triangle.

$\sin\theta = h/s$, or $s = h/\sin\theta = h\csc\theta$.

$$W_F = Fs = Fh\csc\theta \qquad (1)$$

4.B

If the frictional force is constant and denoted by f, find the work done by **f**.

$W_f = \mathbf{f} \cdot \mathbf{s} = -fs = -fh\csc\theta$ (2)

4.C

Find the work done by gravity for the displacement from the bottom to the top.

F_g along the plane $= -mg \sin\theta$, where the positive direction is *up* the plane. Therefore,

$$W_g = F_g s = -mg \sin\theta \frac{h}{\sin\theta} = -mgh \quad (3)$$

4.D

Are there any other forces acting on m that do work?

No. The normal force **N** and the component of weight perpendicular to the plane are at right angles to s, and hence do no work.

4.E

Find the total work done by all forces acting on m. Make use of (1), (2) and (3).

$W = W_F + W_f + W_g = (F - f)h\csc\theta - mgh$ (4)

4.F

Write an expression for the work-energy theorem for the displacement from the bottom to the top.

$$W = \Delta K = \frac{1}{2}mv^2 - \frac{1}{2}mv_0{}^2$$

$$(F - f)\,h\csc\theta - mgh = \frac{1}{2}mv^2 - \frac{1}{2}mv_0{}^2 \quad (5)$$

4.G

What is the *change* in the gravitational potential energy for this displacement?

$$\Delta U = mgh - 0 = mgh \quad (6)$$

4.H

Find a relationship between the change in potential energy and the work done by gravity.

From (3) $W_g = -mgh$

From (6) $\Delta U = mgh$

Therefore, $\Delta U = -W_g$

4.I

Rewrite the work-energy expression in terms of the work done by all forces *other* than gravity, and its relation to the change in kinetic energy and change in potential energy.

$W_F + W_f = \Delta K + \Delta U$

$$(F - f)\,h\csc\theta = \frac{1}{2}mv^2 - \frac{1}{2}mv_0{}^2 + mgh \quad (7)$$

This is equivalent to (5).

4.J

Are there any forces that are conservative in this situation?

Yes. The force of gravity is a conservative force. For this reason, we can define an associated potential energy function U.

4.K

Is the total mechanical energy, E, constant in this problem? Recall that E = K + U.

No. Since there is work done by forces other than gravity, we see from the work-energy theorem that $E - E_0 \neq 0$.

5 *Combined Gravitational and Elastic Potential Energy.* A mass m is attached to a spring hanging vertically in the earth's gravitational field. The mass is released from *rest* while the spring is *unstretched,* as shown below. The maximum extension of the spring is represented by $2y_0$.

5.A

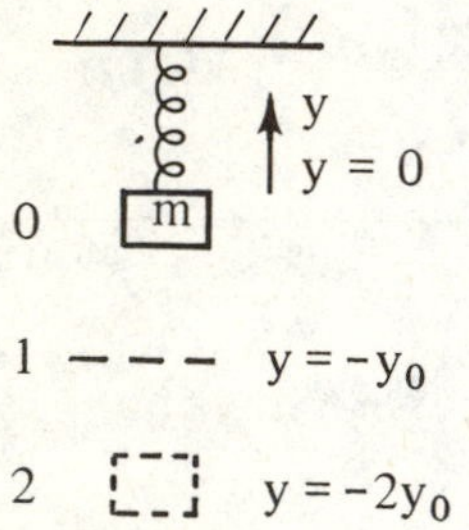

What is the gravitational potential energy at points 0, 1 and 2? Call y = 0 the point at which m is released.

$U_g = 0$ at $y = 0$

$U_g = -mgy_0$ at $y = -y_0$

$U_g = -2\,mgy_0$ at $y = -2y_0$

5.B

Find the kinetic energies of m at points 0, 1 and 2. Let the speed of m be v_m at position 1.

Since m is *released from rest,*

$K_0 = 0$

$K_1 = \frac{1}{2}mv_m{}^2$

$K_2 = 0$ ($v = 0$ at $y = -2y_0$)

5.C

Determine the elastic potential energy in the spring at points 0, 1 and 2.

$U_s = 0$ at $y = 0$

$U_s = \frac{1}{2}ky_0{}^2$ at $y = -y_0$

$U_s = \frac{1}{2}k\,(2y_0)^2$ at $y = -2y_0$

5.D

What is the *total* energy of the system at all times?

$$E = K + U_s + U_g = 0 \qquad (1)$$

Since at y = 0, each term is zero.

5.E

Write an expression for the total energy for *any* extension.

$$E = \frac{1}{2}mv^2 + \frac{1}{2}ky^2 + mgy = 0 \qquad (2)$$

5.F

Determine the maximum extension $2y_0$ in terms of m, g and k.

At point 2, v = 0, $y = -2y_0$, so from (2), the total energy equation, we have

$$\frac{1}{2}k(-2y_0)^2 - 2mgy_0 = 0$$

or,

$$2y_0 = 2mg/k$$

5.G

What is the change in the gravitational potential energy as m moves from 0 to y?

$$\Delta U_g = mgy - 0 = mgy \qquad (3)$$

5.H

What is the change in the elastic potential energy as m moves from 0 to y?

$$\Delta U_s = \frac{1}{2}ky^2 - 0 = \frac{1}{2}ky^2 \qquad (4)$$

5.I

Write an expression for the work-energy theorem applied to the motion from 0 to y.

$$W = \Delta K + \Delta U_g + \Delta U_s = 0$$

$$\frac{1}{2}mv^2 + mgy + \frac{1}{2}ky^2 = 0$$

This is equivalent to (2), that is, E = 0.

6 A particle of mass m slides on a smooth curved track as shown in the frame below. It starts from rest at point A.

6.A

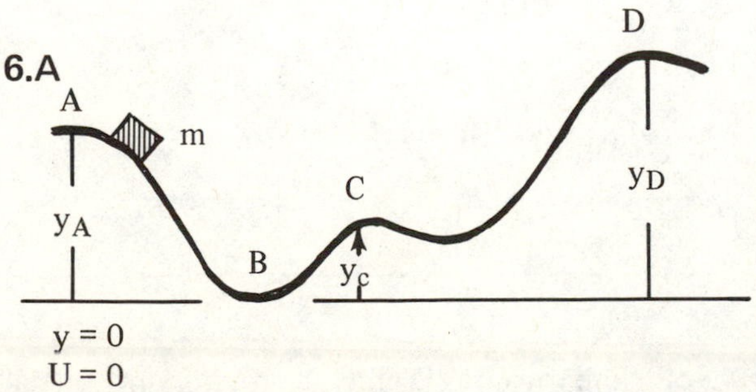

What is the potential energy of m at A? Use the coordinate system shown.

$U_A = mgy_A$ relative to y = 0 where U = 0

6.B

What is the kinetic energy and total energy of m at A?

Since $v_0 = 0$, $K_A = 0$; therefore,

$$E_A = U_A = mgy_A \qquad (1)$$

6.C

What is the potential energy of m at B?

$U_B = 0$ since $y = 0$

6.D

What is the kinetic energy of m at B if its speed is v_B? What is the total energy at B?

$$K_b = \frac{1}{2}mv_B{}^2$$

$$E_B = K_B = \frac{1}{2}mv_B{}^2 \qquad (2)$$

6.E

Is mechanical energy conserved in this situation?

Yes. There are no forces other than gravity that do work on m, since there is no friction, and the normal force is perpendicular to the motion.

6.F

Use the principle of conservation of energy and the results (1) and (2) to find the speed at B.

$$E_B = E_A$$

$$\frac{1}{2}mv_B{}^2 = mgy_A$$

$$v_B = \sqrt{2gy_A}$$

6.G

What is the total energy of m at C? Call the speed v_c at this point.

$$U_c = mgy_c$$

$$K_c = \frac{1}{2}mv_c{}^2$$

$$E_c = K_c + U_c = \frac{1}{2}mv_c{}^2 + mgy_c \qquad (3)$$

6.H

Determine the speed at c in terms of y_c, y_A and g. Make use of (1) and (3).

$$E_c = E_A$$

$$\frac{1}{2}mv_c{}^2 + mgy_c = mgy_A$$

$$v_c = \sqrt{2g(y_A - y_c)}$$

6.I

Will the particle reach the point D? Explain.

No. The point D has a higher potential energy than A. In order for m to reach D, it would have to gain energy. This is impossible.

6.J

If the particle is given an initial velocity v_A at A, can it reach D? If so, what is the condition on v_A in order that m reach D?

Yes, it is possible. By energy conservation, we have

$$K_A + U_A \geqq U_D$$

$$\frac{1}{2}mv_A^2 + mgy_A \geqq mgy_D$$

$$v_A^2 \geqq 2g(y_D - y_A)$$

6.K

Suppose the track is rough. Would the results for v_c and v_B differ? Why?

Yes. When friction is present, mechanical energy is *not* conserved. In fact, mechanical energy decreases as motion commences. Therefore,

$$E_c < E_B < E_A$$

and the values for v_B and v_C would be less than those given in 6.F and 6.H.

†7 A particle moves in the xy plane under the influence of a force given by $\mathbf{F} = \alpha x^2 \mathbf{i}$. The particle moves from the origin to the point P.

7.A

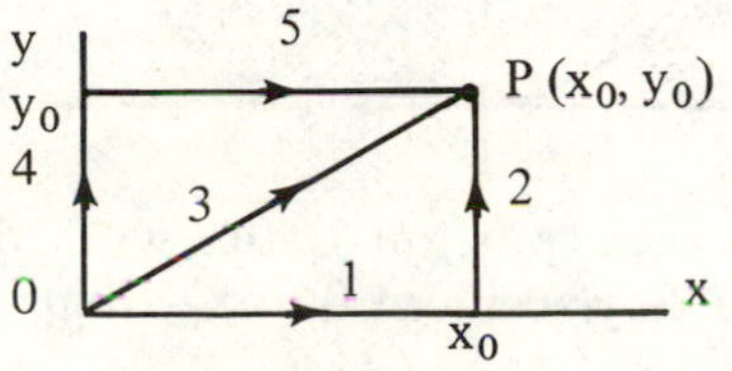

What is the work done by **F** along the path 1?

Recall that $\mathbf{F} \cdot d\mathbf{r} = F_x\,dx + F_y\,dy$. But $F_y = 0$, so

$$W_1 = \int F_x\,dx = \int_0^{x_0} \alpha x^2\,dx$$

$$W_1 = \frac{\alpha x_0^3}{3} \qquad (1)$$

7.B

What is the work done by **F** along the path 2?

$W_2 = 0$ since $x = x_0$, $dx = 0$ along this path.

7.C

What is the net work done by **F** from 0 to P along paths 1 and 2?

$$W_{OP} = W_1 + W_2 = \frac{\alpha x_0^3}{3} \qquad (2)$$

7.D

Calculate the work done by **F** from 0 to P along the paths 4 and 5.

$$W_{OP} = W_4 + W_5$$

$$W_4 = 0 \qquad \text{since } dx = 0$$

$$W_{OP} = W_5 = \int_0^{x_0} \alpha x^2\,dx = \frac{\alpha x_0^3}{3}$$

7.E

Calculate the work done by **F** along the path 3.

Again, $W_{0P} = \int_0^{x_0} \alpha x^2\,dx = \frac{\alpha x_0^3}{3}$ since $F_y = 0$

7.F

Is **F** a conservative force? Explain on the basis of the results to 7.C, 7.D and 7.E.

Yes. **F** is conservative since W is *independent* of the path taken between 0 and P.

7.G

What is the potential energy function corresponding to **F**? Is this function unique?

$U(x) = U_0 - \int F_x\,dx = U_0 - \int \alpha x^2\,dx$

$U(x) = U_0 - \frac{\alpha x^3}{3}$

No. The value of U depends on the choice of the arbitrary constant U_0.

7.H

Draw a rough curve of F *vs* x and interpret the area under the curve bounded by F and the x axis from 0 to x_0. Assume α is positive.

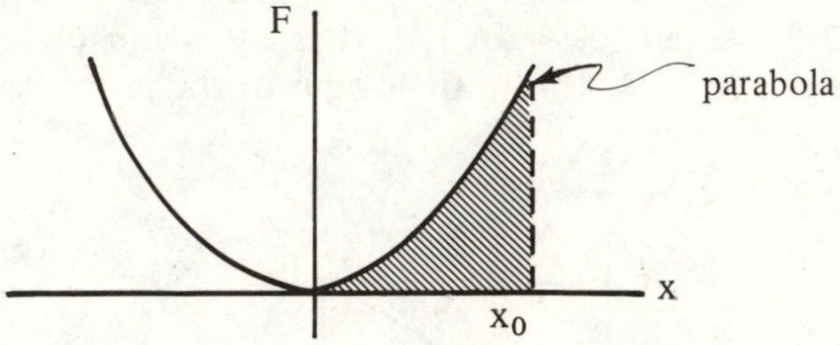

The shaded portion is the work done by F for the displacement 0 to x_0; that is,

$$\text{Area} = W = \int_0^{x_0} \alpha x^2\,dx = \frac{\alpha x_0^3}{3}$$

†8 A particle starts from the origin and follows the paths shown in the frame below. One of the forces acting on the particle is given by $\mathbf{F} = 2y^2\mathbf{i} + 3x\mathbf{j}$, where the force is in newtons and distance is measured in meters.

8.A

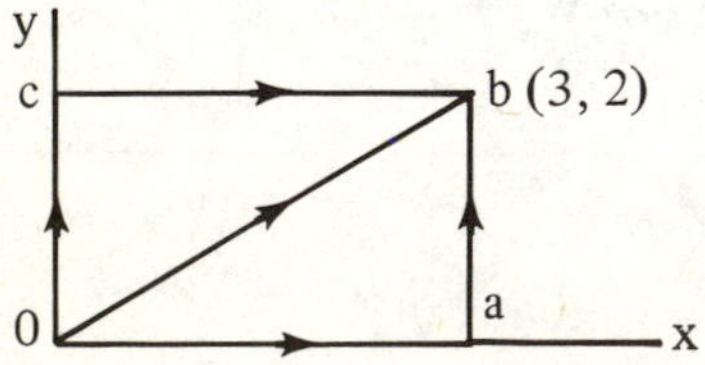

What is the work done by **F** for a general displacement from the origin to a point (x, y)?

From Equation (6.6), we have

$$W = \int_0^x F_x\,dx + \int_0^y F_y\,dy$$

$$W = \int_0^x y^2\,dx + \int_0^y 3\,x dy \qquad (1)$$

8.B

What is the work done by **F** for the path 0a?

Along 0a, y = 0, so dy = 0, and (1) gives

$$W_{0a} = 0 \qquad (2)$$

8.C

What is the work done by **F** for the path ab?

Along ab, x = 3, so dx = 0, and (1) becomes

$$W_{ab} = \int_0^2 3x\,dy = \int_0^2 3\,(3)\,dy$$

$$W_{ab} = 9y\Big|_0^2 = 18\text{ J} \qquad (3)$$

8.D

What is the work done by **F** along 0b? Note that the path 0b is a straight line of the form y = mx.

The path 0b can be represented by the equation $y = \frac{2}{3}x$, so (1) gives

$$W_{0b} = \int_0^3 2\left(\frac{2}{3}x\right)^2 dx + \int_0^2 3\left(\frac{3}{2}y\right) dy$$

$$W_{0b} = 2\left(\frac{4}{9}\right)\int_0^3 x^2\,dx + \frac{9}{2}\int_0^2 y\,dy$$

$$W_{0b} = \frac{8}{27}x^3 \Big|_0^3 + \frac{9}{4}y^2 \Big|_0^2 = 17\text{ J} \qquad (4)$$

8.E

Compare the results for the work done along the path 0ab to the work done along 0b. Why do they differ?

Using the results (2) and (3) gives

$$W_{0ab} = W_{0a} + W_{ab} = 18\text{ J} \qquad (5)$$

$$W_{0b} = 17\text{ J}$$

They differ because **F** is *not* a conservative force. In order that **F** be conservative, the work done should be independent of the path.

8.F

Calculate the work done by **F** along the path 0cb.

$$W_{0cb} = W_{0c} + W_{cb}$$

Along 0c, x = 0, therefore $W_{0c} = 0$.

Along cb, y = 2, dy = 0, so

$$W_{0cb} = \int_0^3 2(2)^2\,dx = 8\,(3) = 24\text{ J} \qquad (6)$$

8.G

Find the work done by F for the *closed* path Ocbo.

$W_{ocbo} = W_{ocb} + W_{bo}$

But $W_{bo} = -W_{ob} = -17$ J

and $W_{ocb} = 24$ J

$\therefore \quad W_{ocbo} = 7$ J (7)

8.H

How do you interpret the result of 8.G?

First, the work done for the closed path is nonzero since F is *not* a conservative force. Second, the result is positive, meaning that the mechanical energy of the particle *increases* by 7 J as it moves along the path Ocb0.

---

9 A particle of mass m is at the top of a smooth sphere of radius R as shown below. The particle starts from rest, and moves to a new point defined by the angle θ, measured from the vertical.

9.A

Take the point of zero potential energy to be at the center of the circle. What is the initial mechanical energy of m at A?

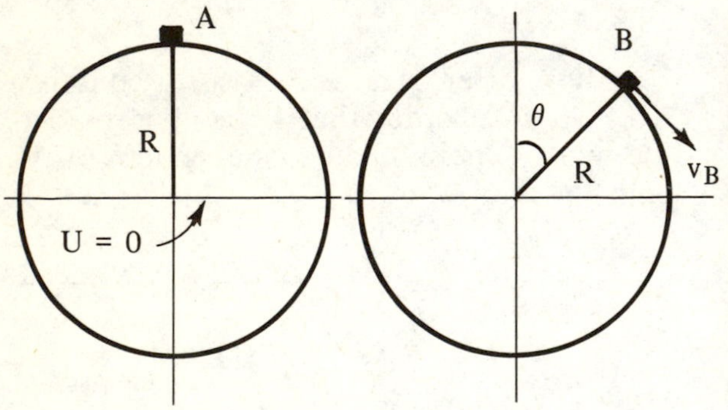

$U_0 = mgR$

$K_0 = 0$

$\therefore \quad E_0 = mgR$ (1)

9.B

What is the gravitational potential energy of m when it has moved to a new point, B, as shown in the frame above?

$U_B = mgR \cos\theta$ (2)

9.C

What is the change in potential energy for this displacement?

$$\Delta U = U_B - U_0 = mgR\cos\theta - mgR$$

$$\Delta U = mgR(\cos\theta - 1) \qquad (3)$$

9.D

Is the change in potential energy positive or negative? Explain.

ΔU is *negative*. Since $\cos\theta$ is always <1 for $0 < \theta < 2\pi$ we see from (3) that $\Delta U < 0$.

9.E

What is the change in kinetic energy as m moves from A to B? Let its velocity be v_B at B.

$$K_A = 0 \quad \text{since } v_A = 0$$

$$K_B = \frac{1}{2}mv_B^2$$

$$\therefore \quad \Delta K = K_B - K_A = \frac{1}{2}mv_B^2 \qquad (4)$$

9.F

Apply the work-energy theorem to this problem, and obtain a relation for the kinetic energy at B in terms of m, R and θ. Make use of Equations (3) and (4).

Since there are no forces other than gravity that do work,

$$W = \Delta K + \Delta U = 0$$

$$K_B + mgR(\cos\theta - 1) = 0$$

$$K_B = mgR(1 - \cos\theta) \qquad (5)$$

9.G

Determine the velocity of the particle at B in terms of m, R and θ.

From (5), we have

$$\frac{1}{2}mv_B^2 = mgR(1 - \cos\theta)$$

$$\therefore \quad v_B = \sqrt{2gR(1 - \cos\theta)} \qquad (6)$$

9.H

What is the magnitude of the radial acceleration of m at B?

Using (6), we get

$$a_r = \frac{v_B^2}{R} = 2g(1 - \cos\theta) \qquad (7)$$

9.I

What is the magnitude of the tangential acceleration of m at B?

$$a_t = g\sin\theta \qquad (8)$$

9.J

Make use of (7) and (8) to find the total acceleration of m at B.

$$a = -g \sin\theta\,\hat{\theta} - 2g\,(1 - \cos\theta)\,\hat{r}$$

$$a = \sqrt{4g^2\,(1 - \cos\theta)^2 + g^2 \sin^2\theta} \qquad (9)$$

9.K

What forces act on m while it is on the sphere? Show them in a free body diagram.

N = Normal Force

mg = weight

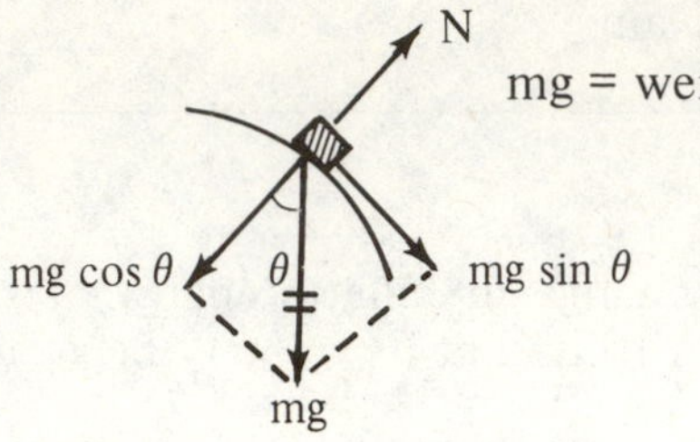

9.L

Write Newton's second law for the motion in the tangential direction and in the radial direction.

$$\Sigma F_t = -mg \sin\theta = ma_t \qquad (10)$$

$$\Sigma F_r = N - mg \cos\theta = -\frac{mv^2}{R} \qquad (11)$$

9.M

What is the condition under which the particle leaves the sphere?

This occurs when N = 0, or from (11), when

$$v^2 = Rg \cos\theta \qquad (12)$$

9.N

Determine the angle θ_c at which the particle leaves the sphere. Make use of (6) and (12).

$$Rg \cos\theta_c = 2gR\,(1 - \cos\theta_c)$$

$$\cos\theta_c = 2 - 2\cos\theta_c$$

$$3\cos\theta_c = 2$$

$$\theta_c = \text{arc cos}\left(\frac{2}{3}\right) \approx 48.2^\circ \qquad (13)$$

9.O

What is the kinetic energy and potential energy of m at the angle θ_c? Use (2), (5) and (13). What is the total energy at θ_c?

$$K = mgR\left(1 - \frac{2}{3}\right) = \frac{mgR}{3} \qquad (14)$$

$$U = \frac{2}{3} mgR \qquad (15)$$

$$E = K + U = mgR = E_0$$

that is, *mechanical energy is conserved.*

9.P

What is the radial acceleration, tangential acceleration and total acceleration at θ_c? Use (7), (8) and (13). Explain your result for the total acceleration.

$$a_r = 2g\left(1 - \frac{2}{3}\right) = \frac{2}{3}g$$

$$a_t = g \sin \theta_c = \sqrt{5/3}\, g$$

$$a = \sqrt{\frac{4}{9}g^2 + \frac{5}{9}g^2} = g$$

At θ_c, the mass is a freely falling body, $\therefore\ a = g$.

9.Q

What is the potential energy of m just before it hits the ground at C?

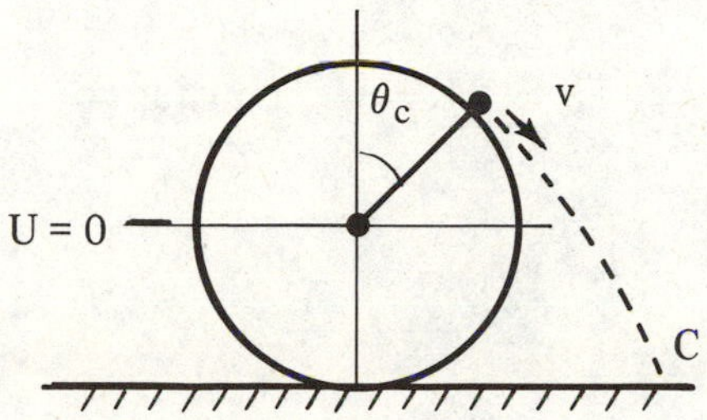

$$U_c = -mgR \qquad (16)$$

relative to the center of the sphere, where $U = 0$.

9.R

What is the speed of m at C? Use energy methods.

$$K_c + U_c = E_0 = mgR$$

$$\frac{1}{2}mv_c^2 - mgR = mgR$$

$$\frac{1}{2}mv_c^2 = 2\,mgR$$

$$v_c = 2\sqrt{Rg}$$

9.S

If R is known, can the trajectory of m be completely predicted? Explain.

Yes. The particles "initial" velocity at θ_c is given by (12), and it falls through a distance of $R \cos \theta_c + R$, so its path is predictable.

6.10 SUMMARY

Work done by a constant force F causing a displacement s:	$W = \mathbf{F} \cdot \mathbf{s} = Fs \cos \theta$
Work done by a variable force F for a displacement from a to b:	$W = \int_a^b \mathbf{F} \cdot d\mathbf{r}$
Kinetic energy of a particle of mass m, speed v:	$K = \frac{1}{2} mv^2$ $(v \ll c)$
Work-energy theorem:	$W = \Delta K = \frac{1}{2} mv^2 - \frac{1}{2} mv_0{}^2$
Potential energy stored in a spring:	$U_s = \frac{1}{2} kx^2$
Gravitational potential energy:	$U_g = mgy$
Conservation of mechanical energy (when only conservative forces act on the system):	$E = K + U = \text{constant}$
Potential energy of two masses separated by a distance r:	$U(r) = -\frac{GMm}{r}$

6.11 PROBLEMS

1. A man lifts a bucket of water at constant speed from the bottom of a well, 100 ft deep. The bucket and its contents weigh 50 lb. How much work does the man do in this chore?

2. A block weighing 25 lb is pulled by on a horizontal rough surface by a force of 20 lb acting at an angle of 37° with the horizontal. The block is displaced 5 ft and the coefficient of kinetic friction is 0.3. (a) How much work is done by the 20 lb force? (b) How much work is done by the force of friction? (c) What is the total work done on the block?

3. A particle whose mass is 0.2 kg moves on a horizontal surface under the action of several external forces. At point A, the speed of the particle is 3 m/sec, and at point B its speed is 5 m/sec. (a) What is the kinetic energy of the particle at point A? (b) What is its kinetic energy at point B? (c) What is the total work done on the particle as it moves from A to B?

4. A man pushes a car from rest to a speed of 3 m/sec with a constant horizontal force. The distance that the car moves in this interval is 30 m and its mass is 2000 kg. Neglecting the frictional force between the car and the road, determine (a) the work done by the man, and (b) the horizontal force he exerts on the car.

5. A 3 kg mass has an initial velocity $\mathbf{v}_0 = (5\mathbf{i} - 3\mathbf{j})$ m/sec. (a) What is its kinetic energy at this time? (b) What is the *change* in kinetic energy if its velocity increases to $\mathbf{v} = (8\mathbf{i} + 4\mathbf{j})$ m/sec?)

†6. A particle situated on the x axis is under the influence of an attractive force given by $\mathbf{F} = -kx^2\mathbf{i}$, where x is in meters, F is in Newtons, and k is a constant having units of N/m². (a) What external force is required to maintain equilibrium of the body 2 m from the origin? (b) How much work is done by an external agent in moving the body from x = 2 m to x = 3 m?

7. A 300 g ball is thrown into the air and reaches a maximum altitude of 50.0 m. (a) What is the initial speed of the ball? (b) What is its initial kinetic energy? (c) What is the ratio of the kinetic energy to the gravitational potential energy when the ball is at an altitude of 10 m. Take the ground to be at zero potential energy.

8. A constant force $\mathbf{F} = (3\mathbf{i} + 5\mathbf{j})$ N acts on a particle whose mass is 2 kg. (a) Calculate the work done by this force if the particle moves from the origin to the point whose vector position is $\mathbf{r} = (2\mathbf{i} - 3\mathbf{j})$ m. Does the result depend on the path? (b) What is the change in kinetic energy of the particle if this is the only force acting on it?

9. A ball of mass m is tied to a string, and is whirled overhead in a circle of radius R, at a constant angular speed ω. (a) What is the kinetic energy of the mass in terms of m, R and ω? (b) Obtain a numerical value for the kinetic energy if m = 500 g, R = 1 m, and the frequency of rotation is 2 rev/sec.

10. A 200 g block starts from rest at **A** on the inside of a rough hemispherical bowl whose radius is 30 cm (Figure 6-9). The speed of the block at the bottom of the bowl is observed to be 150 cm/sec. (a) What is the gravitational potential energy of the block at A relative to the bottom of the bowl? (b) What is the kinetic energy of the block at B? (c) How much energy is lost due to friction?

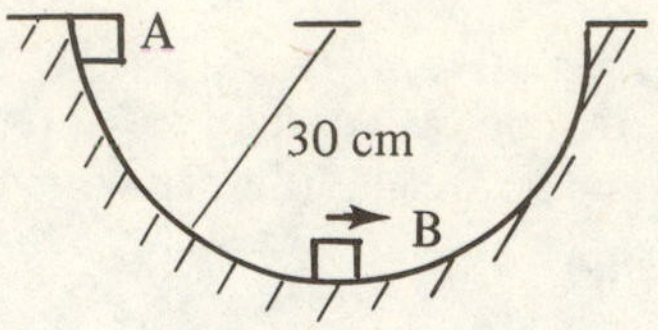

Figure 6-9

11. An object of mass 2 kg is subject to a force in the x direction which varies with position of the object as in Figure 6-10. The particle starts from rest at x = 0. What is the speed of the particle (a) at x = 5 m, (b) at x = 10 m and (c) at x = 15 m?

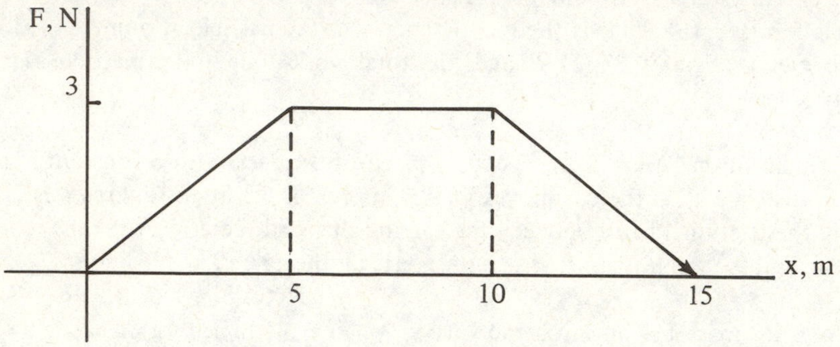

Figure 6-10

†12. An object of mass m is subject to a force in the **x** direction which varies with position according to the expression F = A sin (bx), where A and b are constants. (a) What is the work done by F as the particle moves from x = 0 to x = π/b? (b) If the particle starts with a speed v_0 at x = 0, what is the speed at x = π/b?

†13. A force **F** acting on a particle varies with position according to the expression $\mathbf{F} = (2y\mathbf{i} + 4x^2\mathbf{j})$ N. The particle moves from the origin to the point whose coordinates are x = 2 m, y = 1 m (Figure 6-11). Calculate the work done by **F** (a) along the path 0ac, (b) along the path 0bc, and (c) along the path 0c. (d) Is **F** conservative or nonconservative? Explain.

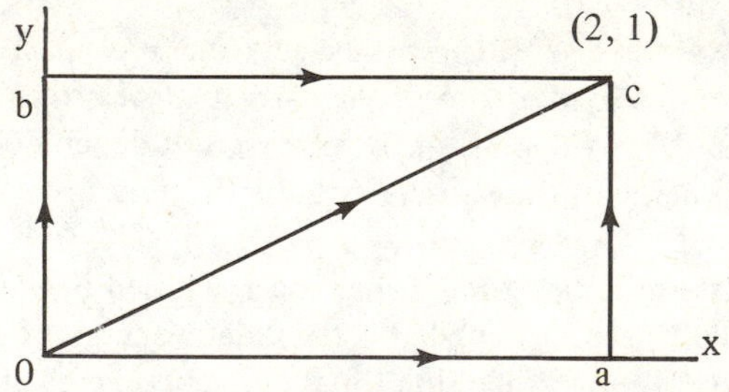

Figure 6-11

14. A block of mass m slides down a frictionless track of height h above a table (Figure 6-12). At the bottom of the track there is a light spring of force constant k. The block strikes and sticks to the spring. (a) How far is the spring compressed? (b) Obtain a numerical value for this distance if m = 2 kg, h = 1 m and k = 490 N/m.

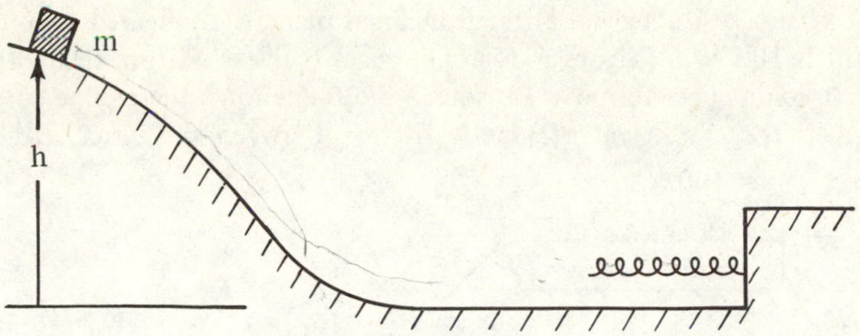

Figure 6-12

15. A particle slides down a frictionless ramp and does the "loop-the-loop" around a track whose radius is 0.5 m (Figure 6-13). If the particle starts from rest at a height h above the bottom, what must the value of h be (a) if the force of the track on the particle at A is six times its weight, (b) if the force of the track on the particle at B is four times its weight and (c) if the particle just completes the loop without leaving the track.

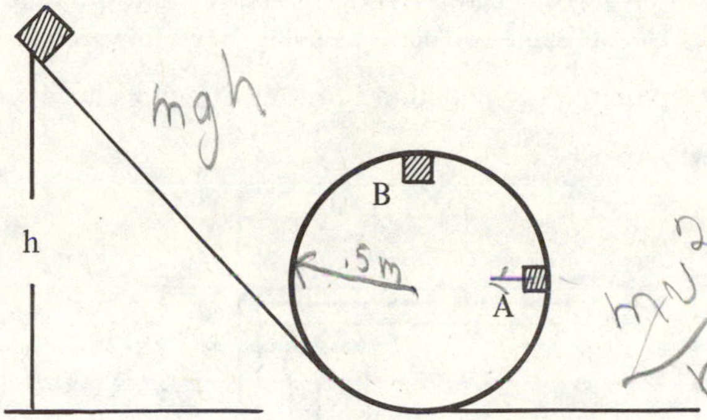

Figure 6-13

16. Let us consider the following track events: the javelin, the discus and the shot put, which have masses of 0.8 kg, 2.0 kg and 7.2 kg, respectively. Record throws for these events are about 89 m, 69 m and 21 m, respectively. Neglect air resistance. (a) Calculate the minimum initial kinetic energies which would produce these throws. (b) Estimate the force exerted on each object during the throw, assuming the force acts over a distance of 2 m. (c) Do any of your results suggest that air resistance plays an important role?

17. A 3 kg mass is connected to a light spring over a frictionless pulley as in Figure 6-14. The mass is released from rest with the spring unstretched initially. If the mass drops a distance of 10 cm before coming to rest, find the force constant k of the spring.

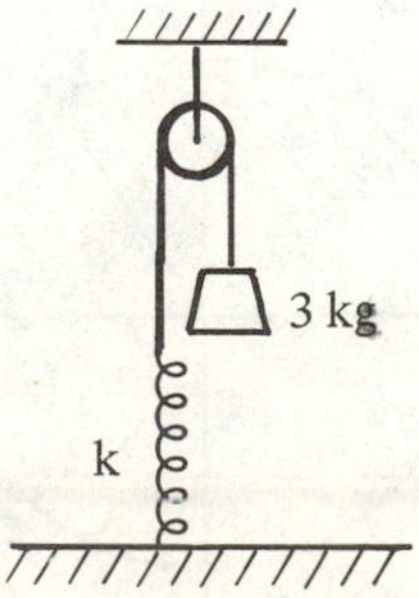

Figure 6-14

18. A 2 kg block situated on a rough inclined plane is connected to a spring whose force constant is 100 N/m (Figure 6-15). The mass is released from rest with the spring unstretched. It commences to move a distance of 20 cm down the incline before coming to rest. What is the coefficient of kinetic friction between the block and the plane?

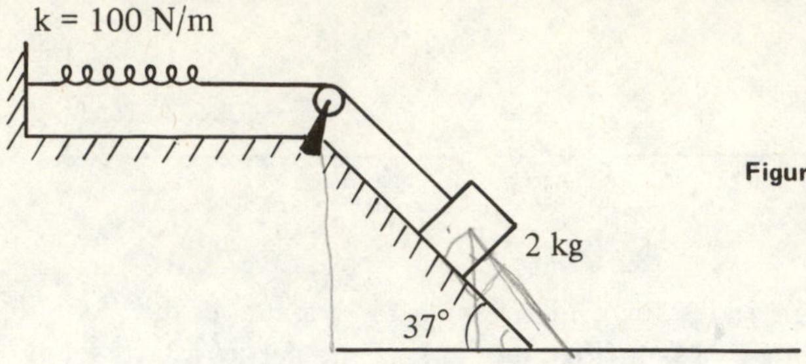

Figure 6-15

†19. A homogeneous rope of length L lies on a horizontal, smooth table whose height is greather than L. Part of the rope, of length y_0, hangs over the table and the rope is released from rest (Figure 6-16). Using energy methods, (a) calculate the velocity of the rope at the instant all of the rope leaves the table, and (b) find the time it takes for this to occur. *Optional:* (c) Do the same calculations using Newton's second law. Note that the acceleration is *not* constant. The equation of motion reduces to $d^2y/dt^2 = \frac{g}{L}y$, whose general solution is $y = Ae^{Bt} + Ce^{-Bt}$.

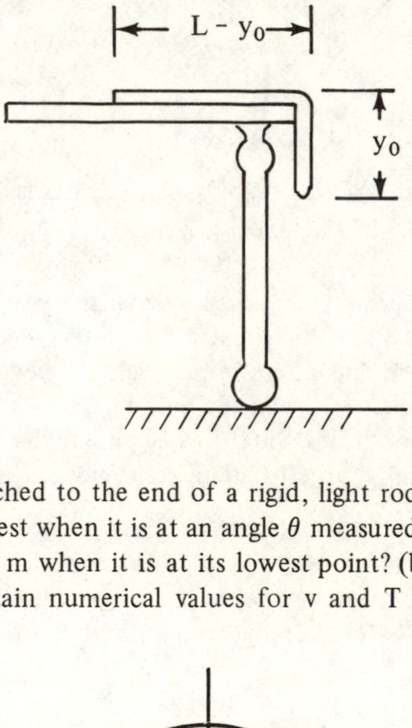

Figure 6-16

20. A mass m is attached to the end of a rigid, light rod of length l (Figure 6-17). The rod is released from rest when it is at an angle θ measured from the +x axis. (a) What is the speed v of the mass m when it is at its lowest point? (b) Find the tension T in the rod at this point. (c) Obtain numerical values for v and T if m = 200 g, l = 1 m and $\theta = 53°$.

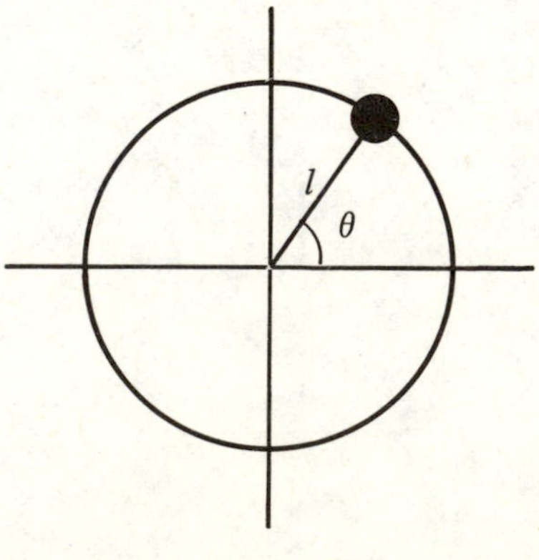

Figure 6-17

21. A 25 kg block is connected to a 30 kg block by a light string which passes over a frictionless pulley. The 30 kg block, in turn, is connected to a spring of force constant 200 N/m (Figure 6-18). The coefficient of kinetic friction between the 25 kg block and inclined plane is 0.10. The spring is unstretched when the system is as shown in the figure. The 25 kg block is then pulled a distance of 20 cm down the incline (so that the 30 kg block is 40 cm above the floor) and is released from rest. (a) What is the tension in the string before the 25 kg block is released? (b) What is the speed of each block when the 30 kg block is 20 cm above the floor (that is, when the spring is unstretched)?

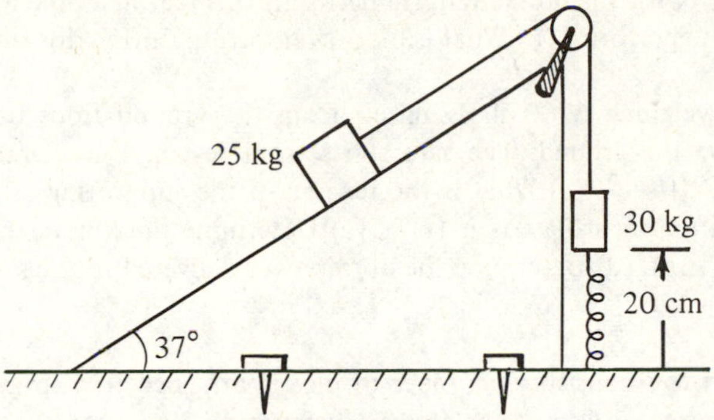

Figure 6-18

†22. A conservative force acts on a particle along the x axis. The force varies, with position according to the expression $\mathbf{F} = (-Ax + Bx^2)\mathbf{i}$, where A and B are constants, x is in meters and F is in Newtons. (a) Calculate the potential energy corresponding to this force, taking $U = 0$ at $x = 0$. (b) Find the change in potential energy as the particle moves from $x = 2$ m to $x = 3$ m.

†23. A particle moves in a region of space where the potential energy varies with position according to the expression

$$U(x) = 3x + 4x^2 - x^3$$

where x is in meters and U is in joules. (a) Determine the force acting on the particle as a function of x. (b) Sketch rough graphs of both U(x) and F(x).

24. Show that Equation (6.30) for the potential energy of a mass m at a distance r from the center of the earth reduces to $U = -mgR + mgy$, when $y \ll R$. R is the radius of the earth and $r = R + y$. Note that relative to the earth's surface, this reduces to $U = mgy$. Hint: Use the binomial expression for $(1 + x)^{-1}$, where $x = y/R$, and retain only the first two terms.

†25. Prove the work-energy theorem, that is, verify Equation (6.8), for a general situation. Note that $\mathbf{F} = m\frac{d\mathbf{v}}{dt}$, and $d\mathbf{r} = \mathbf{v}dt$, so the integrand can be expressed as

$$\mathbf{F} \cdot d\mathbf{r} = m dt \frac{d\mathbf{v}}{dt} \cdot \mathbf{v} = \frac{1}{2} m dt \frac{d}{dt} (\mathbf{v} \cdot \mathbf{v}) = \frac{1}{2} m d(v^2)$$

26. A satellite of mass m is orbiting around the earth in a circular orbit at a distance r from the center of the earth. (a) How much work must be done by forces other than gravity to put the satellite in a new orbit whose radius is 3r? (b) How much work must be done to bring the satellite to ∞, where the earth's gravitational force is zero? Let M be the mass of the earth.

27. (a) Calculate the escape velocity from the moon. Use the data $M_m = 7.35 \times 10^{22}$ kg, $R_m = 1.74 \times 10^6$ m. (b) Calculate the escape velocity from our solar system, that is, the sun. Use the data $M_s = 1.99 \times 10^{30}$ kg. $R_s = 6.96 \times 10^8$ m.

28. What horsepower is required to drag a 2000 lb crate across a rough horizontal surface at a constant speed of 2 ft/sec? Assume the coefficient of kinetic friction is 0.30.

29. An athlete whose mass is 71 kg runs a distance of 500 m up a mountain inclined at an angle of 37° with the horizontal. He performs this feat in a time of 70 sec. (a) How much work does he perform? (b) What is his power output during this run?

30. A safe weighing 3200 lb is raised from the ground floor of a building to a window far above the ground. The safe starts from rest and accelerates upwards at a constant rate of 2 ft/sec^2. (a) What is the tension in the supporting cable? (b) Find the velocity of the safe after it has risen 16 ft. (c) Determine the kinetic energy of the safe after it has risen 16 ft. (d) Determine the horsepower delivered to the safe when its speed is 8 ft/sec.

31. A child's toy consists of a piece of plastic attached to a spring (Figure 6-19). The spring is compressed against the floor a distance of 2 cm, and the toy is released. If the toy has a mass of 100 g and rises to a maximum height of 60 cm above the floor, estimate the force constant of the spring.

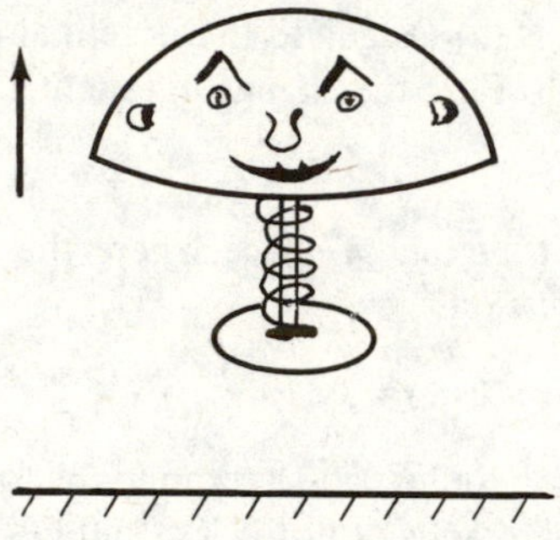

Figure 6–19

7

LINEAR MOMENTUM AND COLLISIONS

In this chapter we will deal with the linear momentum of a particle and a system of particles. Chapter 8 deals with a related topic, angular momentum, a concept useful in the analysis of rotational motion of particles and rigid bodies. As we shall see, the laws of the conservation of momentum and the conservation of energy are the most significant concepts in science and engineering, hence their importance cannot be overemphasized.

7.1 IMPULSE AND LINEAR MOMENTUM

The *linear momentum* of a particle whose mass is m, moving with a velocity **v** relative to an observer in an inertial frame is defined to be

$$\mathbf{p} = m\mathbf{v} \tag{7.1}$$

Therefore, linear momentum is a vector quantity, its direction is that of **v**, and its dimensions are M L/T. Equation (7.1) is a nonrelativistic expression, therefore it is only valid when $v \ll c$, and for particles which have a nonzero rest mass.

Momentum can be related to the force on a particle by the following consideration. If we take the time derivative of **p**, and assume that m remains constant (nonrelativistic case), we see that

$$\frac{d\mathbf{p}}{dt} = m\frac{d\mathbf{v}}{dt}$$

But from Newton's second law,

$$\mathbf{F} = m\mathbf{a} = m\frac{d\mathbf{v}}{dt}$$

therefore we conclude that the *force on m equals the time rate of change of linear momentum,* or

$$\mathbf{F} = \frac{d\mathbf{p}}{dt} \tag{7.2}$$

From Equation (7.2) we see that if $\mathbf{F} = 0$, $\mathbf{p}$ = constant. In other words, the linear momentum (and velocity) of an *isolated* particle remains constant in time, or is conserved. Later, we will see that a similar conservation law applies to a system of particles.

We can rewrite Equation (7.2) as

$$d\mathbf{p} = \mathbf{F}dt \tag{7.3}$$

Suppose that the particle undergoes a change in momentum from $\mathbf{p}_0$ at time t_0 to $\mathbf{p}$ at time t. We can express this change in momentum $\Delta\mathbf{p}$ from Equation (7.3) by integrating:

$$\int_{\mathbf{p}_0}^{\mathbf{p}} d\mathbf{p} = \int_{t_0}^{t} \mathbf{F}dt$$

or,

$$\Delta\mathbf{p} = \mathbf{p} - \mathbf{p}_0 = \int_{t_0}^{t} \mathbf{F}dt \tag{7.4}$$

The quantity on the right side of Equation (7.4) is usually referred to as the *impulse* $\mathbf{I}$ of the force $\mathbf{F}$, or

$$\mathbf{I} = \int_{t_0}^{t} \mathbf{F}dt = \Delta\mathbf{p} \tag{7.5}$$

Therefore, *the impulse of the force* **F** *equals the change in momentum of m.* This is actually an equivalent statement of Newton's second law. Impulse is a vector quantity which has the same direction as **F**, and which has dimensions of force × time, or ML/T. It should be noted that the concept of impulse is *not* a property of the particle or body itself. Impulse is a quantity which measures the degree to which an external force changes the momentum of the particle. Sometimes we say that momentum is transferred from the external agent to the particle or body.

The calculation of the impulse of a force **F** is not always straight-forward as suggested by Equation (7.5). However, the calculation becomes simple if the force acting on m is constant. When this is the case, Equation (7.4) reduces to

$$\mathbf{p} - \mathbf{p}_0 = \mathbf{I} = \mathbf{F}\,(t - t_0) \tag{7.6}$$

or,

$$\Delta\mathbf{p} = \mathbf{F}\Delta t \tag{7.7}$$

Example 7.1

A particle of mass m moves under the action of a constant gravitational force. The particle is given an initial velocity $\mathbf{v}_0$. (a) Find the impulse delivered to the mass by the gravitational force in the time interval Δt. From this result, obtain the final velocity of the mass m. (b) Obtain numerical values for **I** and **v** if $m = 2$ kg, $\Delta t = 3$ sec, and $\mathbf{v}_0 = (4\mathbf{i} + 2\mathbf{j})$ m/sec.

Solution

(a) A particle falling in the earth's gravitational field experiences a force

$$\mathbf{F} = -mg\mathbf{j}$$

Substituting this into Equation (7.7) gives

$$\mathbf{I} = \Delta\mathbf{p} = \mathbf{F}\Delta t = -mg\Delta t\mathbf{j}$$

or,

$$m\mathbf{v} = m\mathbf{v}_0 = -mg\Delta t\mathbf{j}$$

Solving for **v**, we have

$$\mathbf{v} = \mathbf{v}_0 - g\Delta t\mathbf{j}$$

This result could also be obtained from the equations of linear motion under constant acceleration.

(b) If $m = 2$ kg, $\Delta t = 3$ sec and $\mathbf{v}_0 = (4\mathbf{i} + 2\mathbf{j})$ m/sec, we get

$$\mathbf{I} = -2 \text{ kg} \left(9.8 \frac{\text{m}}{\text{sec}^2}\right) 3 \text{ sec } \mathbf{j}$$

$$\mathbf{I} = -59\mathbf{j} \quad \text{kg m/sec}$$

$$\mathbf{v} = \mathbf{v}_0 - g\Delta t\mathbf{j} = [(4\mathbf{i} + 2\mathbf{j}) - 9.8\,(3)\mathbf{j}] \text{ m/sec}$$

$$\mathbf{v} = (4\mathbf{i} - 27\mathbf{j}) \text{ m/sec}$$

Example 7.2

A ball whose mass is 100 g is dropped from a height of 2 m from the floor (Figure 7-1). It rebounds vertically upward after colliding with the floor to a height of 1.5 m. (a) Find the momentum of the ball immediatley before and after colliding with the floor. (b) Determine the *average force* exerted by the floor on the ball. Assume that the collision lasts for 1 msec (10^{-3} sec).

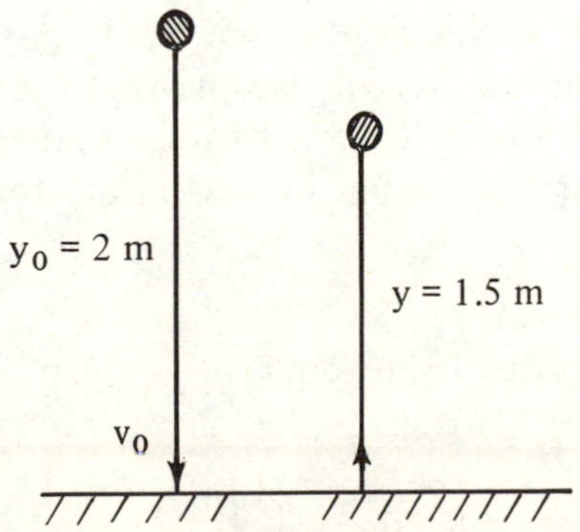

Figure 7-1

Solution

Let us call $\mathbf{v}_0$ the velocity of the ball just before colliding with the floor, and $\mathbf{v}$ its velocity after the collision. From energy considerations we have

$$mgh = \frac{1}{2}mv_0^2 \quad \text{and} \quad mgy = \frac{1}{2}mv^2$$

Substituting into these expressions the values $y_0 = 2$ m and $y = 1.5$ m gives

$$v_0 = \sqrt{2gy_0} = \sqrt{2\,(9.8)\,2}\ \text{m/sec} = 6.2\ \text{m/sec}$$

$$v = \sqrt{2gy} = \sqrt{2(9.8)\,1.5}\ \text{m/sec} = 5.4\ \text{m/sec}$$

Since m = 0.1 kg, the vector expressions for the initial and final linear momenta become

$$\mathbf{p}_0 = m\mathbf{v}_0 = -0.62\mathbf{j}\ \ \text{kg m/sec}$$

$$\mathbf{p} = m\mathbf{v} = 0.54\mathbf{j}\ \ \text{kg m/sec}$$

(b) We can write Equation (7.5) as

$$\mathbf{I} = \mathbf{p} - \mathbf{p}_0 = \mathbf{F}_{av}\,\Delta t$$

$$\mathbf{F}_{av} = \frac{(0.54\mathbf{j} + 0.62\mathbf{j})\ \text{kg m/sec}}{10^{-3}\ \text{sec}}$$

or,

$$\mathbf{F}_{av} = (1.16 \times 10^3)\mathbf{j}\ \ \text{N}$$

Note that this is the *average* force on the ball due to the collision with the floor. It *does not* include the force of gravity on the ball which is much less than this number [$mg \approx 1.0$ N]. In fact, when dealing with collisions occurring over a very short time interval, the impulsive force generally overwhelms the gravitational forces.

7.2 COLLISIONS AND CONSERVATION OF LINEAR MOMENTUM

Let us analyze the change in momentum of two masses m_1 and m_2 which undergo a collision with each other. We will assume that the impulsive force due to the collision is much greater than any external forces that might exist. The actual impulsive force F(t) may vary in time in a complicated fashion, such as in Figure 7–2.

The change in momentum of m_1 due to the collision is given by

$$\Delta\mathbf{p}_1 = \int_{t_0}^{t} \mathbf{F}_1\,dt$$

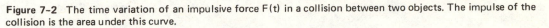

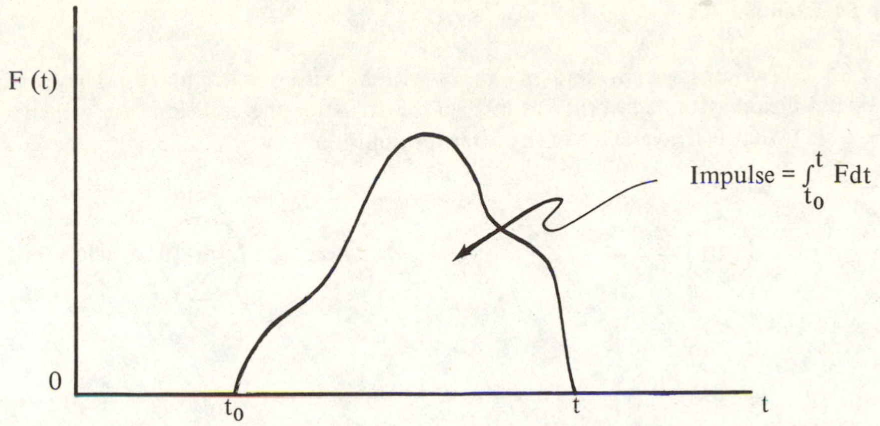

Figure 7-2 The time variation of an impulsive force F(t) in a collision between two objects. The impulse of the collision is the area under this curve.

where $\mathbf{F}_1$ is the force of m_2 on m_1. Likewise, the change in momentum of m_2 due to the collision is

$$\Delta \mathbf{p}_2 = \int_{t_0}^{t} \mathbf{F}_2 \, dt$$

where $\mathbf{F}_2$ is the force of m_1 on m_2. But Newton's third law states that $\mathbf{F}_1 = -\mathbf{F}_2$; that is, the force of m_2 on m_1 is equal and opposite to the force of m_1 on m_2. Hence, we conclude that

$$\Delta \mathbf{p}_1 = -\Delta \mathbf{p}_2$$

If we call the total momentum of the system **p**, then we can write

$$\mathbf{p} = \mathbf{p}_1 + \mathbf{p}_2$$

Consequently, the change in momentum of the system (the two masses) is given by

$$\Delta \mathbf{p} = \Delta \mathbf{p}_1 + \Delta \mathbf{p}_2 = 0$$

In other words, if we disregard external forces in an impulsive collision, the total momentum of the system remains *unchanged.* We can also say that impulsive forces are internal to the system and do not influence the total momentum of the system. This is the *law of conservation of linear momentum.* A mathematical statement of the law of conservation of momentum is the following:

$$\mathbf{p} = \text{constant} \quad (\text{when } \mathbf{F}_{ext} = 0) \tag{7.8}$$

Later, we shall prove that this law also applies to a system of particles. Keep in mind, however, that the conservation of momentum principle assumes that the impulsive force due to the collision is much larger than any external forces. This is a good approximation for collisions whose time of duration is small.

Example 7.3

Two masses m_1 and m_2 move with velocities $\mathbf{v}_{10}$ and $\mathbf{v}_{20}$. They make a head-on collision with each other as in Figure 7-3. After the collision, the velocity of m_1 is $\mathbf{v}_1$. (a) What is the velocity of m_2 after the collision?

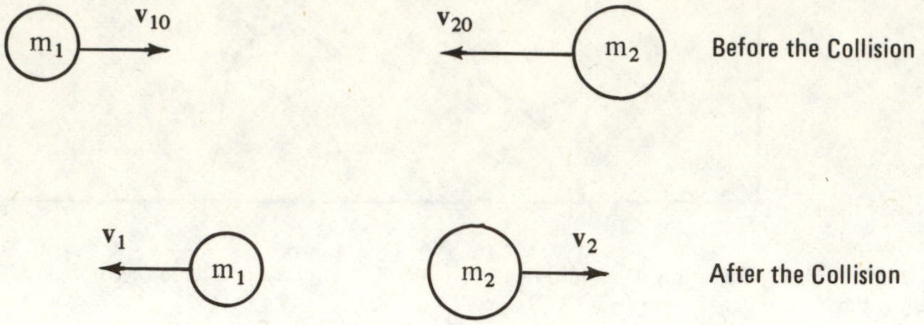

Figure 7-3 Head-on collision between two masses m_1 and m_2.

Solution

(a) Since there are no external forces involved, we consider m_1 and m_2 as an isolated system. Therefore, the total momentum is conserved.

$$\mathbf{p} = \mathbf{p}_{10} + \mathbf{p}_{20} = m_1\mathbf{v}_{10} + m_2\mathbf{v}_{20} \qquad \text{Before the Collision}$$

$$\mathbf{p} = \mathbf{p}_1 + \mathbf{p}_2 = m_1\mathbf{v}_1 + m_2\mathbf{v}_2 \qquad \text{After the Collision}$$

Applying Equation (7.8), we have

$$m_1\mathbf{v}_{10} + m_2\mathbf{v}_{20} = m_1\mathbf{v}_1 + m_2\mathbf{v}_2$$

or,

$$\mathbf{v}_2 = \frac{m_1(\mathbf{v}_{10} - \mathbf{v}_1) + m_2\mathbf{v}_{20}}{m_2}$$

(b) Suppose $m_1 = 2$ kg, $m_2 = 4$ kg, $\mathbf{v}_{10} = 5\mathbf{i}$ m/sec, $\mathbf{v}_{20} = -8\mathbf{i}$ m/sec, and $\mathbf{v}_1 = -2\mathbf{i}$ m/sec. Find $\mathbf{v}_2$.

$$\mathbf{v}_2 = \frac{2\text{ kg}}{4\text{ kg}}(5\mathbf{i} + 2\mathbf{i})\text{ m/sec} + \frac{4\text{ kg}}{4\text{ kg}}(-8\mathbf{i})\text{ m/sec}$$

$$\mathbf{v}_2 = (3.5\mathbf{i} - 8\mathbf{i})\text{ m/sec} = -4.5\mathbf{i}\ \text{ m/sec}$$

In such a situation, we can also write the law of conservation of momentum as

$$m_1 v_{10} + m_2 v_{20} = m_1 v_1 + m_2 v_2$$

where a positive velocity component indicates a velocity vector to the right, and a negative component indicates a velocity vector to the left.

There are two kinds of collisions that can occur between two objects, the difference lying in whether or not kinetic energy is conserved in the collision. The first type is called an *elastic* collision, for which kinetic energy is conserved. The second is called an *inelastic* collision, where kinetic energy is not conserved. Billiard ball collisions and the collision between atomic particles are approximately elastic. The collision of a rubber ball with a hard surface is inelastic, since some of the kinetic energy is lost due to the deformation of the ball. If two particles stick together after colliding with each other, the collision is *perfectly inelastic.* For example, if two pieces of putty collide, they stick together and move with some common velocity. If a meteorite collides with the earth, it becomes buried in the earth's surface, and the collision is inelastic. However, not *all* the kinetic energy is necessarily lost in an inelastic collision. Let us consider the two cases separately by some examples.

Inelastic Collisions

Consider two masses m_1 and m_2 moving with initial velocities $\mathbf{v}_{10}$ and $\mathbf{v}_{20}$, respectively. Suppose the two masses collide, stick together and move with a common velocity $\mathbf{v}$ after the collision. This is a *perfectly inelastic* collision, hence kinetic energy is not conserved. What is the speed of m_1 and m_2 after the collision?

Since momentum is conserved, we have

$$m_1\mathbf{v}_{10} + m_2\mathbf{v}_{20} = (m_1 + m_2)\mathbf{v}$$

or

$$\mathbf{v} = \frac{m_1\mathbf{v}_{10} + m_2\mathbf{v}_{20}}{m_1 + m_2} \qquad \text{Perfectly Inelastic Collision} \qquad (7.9)$$

Example 7.4

Two masses m_1 and m_2 collide in a perfectly inelastic collision. Suppose $m_1 = 0.20$ kg, $m_2 = 0.30$ kg, $\mathbf{v}_{10} = 8\mathbf{i}$ m/sec, and $\mathbf{v}_{20} = -3\mathbf{i}$ m/sec. (a) Find the velocity after the collision.

Solution

Applying Equation (7.9), we have

(a)

$$\mathbf{v} = \frac{0.20\text{ kg }(8\mathbf{i})\text{ m/sec} - 0.30\text{ kg }(3\mathbf{i})\text{ m/sec}}{0.50\text{ kg}}$$

$$\mathbf{v} = \frac{1.6\mathbf{i} - 0.9\mathbf{i}}{0.5}\text{ m/sec} = 1.4\mathbf{i}\ \text{ m/sec}$$

(b) How much kinetic energy is lost in the collision?

Before the collision, $K_0 = \frac{1}{2} m_1 v_{10}{}^2 + \frac{1}{2} m_2 v_{20}{}^2$

$$K_0 = \frac{1}{2}(0.20 \text{ kg})(64 \text{ m}^2/\text{sec}^2) + \frac{1}{2}(0.30 \text{ kg})(9 \text{ m}^2/\text{sec}^2)$$

$$K_0 = 7.75 \text{ J}$$

After the collision, $K = \frac{1}{2}(m_1 + m_2)v^2 = \frac{1}{2}(0.50 \text{ kg})(1.4)^2 \text{ m}^2/\text{sec}^2 = 0.49 \text{ J}$

Hence, the loss in kinetic energy is

$$K_0 - K = 7.26 \text{ J}$$

Elastic Collisions

Now suppose m_1 and m_2 undergo a head-on, perfectly elastic collision. In this case, we can apply conservation of momentum *and* conservation of kinetic energy to the collision. These conditions can be written as

$$m_1 v_{10} + m_2 v_{20} = m_1 v_1 + m_2 v_2$$

$$\frac{1}{2} m_1 v_{10}{}^2 + \frac{1}{2} m_2 v_{20}{}^2 = \frac{1}{2} m_1 v_1{}^2 + \frac{1}{2} m_2 v_2{}^2$$

Usually, the initial velocities and masses are known. The above equations would therefore represent two equations with two unknowns. It is left as an exercise to show that the velocities of the particles *after* the collision are given by

Perfectly Elastic Collision

$$v_1 = \left(\frac{m_1 - m_2}{m_1 + m_2}\right) v_{10} + \left(\frac{2 m_2}{m_1 + m_2}\right) v_{20} \quad (7.10)$$

$$v_2 = \left(\frac{2m_1}{m_1 + m_2}\right) v_{10} + \left(\frac{m_2 - m_1}{m_1 + m_2}\right) v_{20} \quad (7.11)$$

If m_2 is *initially at rest,* $v_{20} = 0$, the velocities after the collision reduce to

$$v_1 = \left(\frac{m_1 - m_2}{m_1 + m_2}\right) v_{10}\,; \quad v_2 = \left(\frac{2 m_1}{m_1 + m_2}\right) v_{10}$$

Furthermore, if $m_1 = m_2$, $v_1 = 0$ and $v_2 = v_{10}$. That is, the incoming particle comes to rest and the second one moves off with the velocity of the first particle. You may have seen this occur (approximately) for billiard ball collisions. If $m_1 = m_2$, but both particles have finite velocities before the collision, then $v_1 = v_{20}$ and $v_2 = v_{10}$. That is, the particles exchange velocities after the collision.

Second, if $m_1 \gg m_2$, $v_1 \approx v_{10}$ and $v_2 \approx 2v_{10}$. That is, the incoming heavy particle m_1 "hardly knows" m_2 is present; therefore, it continues its motion unaltered by the collision. On the other hand, the light particle m_2 rebounds with a velocity equal to about twice the initial velocity of m_1.

Finally, if $m_2 \gg m_1$, $v_1 \approx -v_{10}$ and $v_2 \approx 0$. That is, if a very light particle collides with a very heavy one, initially at rest, the light particle will have its velocity reversed, while the massive particle will remain approximately at rest.

Example 7.5

A hard sphere whose mass is 5 g makes a head-on, perfectly elastic collision with a mass of 10 g. The 5 g mass is initially moving to the right with a speed of 20 cm/sec, while the 10 g mass is initially at rest. (a) Find the velocities of the two masses after the collision. (b) Determine the kinetic energies of the two masses after the collision. What fraction of the total kinetic energy is transferred to the 10 g mass?

Solution

(a) Since energy and momentum are conserved, we can apply Equations (7.10) and (7.11) to get the velocities after the collision. Letting $m_1 = 5$ g, $m_2 = 10$ g, $v_{10} = 20$ cm/sec and $v_{20} = 0$, we have

$$v_1 = \left(\frac{5 - 10}{5 + 10}\right) 20 \text{ cm/sec} = -\frac{20}{3} \text{ cm/sec}$$

$$v_2 = \left(\frac{2\,(5)}{5 + 10}\right) 20 \text{ cm/sec} = \frac{40}{3} \text{ cm/sec}$$

(b) The initial kinetic energy is equal to $\frac{1}{2} m_1 v_{10}{}^2$, or

$$K_0 = \frac{1}{2}(5)\,(20)^2 \text{ ergs} = 10^3 \text{ ergs}$$

The kinetic energies of the 5 g and 10 g masses after the collision are

$$K_1 = \frac{1}{2} m_1 v_1{}^2 = \frac{1}{2}(5)\left(-\frac{20}{3}\right)^2 = \frac{10^3}{9} \text{ ergs}$$

$$K_2 = \frac{1}{2} m_2 v_2{}^2 = \frac{1}{2}(10)\left(\frac{40}{3}\right)^2 = \frac{8}{9} \times 10^3 \text{ ergs}$$

Note that $K_1 + K_2 = 10^3$ ergs $= K_0$, so our calculation is consistent with energy conservation. The fraction of energy transferred to the 10 g mass is

$$K_2/K_0 = 8/9$$

7.3 TWO DIMENSIONAL ELASTIC COLLISIONS

In the previous section, it was shown that the total momentum of a system is conserved when there are no external forces acting on the system. Therefore, this implies that the total momentum in each of the x, y and z directions is conserved. Let us consider a two-dimensional problem, where m_1 collides with m_2, where m_2 is initially at rest as shown in Figure 7-4.

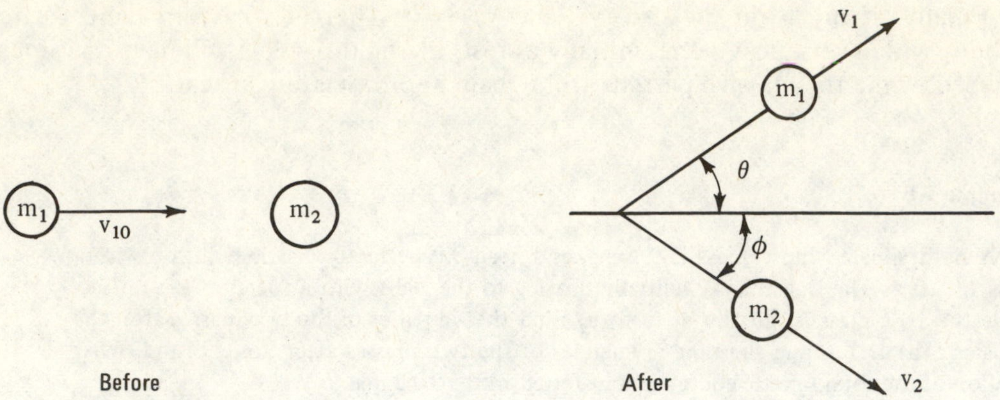

Figure 7-4 A collision in two dimensions. The mass m_2 is initially at rest.

Again, we can apply Equation (7.8) to the problem, since momentum is conserved. Let us write the component form of the momentum conservation law.

$$p_i \text{ before the collision} = p_i \text{ after the collision}$$

$$p_x = m_1 v_{10} = m_1 v_1 \cos\theta + m_2 v_2 \cos\phi \qquad (7.12)$$

$$p_y = 0 = m_1 v_1 \sin\theta - m_2 v_2 \sin\phi \qquad (7.13)$$

Furthermore, let us assume that the collision is perfectly elastic, so kinetic energy is conserved.

$$\frac{1}{2} m_1 v_{10}{}^2 = \frac{1}{2} m_1 v_1{}^2 + \frac{1}{2} m_2 v_2{}^2 \qquad (7.14)$$

The last three equations contain v_{10}, v_1, v_2, θ and ϕ as unknowns. If two of these are given, the remaining three can be found, since there are three independent equations available in the analysis.

Example 7.6

Two billiard balls collide in a perfectly elastic fashion. The incoming ball has an initial speed of 50 cm/sec, and makes a glancing collision with a second ball initially at rest. After the collision, the first ball is observed to move at an angle of +37° with its initial direction of motion, and the second moves at an angle of ϕ with the same axis. What are the final speeds of the two billiard balls, and what is ϕ?

Solution

Since $m_1 = m_2$, $\theta = 37°$, and $v_{10} = 50$ cm/sec, Equations (7.12), (7.13), and (7.14) become

$$v_1 \cos(37) + v_2 \cos\phi = 50$$

$$v_1 \sin(37) - v_2 \sin\phi = 0$$

$$v_1{}^2 + v_2{}^2 = (50)^2$$

These represent three equations, with three unknowns, v_1, v_2 and ϕ. Solving them simultaneously gives

$$v_1 = 40 \text{ cm/sec}$$

$$v_2 = 30 \text{ cm/sec}$$

$$\phi = 53^\circ$$

It is interesting to note that $\theta + \phi = 90^\circ$. This result is not fortuitous. Whenever two equal masses collide *elastically*, and one is initially at rest, the recoiling masses move away at *right angles* to each other. Of course, this is only true if the collision is *not* head-on.

Conservation of momentum in two or three dimensions is also a useful principle in other areas of physics. For example, in the preceding problem, the stationary particle may have been a stationary atom in a target, and the incident particle might be a nuclear particle from an accelerator. In this case, although the nuclear particle may not actually be on a direct collison course with the target atom, the particles could recoil due to a strong repulsive force between them which is large at small separation.

7.4 THE MOTION OF A SYSTEM OF PARTICLES–CENTER OF MASS

A system of particles consists of a collection of two or more objects whose masses, position vectors and velocities may be known. Suppose the mass m_1 has a position vector $\mathbf{r}_1$, m_2 has a position vector $\mathbf{r}_2$, and so on, as shown in Figure 7–5. These position vectors are measured with respect to an inertial frame of reference.

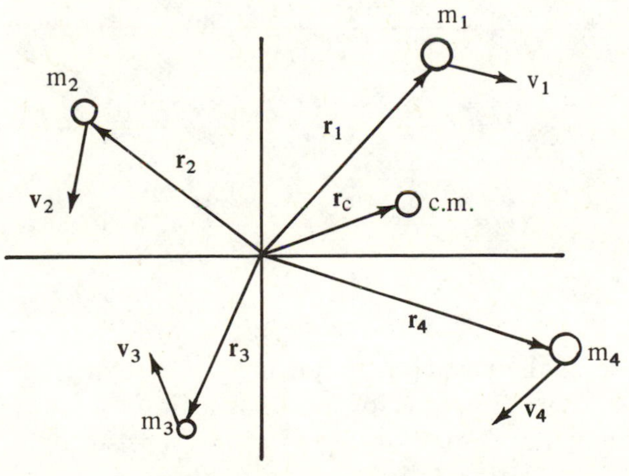

Figure 7–5 A system of particles. The position vector of the center of mass, $\mathbf{r}_c$, is defined by Equation (7.15).

The center of mass of the system of particles is defined by the position vector

$$\mathbf{r}_c = \frac{m_1 \mathbf{r}_1 + m_2 \mathbf{r}_2 + \ldots}{m_1 + m_2 + \ldots} = \frac{\Sigma_i m_i \mathbf{r}_i}{\Sigma_i m_i} \qquad (7.15)$$

The denominator of Equation (7.15) is the *total mass* of the system, M, given by

$$M = \Sigma_i m_i \tag{7.16}$$

The center of mass as defined by Equation (7.15) has x, y and z coordinates given by

$$x_c = \frac{\Sigma m_i x_i}{\Sigma m_i} \tag{7.17}$$

$$y_c = \frac{\Sigma m_i y_i}{\Sigma m_i} \tag{7.18}$$

$$z_c = \frac{\Sigma m_i z_i}{\Sigma m_i} \tag{7.19}$$

Example 7.7

A system consists of three particles. Particle 1 has a mass of 2 kg, and its position vector is $\mathbf{r}_1 = 3\mathbf{i}$ m. Particle 2 has a mass of 5 kg, and its position vector is $\mathbf{r}_2 = (2\mathbf{i} + 3\mathbf{j})$ m. Particle 3 has a mass of 3 kg, and its position vector is $\mathbf{r}_3 = (4\mathbf{i} - 3\mathbf{j})$ m. What is the position vector of the center of mass?

Solution

We can apply Equation (7.15) directly.

$$\mathbf{r}_c = \frac{m_1\mathbf{r}_1 + m_2\mathbf{r}_2 + m_3\mathbf{r}_3}{m_1 + m_2 + m_3}$$

$$\mathbf{r}_c = \frac{[2\,(3\mathbf{i}) + 5\,(2\mathbf{i} + 3\mathbf{j}) + 3\,(4\mathbf{i} - 3\mathbf{j})]\ \cancel{\text{kg}}\ \text{m}}{(2 + 5 + 3)\ \cancel{\text{kg}}}$$

$$\mathbf{r}_c = \frac{6\mathbf{i} + 10\mathbf{i} + 12\mathbf{i} + 15\mathbf{j} - 9\mathbf{j}}{10} = (2.8\,\mathbf{i} + 0.6\,\mathbf{j})\ \text{m}$$

If one or more of the particles comprising the system is moving with respect to the inertial frame, then the center of mass has a finite velocity $\mathbf{v}_c$. This velocity is obtained by differentiating Equation (7.15) with respect to time.

$$\mathbf{v}_c = \frac{d\mathbf{r}_c}{dt} = \frac{1}{M}\Sigma_i m_i \frac{d\mathbf{r}_i}{dt} = \frac{1}{M}\Sigma_i\, m_i\mathbf{v}_i \tag{7.20}$$

We can rewrite this expression as

$$M\mathbf{v}_c = \Sigma\ m_i\mathbf{v}_i = \Sigma_i\ \mathbf{p}_i \tag{7.21}$$

where $\mathbf{p}_i = m_i\mathbf{v}_i$, is the linear momentum of the ith particle. Therefore, $M\mathbf{v}_c$ is the *total momentum* of the system of particles, **P**.

$$\mathbf{P} = M\mathbf{v}_c \qquad (7.22)$$

We conclude that the system of particles has a total momentum which can be described by a single particle of mass M located at the center of mass, moving with a velocity $\mathbf{v}_c$.

Differentiating Equation (7.22) with respect to time gives the time rate of change of total momentum.

$$\frac{d\mathbf{P}}{dt} = M\frac{d\mathbf{v}_c}{dt} = M\mathbf{a}_c \qquad (7.23)$$

The vector $\mathbf{a}_c$ is the acceleration of the center of mass. Using Equation (7.20), we can write $\mathbf{a}_c$ as

$$\mathbf{a}_c = \frac{d\mathbf{v}_c}{dt} = \frac{1}{M}\Sigma m_i\frac{d\mathbf{v}_i}{dt} = \frac{1}{M}\Sigma m_i\mathbf{a}_i \qquad (7.24)$$

Example 7.8

A ball whose mass is 1 kg is thrown with an initial velocity $\mathbf{v}_{10} = (3\mathbf{i} + 5\mathbf{j})$ m/sec. One second later a second ball of the same mass is thrown from the same point with a velocity of $\mathbf{v}_{20} = (5\mathbf{i} + 3\mathbf{j})$ m/sec. What is the velocity and acceleration of the center of mass after the second ball is thrown?

Solution

We can use our expression for the velocity of a projectile moving in the earth's gravitational field, where $\mathbf{a} = -g\mathbf{j}$.

$$\mathbf{v} = \mathbf{v}_0 - gt\mathbf{j}$$

Applying this to the first ball gives

$$\mathbf{v}_1 = \mathbf{v}_{10} - gt\mathbf{j}$$

$$\mathbf{v}_1 = (3\mathbf{i} + 5\mathbf{j}) - gt\mathbf{j} \qquad (1)$$

Since the second ball is thrown one second later, we can write for its velocity

$$\mathbf{v}_2 = \mathbf{v}_{20} - g(t-1)\mathbf{j} = (5\mathbf{i} + 3\mathbf{j}) - g(t-1)\mathbf{j} \qquad (2)$$

where the time $t = 0$ corresponds to the time at which the *first* ball is thrown. Note that (2) is not physically meaningful for times $t < 1$ sec, since the second ball is not yet in motion. We are really interested in $\mathbf{v}_c$ and $\mathbf{a}_c$ for times $t > 1$ sec, since this corresponds to both balls in motion. Applying Equation (7.20), and making use of (1) and (2) gives

$$\mathbf{v}_c = \frac{1}{M}\Sigma_i m_i\mathbf{v}_i = \frac{1}{2}[(3\mathbf{i} + 5\mathbf{j}) - gt\mathbf{j} + (5\mathbf{i} + 3\mathbf{j}) - g(t-1)\mathbf{j}]$$

or

$$\mathbf{v}_c = [4\mathbf{i} + 4\mathbf{j} - g\left(t - \frac{1}{2}\right)\mathbf{j}] \text{ m/sec} \quad t > 1 \text{ sec} \tag{3}$$

The acceleration of the center of mass can be obtained from Equation (7.24) and (3).

$$\mathbf{a}_c = \frac{d\mathbf{v}_c}{dt} = \frac{d}{dt}\ [(4\mathbf{i} + 4\mathbf{j}) - g\left(t - \frac{1}{2}\right)\mathbf{j}] = -g\,\mathbf{j}$$

We could have guessed this without carrying through with the calculation, since both particles are freely falling bodies and have an acceleration $-g\mathbf{j}$; therefore, the center of mass must have the same acceleration.

7.5 NEWTON'S SECOND LAW APPLIED TO A SYSTEM OF PARTICLES

The total force on a system of particles may be written as the vector sum of all the individual forces on each part of the system. That is,

$$\mathbf{F} = \mathbf{F}_1 + \mathbf{F}_2 + \mathbf{F}_3 + \ldots$$

where $\mathbf{F}_1 = m_1\mathbf{a}_1$, $\mathbf{F}_2 = m_2\mathbf{a}_2$, and so on. In other words, each particle obeys Newton's second law, hence the system as a whole obeys Newton's second law. In the previous section, we found that

$$M\mathbf{a}_c = \sum_i m_i\mathbf{a}_i$$

Therefore, the total force acting on the system is given by

$$\mathbf{F} = M\mathbf{a}_c = \sum_i m_i\mathbf{a}_i \tag{7.25}$$

The individual forces acting on each particle of the system consist of both internal forces and external forces. The internal forces may arise, for example, from gravitational interactions, collisions or electromagnetic interactions. However, by Newton's third law, these forces occur in pairs which are equal in magnitude, opposite in direction. Therefore, when we take the vector sum to get **F**, the internal forces cancel, and the sum reduces to a sum over *external* forces. Therefore, we can write Equation (7.25) as

$$\mathbf{F}_{ext} = M\mathbf{a}_c = \frac{d\mathbf{P}}{dt} \tag{7.26}$$

This result implies that the motion of a system of particles can be represented by a single particle of mass M located at the center of mass, with all the external forces applied at that point. In other words, the center of mass is a special point whose motion represents the average motion of the system of particles.

If $\mathbf{F}_{ext} = 0$, we see from Equation (7.26) that the total momentum of the system remains constant, or is conserved.

$$\mathbf{P} = \text{constant} \qquad (\text{when } \mathbf{F}_{ext} = 0) \tag{7.27}$$

This result is similar to Equation (7.8) obtained for a single particle. Therefore, Equation (7.27) is a statement of the *law of conservation of linear momentum* for a system of particles. From this, we can also conclude that the *center of mass of an isolated system of particles moves with constant velocity.* In other words,

$$\mathbf{v}_c = \text{constant} \qquad (\text{when } \mathbf{F}_{ext} = 0) \tag{7.28}$$

Example 7.9

Two carts resting on a horizontal frictionless surface are connected by a spring as in Figure 7-6. The blocks are pulled some distance apart so that the spring is stretched and are then released from rest. The cart, whose mass is 300 g, is later observed to have a velocity of 50 cm/sec in the +x direction. What is the velocity of the 200 g mass at this time?

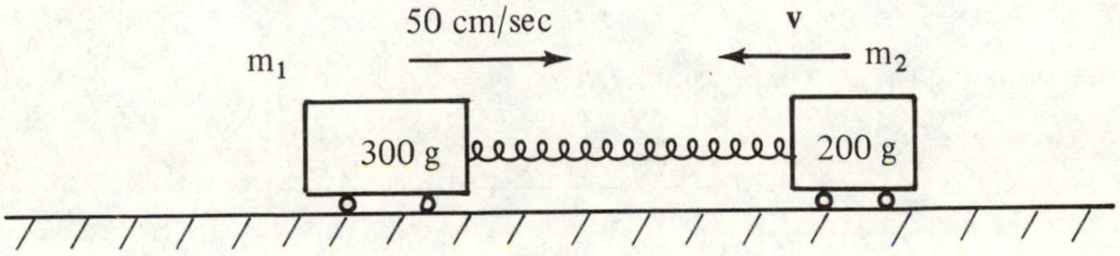

Figure 7-6

Solution

Since the system is released from rest, the *total* momentum of the system is zero.

$$\mathbf{P} = 0$$

The total momentum of the system for the subsequent motion will remain zero, since there are *no* external forces acting on the system. (The spring force is internal to the system.) Therefore, when the velocity of the 300 g mass is 50 **i** cm/sec, we can write

$$\mathbf{P} = m_1\mathbf{v}_1 + m_2\mathbf{v}_2 = 300\,(50)\,\mathbf{i}\text{ g cm/sec} + 200\,\mathbf{v} = 0$$

or,

$$\mathbf{v} = -\frac{300\,(50)\,\mathbf{i}}{200}\text{ cm/sec}$$

$$\mathbf{v} = -75\,\mathbf{i}\ \text{ cm/sec}$$

That is, the 200 g mass is moving to the left with a speed of 75 cm/sec. In this situation, the center of mass remains *stationary,* since $\mathbf{v}_c = 0$ initially, and must remain zero, since $\mathbf{F}_{ext} = 0$.

Example 7.10

A rocket is fired vertically upwards. It reaches an altitude of 1000 m, with an upward velocity of 300 m/sec, at which time an explosion occurs, separating the rocket into three equal fragments. One fragment continues to move upwards with a speed of 400 m/sec right after the explosion. The second fragment has a speed of 200 m/sec at a 45° angle to the original motion right after the explosion. (a) What is the velocity of the third fragment right after the explosion?

Solution

Let us call the total mass of the rocket M, and the mass of each fragment M/3. The total initial momentum before the explosion must equal the total momentum of the fragments after the explosion since the forces of the explosion represent *internal* forces, and cannot change the total momentum of the system.

$$\mathbf{P} = M\mathbf{V} = M\,300\,\mathbf{j}\ \text{m/sec} \qquad \textbf{Before the Explosion}$$

$$\mathbf{P} = \frac{M}{3}(400)\,\mathbf{j} + \frac{M}{3}100\sqrt{2}\,(\mathbf{i}+\mathbf{j}) + \frac{M}{3}\mathbf{v} \qquad \textbf{After the Explosion}$$

Therefore, equating the two expressions gives

$$\frac{\mathbf{v}}{3} = \left(300 - \frac{400}{3} - \frac{100\sqrt{2}}{3}\right)\mathbf{j} - \frac{100\sqrt{2}}{3}\mathbf{i}$$

or,

$$\mathbf{v} = (-141\,\mathbf{i} + 359\,\mathbf{j})\ \ \text{m/sec}$$

(b) What is the position of the center of mass 5 sec after the explosion relative to the ground? Assume the rocket engine is nonoperative immediately after the explosion.

Solution

Immediately after the explosion, we can treat the center of mass of the system as a freely falling body with $\mathbf{v}_0 = 300\,\mathbf{j}$. In other words, we can ignore the explosion, since it doesn't affect the motion of the center of mass. Since the rocket has an initial coordinate $y_0 = 1000$ m at $t = 0$ (the time of the explosion), we have

$$y_c = y_0 + v_0\,t - \frac{1}{2}gt^2 = 1000 + 300\,t - \frac{1}{2}gt^2$$

At $t = 5$ sec,

$$y_c = [1000 + 300\,(5) - 4.9\,(5)^2]\ \text{m} \cong 2377\ \text{m}$$

7.6 ROCKET PROPULSION

When ordinary vehicles, such as automobiles, boats, locomotives or airplanes are propelled, the driving force for the acceleration is one of friction. An automobile pushes against the road, so the driving force is the force of the road on the car. A locomotive pushes against the tracks, hence the driving force is the force of the track on the locomotive. However, rockets moving in space have no air, tracks or water to push against; therefore, the sources of their propulsion must be different. The principle of rocket propulsion lies in the law of conservation of momentum. If the rocket moves in free space, its velocity will only change if some of its mass is ejected. The mass ejected is the fuel carried by the rocket. If the exhaust velocity of the fuel particles is large, the exhaust particles have a large momentum. The rocket must receive a compensating momentum in the opposite direction of the exhaust particles, since the fuel + rocket make up the system, and the total momentum of the system remains constant. Therefore, the rocket is accelerated due to the "push" of the exhaust-fuel on the rocket. Likewise, by Newton's third law, the rocket pushes back on the exhaust-fuel with an equal and opposite force. During this time, the center of mass moves uniformly, independent of the propulsion process.

To be more quantitative, let us consider the rocket moving upwards in the earth's gravitational field as in Figure 7–7. We will assume that the acceleration of gravity is constant and equal to $-g\mathbf{j}$, and we shall neglect air resistance. The initial momentum of the rocket, whose mass is m, velocity $v\mathbf{j}$, is given by

$$\mathbf{p}_0 = mv\mathbf{j}$$

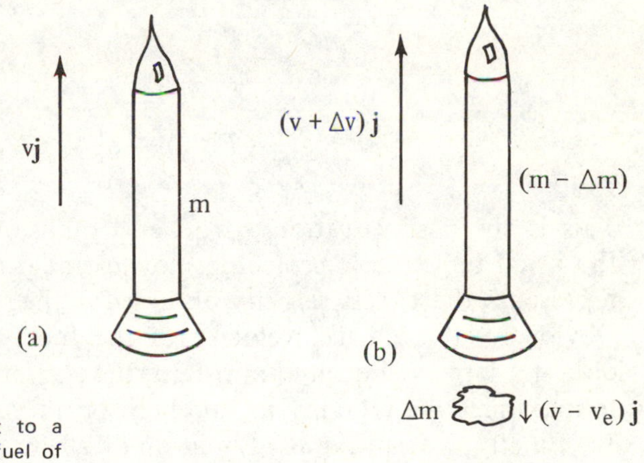

Figure 7–7 The rocket system with respect to a stationary inertial frame *(a)* before it ejects fuel of mass Δm and *(b)* after it ejects some fuel.

A short time later, it ejects some fuel Δm, and its speed increases to $v + \Delta v$. The fuel is ejected at a velocity $-v_e\mathbf{j}$ *relative to the rocket,* hence its velocity relative to a stationary frame is $(v - v_e)\mathbf{j}$. At the end of the short time interval, the momentum of the system is

$$\mathbf{p} = (m - \Delta m)(v + \Delta v)\,\mathbf{j} + \Delta m\,(v - v_e)\,\mathbf{j}$$

Since the only external force acting on the system during this interval is that of gravity, $-mg\mathbf{j}$, we can apply Equation (7.7) to get the impulse of this force in terms of the change in momentum of the system. In a time interval Δt, the impulse of this force becomes

$$\mathbf{I} = -\,mg\Delta t\,\mathbf{j} = \mathbf{p} - \mathbf{p}_0$$

or,

$$-mg\Delta t = (m - \Delta m)(v + \Delta v) + \Delta m (v - v_e) - mv$$

Simplifying this expression, and neglecting the higher order term $\Delta m \Delta v$ gives

$$-mg\Delta t = m\Delta v - v_e \Delta m$$

Taking the limit as Δm and Δv approach zero, and noting that an increase in v corresponds to a decrease in m (so as to effectively change signs of the last term) gives

$$dv = -v_e \frac{dm}{m} - gdt \tag{7.29}$$

Letting the mass and velocity of the rocket be m_0 and v_0 at $t = 0$, and integrating Equation (7.29) once gives

$$v - v_0 = -v_e \, ln\left(\frac{m}{m_0}\right) - gt$$

or,

$$v = v_0 - gt + v_e \, ln\left(\frac{m_0}{m}\right) \tag{7.30}$$

This is the basic equation of rocket propulsion. The first two terms correspond to the usual terms for a projectile moving under the influence of gravity. The last term represents the excess velocity obtained as the result of ejecting fuel of mass $m_0 - m$. Obviously, the exhaust velocity of the fuel, v_e, should be as large as possible to obtain a large v. Equally important, the rocket should carry as much fuel as possible to maximize the ratio m_0/m, and hence maximize v.

Finally, we can express Equation (7.29) in the following equivalent form:

$$-v_e \frac{dm}{dt} - mg = m\frac{dv}{dt} \tag{7.31}$$

This is simply Newton's second law applied to the rocket. The left side is the total external force on the rocket due to the exhaust (or so-called *thrust*), and the right side is simply ma.

Example 7.11

A rocket moving in *free space* has a speed of 5×10^3 m/sec. Suddenly, its engines are turned on, and fuel is ejected in the reverse direction at a speed of 2×10^3 m/sec relative to the rocket. (a) What is the speed of the rocket when its mass is reduced to 0.1 that of the rocket before its engines were ignited?

Solution

In free space, $g = 0$, so Equation (7.30) reduces to

$$v = v_0 + v_e \ln\left(\frac{m_0}{m}\right)$$

In this example, $v_0 = 5 \times 10^3$ m/sec, $v_e = 2 \times 10^3$ m/sec, and $m = 0.1\ m_0$. Therefore, v becomes

$$v = [5 \times 10^3 + 2 \times 10^3 \ln(10)] \text{ m/sec}$$

$$v = 9.6 \times 10^3 \text{ m/sec.}$$

(b) Suppose the rocket ejects the fuel at the rate of 30 kg/sec. Determine the thrust of the rocket, and its acceleration when its mass is 5000 kg.

Solution

From Equation (7.31), the thrust is given by

$$F_t = v_e \left|\frac{dm}{dt}\right| = 2 \times 10^3 \text{ m/sec}\left(\frac{30 \text{ kg}}{\text{sec}}\right) = 6 \times 10^4 \text{ N}$$

Since $g = 0$, we can use Equation (7.31) and the result above to obtain the acceleration of the vehicle.

$$a = \frac{dv}{dt} = \frac{v_e \left|\frac{dm}{dt}\right|}{m} = \frac{6 \times 10^4 \text{ N}}{5 \times 10^3 \text{ kg}} = 12 \text{ m/sec}^2$$

7.7 PROGRAMMED EXERCISES

1.A

What is the linear momentum of a particle whose mass is m and whose velocity is **v**?

$\mathbf{p} = m\mathbf{v}$

1.B

Under what condition will **p** remain constant?

The linear momentum **p** of the particle remains constant when $\mathbf{F}_{ext} = 0$, since $\mathbf{F} = d\mathbf{p}/dt = 0$ only when $\mathbf{p}$ = constant.

1.C

Under what condition will **p** change in time?

When $\mathbf{F}_{ext} \neq 0$, $d\mathbf{p}/dt \neq 0$, so the linear momentum varies in time.

1.D

A particle of mass m has an initial velocity $\mathbf{v}_0$. A force **F** acts on the particle in a short time interval Δt. If the final velocity of the particle is **v**, what is the *change* in momentum of the particle?

$\mathbf{p}_0 = m\mathbf{v}_0$ initially

$\mathbf{p} = m\mathbf{v}$ finally

$\Delta\mathbf{p} = \mathbf{p} - \mathbf{p}_0 = m(\mathbf{v} - \mathbf{v}_0)$

1.E

What is the impulse of the force **F**?

$\mathbf{I} = \mathbf{F}\Delta t = \Delta\mathbf{p}$

1.F

Suppose the particle is a baseball weighing 1/2 lb. If its initial velocity is 100 **i** ft/sec, and it is struck by a batter giving it a final velocity of (−100 **i** + 50 **j**) ft/sec, what is the Impulse delivered by the bat?

$$m\mathbf{v} - m\mathbf{v}_0 = \frac{1/2 \text{ lb}}{32 \text{ ft/sec}^2}[-100\mathbf{i} + 50\mathbf{j} - 100\mathbf{i}] \text{ ft/sec}$$

$$\therefore \mathbf{I} = \Delta\mathbf{p} = (-3.1\,\mathbf{i} + 0.78\,\mathbf{j}) \text{ lb sec}$$

1.G

If the bat is in contact with the ball for 10^{-3} sec, what is the average impulsive force exerted on the ball?

$$\mathbf{F}_{av} = \frac{\mathbf{I}}{\Delta t} = (-3.1\,\mathbf{i} + 0.78\,\mathbf{j}) \times 10^3 \text{ lb}$$

or

$$|\mathbf{F}_{av}| \cong 3.2 \times 10^3 \text{ lb}$$

2 Batman drops from a height h onto a platform which is supported by a sturdy spring of force constant k. Since Batman is wearing suction cups on his feet, he sticks to the platform upon colliding with it. Assume that Batman has a mass M, and the platform has a mass m.

2.A

What is Batman's speed *just before* he collides with the platform?

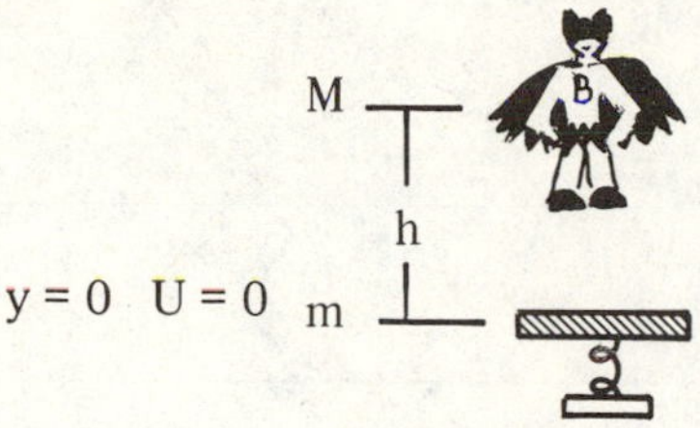

At $y = h$, $K = 0$ and $U = Mgh$.

At $y = 0$, $K = \frac{1}{2}Mv^2$ and $U = 0$.

$$\therefore \frac{1}{2}Mv^2 = Mgh$$

$$v = \sqrt{2gh} \tag{1}$$

2.B

Since Batman sticks to the platform, the collision is perfectly inelastic. Find the speed V of Batman + platform right after the collision.

Momentum is conserved, so

$$Mv = (M + m)\,V$$

$$V = \left(\frac{M}{M+m}\right)\sqrt{2\,gh} \tag{2}$$

2.C

Is the initial energy of the system equal to the energy after the collision? Explain.

No. Mechanical energy is lost in an inelastic collision. From (1) and (2), we see that $\frac{1}{2}Mv^2 > \frac{1}{2}(M + m)\,V^2$ or,

$$Mgh > \frac{1}{2}(M + m)\,V^2$$

2.D

What fraction of the energy is *lost* in the collision?

$$f = 1 - \frac{E}{E_0} = 1 - \frac{\frac{1}{2}(M + m)\,V^2}{Mgh}$$

$$f = 1 - \left(\frac{M}{M+m}\right) = \frac{m}{M+m} \tag{3}$$

2.E

Find the maximum compression of the spring y. Use energy methods. Neglect the initial compression of the spring due to the platform, and assume Batman is a point mass.

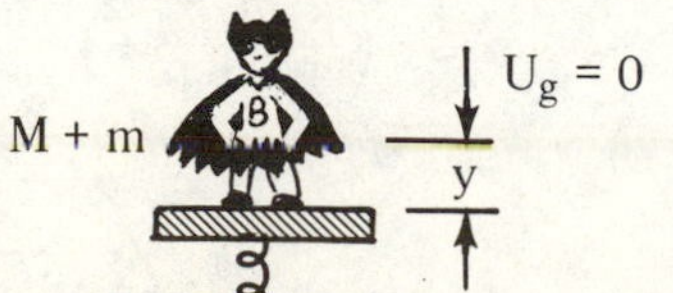

At $y = 0$, $E = \frac{1}{2}(m + M)\,V^2$ right after the collision. When the spring is compressed, $K = 0$, $U_s = \frac{1}{2}ky^2$ and $U_g = -(m + M)\,gy$. Therefore,

$$\frac{1}{2}(m + M)\,V^2 = \frac{1}{2}ky^2 - (m + M)\,gy \tag{4}$$

Now use (2), and solve the quadratic equation for y.

3 Two carts of masses m_1 and m_2 are placed on a horizontal, *frictionless* air track. The mass m_2 is connected to a spring of force constant k, which in turn is fastened to a rigid support. m_1 is given an initial velocity v_0 to the right. It collides with m_2 and the masses *stick* together.

3.A

What kind of collision is this? Is any mechanical quantity conserved for this system? Explain.

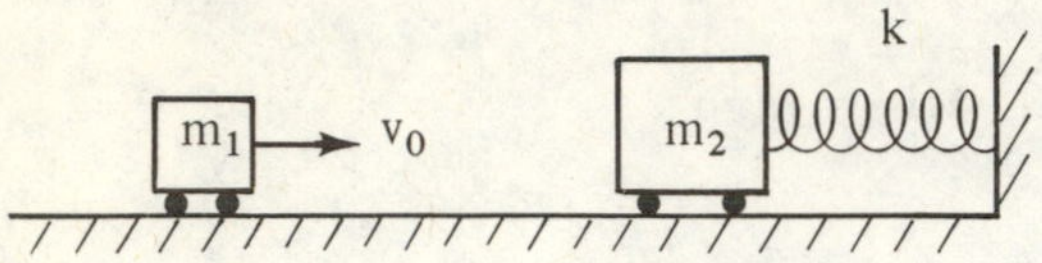

This is a completely inelastic collision. Yes, linear momentum is conserved for all collisions.

3.B

Calculate the velocity v of the system *immediately after* the collision.

The initial momentum must equal the final momentum.

$$\therefore m_1 v_0 = (m_1 + m_2)v$$

or,

$$v = \left(\frac{m_1}{m_1 + m_2}\right) v_0 \qquad (1)$$

3.C

What is the kinetic energy of the system before and after the collision? Is kinetic energy conserved?

$$K_0 = \frac{1}{2} m_1 v_0^2 \qquad \text{(before)} \qquad (2)$$

$$K = \frac{1}{2}(m_1 + m_2) v^2$$

$$K = \frac{1}{2}\left(\frac{m_1^2}{m_2 + m_2}\right) v_0^2 \qquad \text{(after)} \qquad (3)$$

No. $K < K_0$

3.D

Find the maximum distance x the spring will be compressed. Use energy methods.

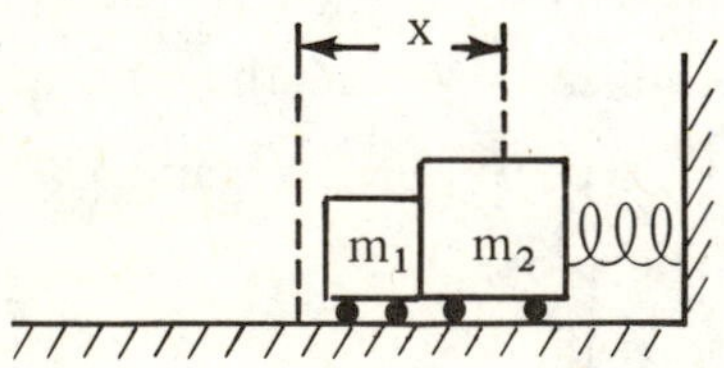

The initial energy of the system after the collision is K given by (3). When the spring is completely compressed, K = 0, and $U_s = \frac{1}{2} kx^2$. Since there is no friction, mechanical energy is conserved. Therefore,

$$\frac{1}{2}\left(\frac{m_1^2}{m_1 + m_2}\right) v_0^2 = \frac{1}{2} kx^2$$

or

$$x = v_0 \sqrt{\frac{m_1^2}{k(m_1 + m_2)}}$$

3.E

Obtain a numerical value for x if $v_0 = 50$ cm/sec, $m_1 = 100$ g, $m_2 = 200$ g and $k = 10^4$ dyne/cm.

$$x = 50\,\frac{\text{cm}}{\text{sec}}\sqrt{\frac{10^4 \qquad \text{g}^2}{10^4\ (300)\ \frac{\text{dyne}}{\text{cm}}\ \text{g}}}$$

$$x = 2.9\ \text{cm}$$

4 Two masses undergo a *perfectly elastic head-on* collision. The first mass m moves with an initial velocity of 50 m/sec in the +x direction. The second mass 2m moves with an initial speed of 40 m/sec in the −x direction as shown.

4.A

What is the final velocity of the mass m? Make use of Equation (7.10).

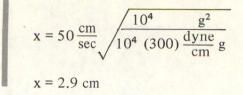

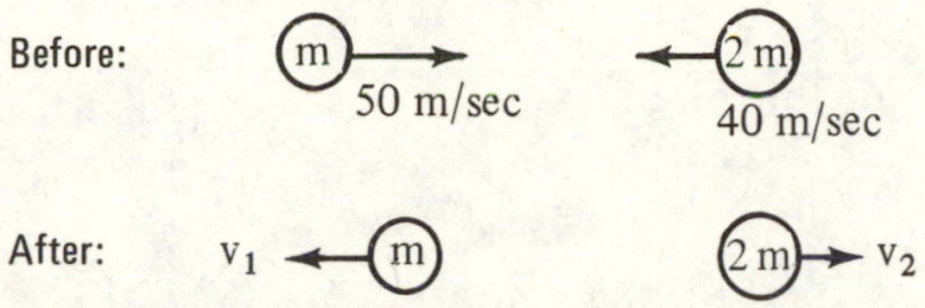

Since $v_{10} = 50$ m/sec and $v_{20} = -80$ m/sec,

$$v_1 = \left(\frac{m - 2m}{m + 2m}\right) 50 - \left(\frac{4m}{m + 2m}\right) 40$$

$$v_1 = -\frac{50}{3} - \frac{4}{3}(40) = -70 \quad \text{m/sec} \qquad (1)$$

4.B

Using Equation (7.11), find the final velocity of the second mass of 2m.

$$v_2 = \left(\frac{2m}{m + 2m}\right) 50 - \left(\frac{2m - m}{m + 2m}\right) 40$$

$$v_2 = \frac{2}{3}(50) - \frac{40}{3} = 20 \quad \text{m/sec} \qquad (2)$$

4.C

What is the kinetic energy of the *system before* the collision? Take m = 1 kg, 2m = 2 kg.

$$K = \frac{1}{2}\,m v_{10}^2 + \frac{1}{2}(2m)\,v_{20}^2$$

$$K = \left[\frac{1}{2}(1)\,(50)^2 + \frac{1}{2}(2)\,(40)^2\right]\ \text{J}$$

$$K = 2.85 \times 10^3\ \text{J} \qquad (3)$$

4.D

What is the kinetic energy of the *system after* the collision?

$K = 2.85 \times 10^3$ J since kinetic energy is *conserved* for an elastic collision. This can also be shown by calculating the final kinetic energy using (1) and (2).

$$K = \frac{1}{2}\,m v_1{}^2 + \frac{1}{2}(2m)\,v_2{}^2$$

$$= \frac{1}{2}(1)\,(70)^2 + (20)^2$$

$$K = 2.85 \times 10^3\ \text{J}$$

4.E

What is the ratio of the final kinetic energy to the initial kinetic energy for m?

$$K_0 = \frac{1}{2} m v_{10}^{\ 2} \quad K = \frac{1}{2} m v_1^{\ 2}$$

$$\frac{K}{K_0} = \frac{v_1^{\ 2}}{v_{10}^{\ 2}} = \frac{(-70)^2}{(50)^2}$$

$$\frac{K}{K_0} = 1.96$$

4.F

Calculate the *change* in linear momentum for the 1 kg mass. Likewise, calculate the change in linear momentum for the 2 kg mass.

$$p_0 = mv_{10} = 1\ (50) \quad \text{kg m/sec}$$

$$p \ = mv_1 = -1\ (70) \quad \text{kg m/sec}$$

$$\Delta p_1 = p - p_0 = -120 \quad \text{kg m/sec}$$

For the 2 kg mass,

$$\Delta p_2 = 2\ mv_2 - 2\ mv_{20}$$

$$\Delta p_2 = 2\ (20) + 2\ (40) = 120 \quad \text{kg m/sec}$$

4.G

What is the change in momentum for the system? How do you interpret this result?

$$\Delta p = \Delta p_1 + \Delta p_2 = 0$$

This must be true, since there are *no* external forces, hence the momentum of the system is conserved.

5 Consider two particles of *equal mass* m making a *perfectly elastic* glancing collision as in the figure below. One mass is initially at rest and the incoming particle has an initial momentum $\mathbf{p}_0$. The final momenta of the two particles are $\mathbf{p}_1$ and $\mathbf{p}_2$. We will show that the particles scatter at right angles, that is, $\theta + \phi = \pi/2$.

5.A

Write a vector expression for the conservation of linear momentum.

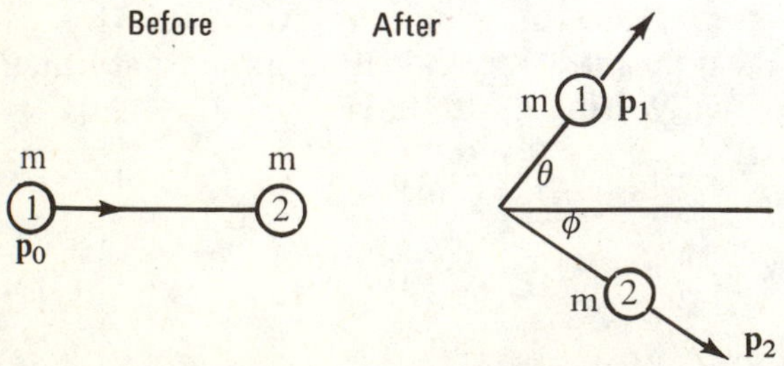

$$\mathbf{p}_0 = \mathbf{p}_1 + \mathbf{p}_2 \tag{1}$$

where $\mathbf{p}_0 = m\mathbf{v}_0$, $\mathbf{p}_1 = m\mathbf{v}_1$, and

$$\mathbf{p}_2 = m\mathbf{v}_2$$

5.B

Obtain an expression for the initial kinetic energy of the system in terms of the momentum of particle (1) and its mass m.

$$K_0 = \frac{1}{2} m v_0^2$$

Since $p_0 = m v_0$

$$K_0 = \frac{p_0^2}{2m} \qquad (2)$$

5.C

Obtain an expression for the final kinetic energy of the system in terms of the final momenta of the particles and their masses.

$$K = \frac{1}{2} m_1 v_1^2 + \frac{1}{2} m_2 v_2^2$$

Since $p_1 = m v_1$ and $p_2 = m v_2$

$$K = \frac{p_1^2}{2m} + \frac{p_2^2}{2m} \qquad (3)$$

5.D

Since this is assumed to be an elastic collision, kinetic energy is conserved. Use this fact, together with (2) and (3). to obtain a second relation between the initial and final momenta.

$$K_0 = K$$

$$\frac{p_0^2}{2\,m} = \frac{p_1^2}{2\,m} + \frac{p_2^2}{2\,m}$$

or,

$$p_0^2 = p_1^2 + p_2^2 \qquad (4)$$

5.E

According to (1) and (4), what can you conclude about the sum of the angles $\theta + \phi$? *Hint:* Draw a *vector* diagram corresponding to the condition required by (1). *Note:* The result is valid *only* for elastic collisions between two equal masses.

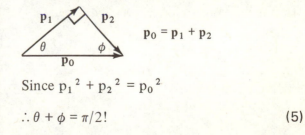

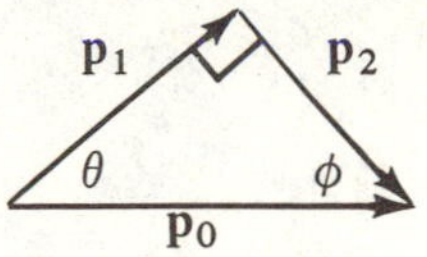

$$p_0 = p_1 + p_2$$

Since $p_1^2 + p_2^2 = p_0^2$

$$\therefore \theta + \phi = \pi/2! \qquad (5)$$

6 A 5 g mass and a 3 g mass move along the x axis as shown below. At some time t, the 5 g mass is located at x = 2 cm, it has a velocity of +10 cm/sec and an acceleration of +20 cm/sec^2. At the same instant, the 3 g mass is located at x = 10 cm, has a velocity of −6 cm/sec and an acceleration of −30 cm/sec^2.

6.A

Write vector expressions for the positions of the masses at this time.

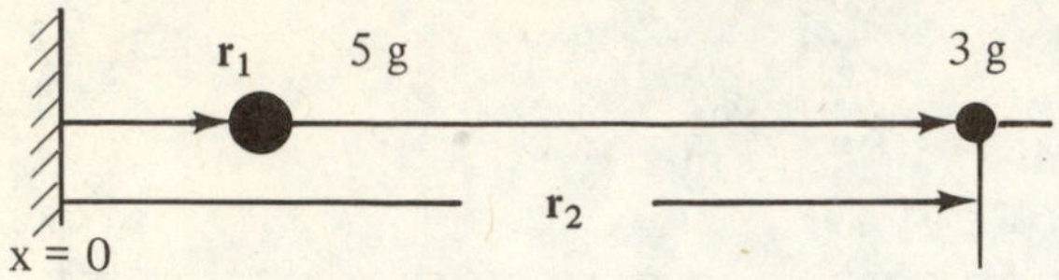

$\mathbf{r}_1 = 2\mathbf{i}$ cm

$\mathbf{r}_2 = 10\mathbf{i}$ cm

6.B

Find the position vector of the center of mass measured from x = 0. Use Equation (7.15).

By definition, $\mathbf{r}_c = \dfrac{m_1 \mathbf{r}_1 + m_2 \mathbf{r}_2}{m_1 + m_2}$

$$\mathbf{r}_c = \frac{5\,(2\,\mathbf{i}) + 3\,(10\,\mathbf{i})}{5 + 3} = 5\mathbf{i} \text{ cm}$$

6.C

Write vector expressions for the velocities of the two particles at this time.

$\mathbf{v}_1 = 10\mathbf{i}$ cm/sec (1)

$\mathbf{v}_2 = -6\mathbf{i}$ cm/sec (2)

6.D

What is the velocity of the center of mass? Use Equation (7.20) and (1) and (2).

$$\mathbf{v}_c = \frac{\Sigma m_i \mathbf{v}_i}{\Sigma m_i} = \frac{m_1 \mathbf{v}_1 + m_2 \mathbf{v}_2}{m_1 + m_2}$$

$$\mathbf{v}_c = \frac{5\,(10\,\mathbf{i}) - 3\,(6\mathbf{i})}{5 + 3} = 4\mathbf{i} \text{ cm/sec} \qquad (3)$$

6.E

What is momentum of the center of mass?

$\mathbf{p} = M\mathbf{v}_c = (m_1 + m_2)\,\mathbf{v}_c$

$$\mathbf{p} = 8\,(4\,\mathbf{i}) = 32\mathbf{i}\,\frac{\text{g cm}}{\text{sec}} \qquad (4)$$

6.F

Calculate the total momentum of the system using (1) and (2). Your result should be equal to (4).

$\mathbf{p} = \mathbf{p}_1 + \mathbf{p}_2 = m_1 \mathbf{v}_1 + m_2 \mathbf{v}_2$

$\mathbf{p} = 5\,(10\mathbf{i}) - 3\,(6\mathbf{i})$

$$\mathbf{p} = 32\mathbf{i}\,\frac{\text{g cm}}{\text{sec}}$$

6.G

Determine the kinetic energy of the system using (1) and (2).

$K = K_1 + K_2 = \frac{1}{2} m_1 v_1{}^2 + \frac{1}{2} m_2 v_2{}^2$

$K = \left[\frac{1}{2}(5)\,(10)^2 + \frac{1}{2}(3)\,(6)^2\right]$ ergs

$K = 304$ ergs (5)

6.H

What is the *relative* speed of the two particles?

5 g
3 g
10 cm/sec
-6 cm/sec

$v_r = v_1 - v_2 = 10 - (-6)$

$v_r = 16$ cm/sec

That is, one particle "sees" the other approaching it at a speed of 16 m/sec.

6.I

What are the velocities of the two particles as "seen" by an observer in the center of mass frame?

$v_1{}' = v_1 - v_c = 10\mathbf{i} - 4\mathbf{i}$

$v_1{}' = 6\mathbf{i}$ cm/sec (6)

$v_2{}' = v_2 - v_c = -6\mathbf{i} - 4\mathbf{i}$

$v_2{}' = -10\mathbf{i}$ cm/sec (7)

6.J

Calculate the kinetic energy of the system as measured by the observer in the center of mass.

$K' = \frac{1}{2} m_1 v_1{}'^2 + \frac{1}{2} m_2 v_2{}'^2$

$K' = \frac{1}{2}(5)\,(6)^2 + \frac{1}{2}(3)\,(10)^2$

$K' = 240$ ergs (8)

6.K

Now calculate the kinetic energy of the center of mass, $K_c = \frac{1}{2} M v_c{}^2$, where $M = m_1 + m_2$.

$K_c = \frac{1}{2} M v_c{}^2 = \frac{1}{2}(5 + 3)\,(4)^2$

$K_c = 64$ ergs (9)

6.L

Add the results (8) and (9) and compare this to the total kinetic energy (5) measured by the stationary observer. What general expression does this result suggest?

$K' + K_c = (240 + 64)$ ergs

$K' + K_c = 304$ ergs

This equals K given by (5). This suggests that $K = K' + K_c$, or $\frac{1}{2} m_1 v_1{}^2 + \frac{1}{2} m_2 v_2{}^2 = \frac{1}{2} m_1 v_1{}'^2 + \frac{1}{2} m_2 v_2{}'^2 + \frac{1}{2} M v_c{}^2$

6.M

Find the acceleration of the center of mass of the system. Make use of Equation (7.24).

$\mathbf{a}_1 = 20\mathbf{i}$ cm/sec²

$\mathbf{a}_2 = -30\mathbf{i}$ cm/sec²

$$\mathbf{a}_c = \frac{m_1\mathbf{a}_1 + m_2\mathbf{a}_2}{m_1 + m_2}$$

$$\mathbf{a}_c = \frac{5\,(20\mathbf{i}) - 3\,(30\mathbf{i})}{8} = 1.25\mathbf{i} \quad \text{cm/sec}^2 \qquad (10)$$

6.N

What is total external force on the system?

$$F_{ext} = Ma_c = 8\,(1.25)\,\mathbf{i}\ \frac{\text{g cm}}{\text{sec}^2}$$

$$F_{ext} = 10\mathbf{i} \quad \text{dynes} \qquad (11)$$

This is also equal to $m_1\mathbf{a}_1 + m_2\mathbf{a}_2$

6.O

What would you conclude about the behavior of the total momentum as a function of time? What about the velocity of center of mass?

Since $F_{ext} = \dfrac{d\mathbf{p}}{dt} \neq 0$, $\mathbf{p}$ changes with time; therefore, $\mathbf{v}_c$ changes with time.

7 This exercise demonstrates the utility of the *center of mass* concept for an isolated system. Suppose a fisherman situated on a frozen lake catches a large fish and wishes to know the weight of the fish. Other than his fishing equipment, he carries a measuring tape and a rope. If he knows his weight to be 150 lb, how can he estimate the weight of the fish, W?

7.A

Suppose the fisherman ties the rope to the fish and stands 10 ft from the fish. Where is the center of mass of the system (man + fish) relative to a point midway between the fish and man?

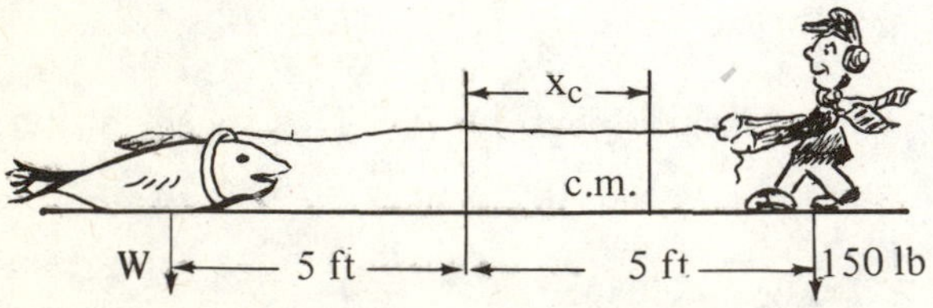

Let W represent the weight of the fish. By definition,

$$x_c = \frac{150\,(5) - 5W}{W + 150}$$

$$x_c = \frac{750 - 5W}{W + 150} \ \text{ft} \qquad (1)$$

Note: We should use masses in these equations, but g cancels in every term.

7.B

The man is initially at rest at the position shown above. Now he begins to pull on the rope. Does the position of the center of mass change? Explain.

No. There are *no* external forces acting on the system, and the total momentum is zero, initially. Since the forces on the rope are internal, the momentum remains zero, and the *center of mass does not move.*

7.C

If the man continues to pull the fish, where will they meet?

They will meet at the *center of mass.*

7.D

Suppose the man measures the distance from $x = 0$ to the center of mass to be $x_c = 2$ ft. What is the weight of the fish?

From (1), we have

$$\frac{750 - 5W}{150 + W} = 2$$

$$750 - 5W = 300 + 2W$$

$$7W = 450$$

$$W \cong 64 \text{ lb}$$

Quite a catch!

7.8 SUMMARY

Linear momentum of a particle:	$\mathbf{p} = m\mathbf{v}$	(7.6)
Impulse of a force **F**:	$\mathbf{I} = \int \mathbf{F}dt = \Delta\mathbf{p}$	(7.5)
Conservation of linear momentum due to a collision between two particles:	$m_1\mathbf{v}_{10} + m_2\mathbf{v}_{20} = m_1\mathbf{v}_1 + m_2\mathbf{v}_2$	
Position vector of the center of mass for a *system* of particles:	$\mathbf{r}_c = \dfrac{\Sigma m_i\mathbf{r}_i}{\Sigma m_i}$	(7.15)
Velocity of the center of mass:	$\mathbf{v}_c = \dfrac{\Sigma m_i\mathbf{v}_i}{\Sigma m_i}$	(7.20)
Momentum of the center of mass:	$\mathbf{P} = M\mathbf{v}_c$	(7.22)
Acceleration of the center of mass:	$\mathbf{a}_c = \dfrac{\Sigma m_i\mathbf{a}_i}{\Sigma m_i}$	(7.24)

7.9 PROBLEMS

1. A golfball, whose mass is 25 g, is hit off a tee with an initial speed of 50 m/sec. (a) What is the magnitude of the initial momentum of the ball? (b) What is its kinetic energy? (c) What is the impulse imparted to the ball?

2. An automobile weighing 3200 lb is traveling Eastward at a speed of 30 mi/hr. The car makes a 90 degree turn to the North in a time of 3 sec, and continues with the same speed. (a) What is the impulse delivered to the car during the turn? (b) Calculate the average force exerted on the car during the turn.

3. A ball of mass 200 g is thrown horizontally against a brick wall with an initial velocity of 20 m/sec to the right. It rebounds with a velocity of 15 m/sec to the left. (a) Calculate the change in momentum of the ball. (b) Find the average force of the wall on the ball during the collision, if the ball is in contact with the wall for 10^{-2} sec. (c) How much mechanical energy is lost in this collision?

4. A 2000 lb car traveling initially at a speed of 60 mi/hr in the Easterly direction crashes into the rear of a 16,000 lb truck moving in the same direction as the car, with a speed of 20 mi/hr before the collision (Figure 7-8). The speed of the car after the collision is 20 mi/hr. (a) What is the velocity of the truck immediately after the collision? (b) How much mechanical energy is lost in this collision? How do you account for this loss in energy?

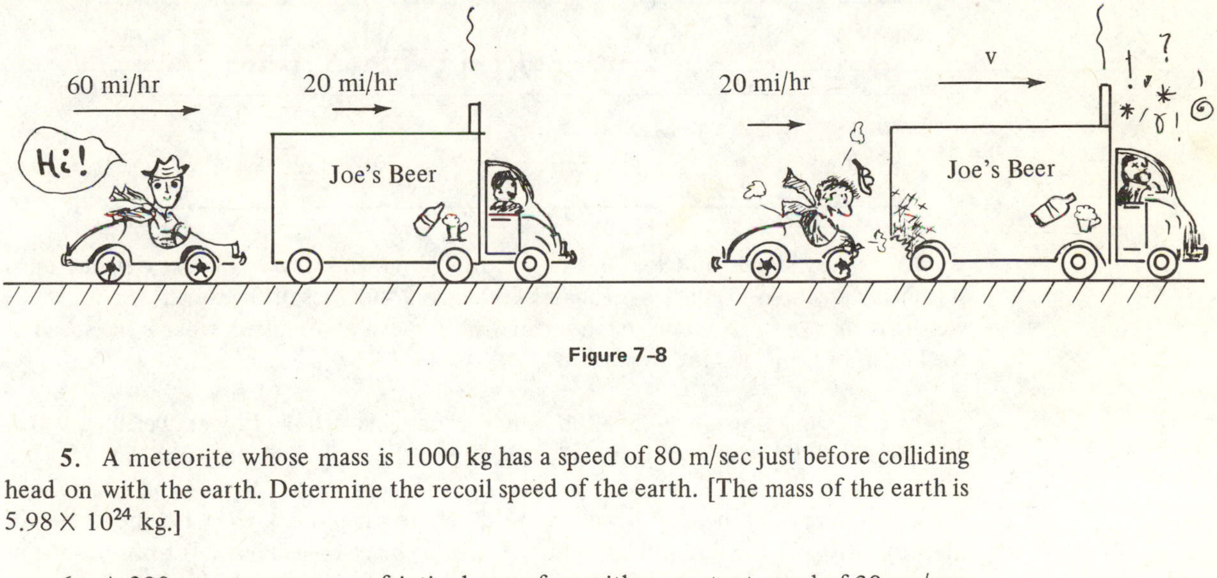

Figure 7-8

5. A meteorite whose mass is 1000 kg has a speed of 80 m/sec just before colliding head on with the earth. Determine the recoil speed of the earth. [The mass of the earth is 5.98×10^{24} kg.]

6. A 200 g cart moves on a frictionless surface with a constant speed of 30 cm/sec. A piece of putty is dropped vertically onto the cart. If the putty has a mass of 50 g, what is the final speed of the cart?

7. Two boys in a boat are drifting in the Southerly direction with a constant speed of 20 ft/sec. Each boy weighs 120 lb and the boat weighs 150 lb. What is the velocity of the boat immediately after (a) one of the boys *falls* off the rear of the boat, (b) one of the doys dives off the rear in the Northerly direction with a speed of 20 ft/sec relative to a stationary land observer, (c) one of the boys dives Eastward (perpendicular to the boat) with a speed of 20 ft/sec?

8. A 5 g bullet is fired into a 1 kg block as in Figure 7-9. The block is initially at rest on a frictionless horizontal surface, and is connected to a spring whose force constant is 900 N/m. The bullet has an initial speed of 400 m/sec. If the block moves a distance of 5 cm after impact, determine the speed at which the bullet emerges from the block.

Figure 7-9

9. Consider a perfectly elastic head-on collision between two hard spheres of masses m_1 and m_2 as in Figure 7-2. Verify Equations (7.10) and (7.11) for this collision.

10. Show that Equations (7.34) and (7.35) follow from solving Equations (7.32) and (7.33), simultaneously.

11. Two particles each having a mass of 0.5 kg move in the xy plane under the action of some external forces. At some instant of time, the positions, velocity components and acceleration components are known and are tabulated in the table below. At this time, find (a) the vector position of the center of mass, (b) the velocity of the center of mass and (c) the acceleration of the center of mass.

	x(m)	y(m)	$v_x\left(\frac{m}{sec}\right)$	$v_y\left(\frac{m}{sec}\right)$	$a_x\left(\frac{m}{sec^2}\right)$	$a_y\left(\frac{m}{sec^2}\right)$
particle 1	2	3	5	-4	4	0
particle 2	-2	3	3	8	2	-2

12. A boy stands on ice skates on a frozen pond, and throws a heavy weight from a stationary position. The boy has a mass of 30 kg and the mass of the weight is 2 kg. If the weight is thrown horizontally to the right with a speed of 40 m/sec, what is the speed of the boy after he throws the weight?

13. A young boy weighing 80 lb stands in one end of a 14 ft boat, weighing 100 lb as in Figure 7-10. The boat is initially 10 ft from the dock. Suddenly, the boy notices a turtle on a rock at the far end of the boat and proceeds to walk to that end to catch the turtle. (a) Where will the boat be relative to the shore when the boy reaches the far end of the boat? (b) Will the boy catch the turtle? Assume he can reach out 2 ft from the end of the boat.

Figure 7-10

14. The mass of the moon is about 0.0123 times the mass of the earth, and the earth-moon separation (measured from their centers) is about 3.84×10^8 m. Determine the location of the center of mass of the earth-moon system as measured from the center of the earth.

†15. A 1 g mass moves in the xy plane, and its vector position varies in time according to the expression $\mathbf{r}_1 = (3\mathbf{i} + 3\mathbf{j})t + (2\mathbf{j})t^2$. At the same time, a 2 g mass moves in the xy plane, and its vector position varies as $\mathbf{r}_2 = 3\mathbf{i} - (6\mathbf{j})t - (2\mathbf{i})t^2$, where the units are in centimeters. At t = 2 sec, determine (a) the vector position of the center of mass, (b) the velocity of each particle, (c) the linear momentum of the system, (d) the velocity of the center of mass, (e) the acceleration of each particle and (f) the acceleration of the center of mass.

16. A fireman whose mass is 60 kg slides down a pole while a constant frictional force of 300 N retards his motion. A horizontal platform of negligible mass is supported by a light spring at the bottom of the pole to "cushion" his fall. The fireman starts from rest 5 m above the platform, and the spring constant is 2000 N/m. Find (a) the speed of the fireman just before he collides with the platform and (b) the maximum distance the spring will be compressed. Assume that friction acts on the fireman during his entire motion.

17. A 1 kg block rests on the edge of a table which is 1 m high (Figure 7-11). A 5 g bullet is fired into the block with an initial speed v_0. The bullet remains in the block, and the block is observed to land 3 m from the bottom of the table. Determine the initial speed of the bullet.

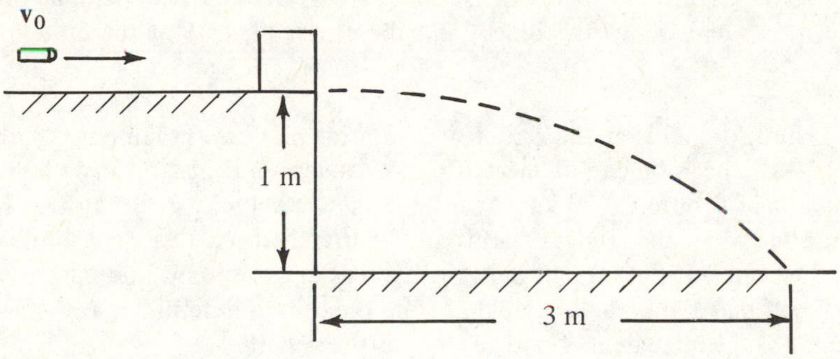

Figure 7-11

18. Two pucks lie on a horizontal frictionless surface. A 300 g puck is initially at rest and is struck by a second 200 g puck moving initially along the +x axis with a velocity of 200 cm/sec. After the collision, the 200 g puck has a velocity of 100 cm/sec at an angle of 53° with the x axis. (a) Determine the velocity of the 300 g puck after the collision. (b) Find the fraction of kinetic energy lost in this collision.

19. A billiard ball moving at a speed of 40 cm/sec collides head-on with a second billiard ball, initially at rest. The collision is inelastic and the coefficient of restitution is 0.70. (a) Determine the velocity of each ball after the collision. (b) Find the fraction of kinetic energy lost in the collision. *Note:* The coefficient of restitution e is defined as $e = -\dfrac{(v_1 - v_2)}{(v_{10} - v_{20})}$, where the v_i are defined in Section 7.2.

20. A proton having a mass of 1 amu (atomic mass unit), and initial velocity 2×10^6 m/sec, makes a head-on elastic collision with a stationary He atom nucleus of

4 amu. (a) Calculate the final velocities of each particle. (b) Determine the fraction of the initial kinetic energy transferred to the He nucleus.

21. A rocket has a total mass of 7×10^5 kg on the launching pad. The rocket engines start and burn fuel at the rate of 100 kg/sec, ejecting gas at an exhaust velocity of 9×10^4 m/sec. (a) Find the thrust exerted on the rocket as it leaves the pad. (b) Determine the initial acceleration of the rocket. (c) Estimate the speed of the rocket after it has lost 90% of its mass in the form of burned fuel (neglect gravity for this last calculation).

†22. (a) Neglecting all external forces on a rocket, including gravity, show that if the rocket has a mass m_0 when it starts from rest, its mass m when its velocity is v is given by

$$m = m_0 e^{-v/v_e}$$

where v_e is the exhaust velocity relative to the rocket. (b) Make a plot of the final speed v versus m_0/m if the exhaust velocity of the fuel is 3×10^3 m/sec. (c) The escape velocity of an object from the earth is about 1.1×10^4 m/sec. A typical rocket attempting to travel from the earth to the moon has a payload mass of 5000 kg. The exhaust velocity relative to the rocket is 3×10^3 m/sec. What must the mass of the fuel be in order for the rocket to reach the required escape speed? (Neglect external forces on the rocket. *Note:* Payload mass means the mass of the rocket without fuel.)

23. A ballistic pendulum is a system used to measure the velocity of projectiles, such as a rifle bullet. A large piece of wood of mass 1 kg is suspended from a light wire. A bullet of mass 5 g is fired into the block in the horizontal direction with an initial speed v_0. If the bullet remains in the block, and the center of mass of the block rises 5 cm, determine v_0.

24. A student stands at the center of a rotating platform holding two barbells, each of mass 10 kg. The moment of inertia of the student and platform (without barbells) about the axis of rotation is 25 kg m^2, and the angular velocity of the system is observed to be 3 rad/sec when the student's arms are outstretched. At this time, the barbells are 1 m from the center. If the student pulls the barbells in close to his body in the horizontal direction, such that the barbells are 0.3 m from the center, determine (a) the new angular velocity and (b) the increase in kinetic energy of the system.

25. A 1 kg mass moving with an initial velocity of 5 m/sec collides and sticks to a 6 kg mass initially at rest. They then proceed to collide and stick to a 3 kg mass also initially at rest. If the collisions are all head-on, (a) what is the final velocity of the system and (b) how much kinetic energy is lost?

8

ANGULAR MOMENTUM AND ROTATIONAL DYNAMICS

In this chapter we will deal with the concept of angular momentum–its application to a rotating system of particles and to rotating rigid bodies. We have already learned how to deal with a system of particles. Therefore, since a rigid body can be thought of as a collection of particles, there are no new physical features in our description of rotational dynamics. We will also show that the angular momentum of a system is conserved when the net torque acting on the system is zero.

8.1 THE ANGULAR MOMENTUM OF PARTICLES

The angular momentum of a particle of mass m is a *vector* defined by the equation

$$\mathbf{L} = m\mathbf{r} \times \mathbf{v} \tag{8.1}$$

where $\mathbf{r}$ is the vector position of the particle and $\mathbf{v}$ is its velocity. Since the linear momentum of the particle is given by $\mathbf{p} = m\mathbf{v}$, the angular momentum is sometimes written as

$$\mathbf{L} = \mathbf{r} \times \mathbf{p} \tag{8.2}$$

The dimension of angular momentum in SI units, by definition, is kg m^2/sec, while in cgs units it has dimension g cm^2/sec. The origin of the inertial reference frame for $\mathbf{r}$ is *arbitrary;* therefore, both the magnitude and direction of $\mathbf{L}$ depend on the choice of the origin. The direction of $\mathbf{L}$ is perpendicular to the plane formed by $\mathbf{r}$ and $\mathbf{v}$, and its sense is determined by the right-hand rule as shown in Figure 8-1. The choice of $\mathbf{r} \times \mathbf{p}$ to define $\mathbf{L}$ is consistent with the conventional choice of a right-handed coordinate system. Therefore, we swing $\mathbf{r}$ into $\mathbf{p}$ by using the curled four fingers of our right hand, always through the *smaller angle* between **r** and **v**. The right thumb points in the direction of **L**. If **r** and **p** are in the xy plane as shown in Figure 8-1, then **L** points in the z direction. The magnitude of **L** is

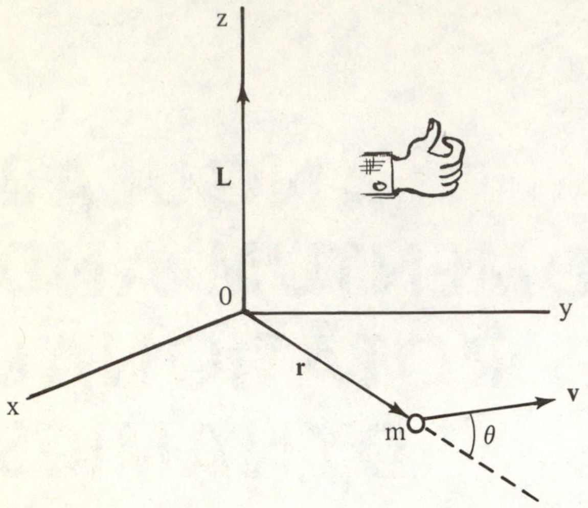

Figure 8-1 The direction of the angular momentum vector **L** for a particle, as defined by **L** = mr × **v**.

$$L = mrv \sin \theta = rp \sin \theta \tag{8.3}$$

where θ is the angle between **r** and **p**. Therefore, L is zero when $\theta = 0$ or π, corresponding to **r**||**v**. In other words, when the particle moves along a line which passes through the origin (either towards or away from the origin), it has no angular momentum with respect to that point. That is to say, it has *no* tendency to rotate about that origin. However, if $\theta = \pi/2$, we see that L is a maximum equal to mrv (or rp). Therefore, we can say that there is a maximum tendency for m to rotate about the origin when **r**⊥**v**. Another equivalent description of angular momentum is to think of the behavior of the position vector **r** as the particle moves. If this vector rotates about the origin, the particle has angular momentum. However, if **r** simply increases or decreases in length, the particle has zero angular momentum with respect to that origin.

With this in mind, we conclude that a particle moving in a straight line possesses angular momentum, providing the origin of **r** is not along the line of motion. If the particle has a *constant* velocity **v**, and its distance of closest approach to the origin is d, as in Figure 8-2(*a*), the angular momentum is *constant* and has a magnitude

$$L = mvr \sin \theta = mvd \tag{8.4}$$

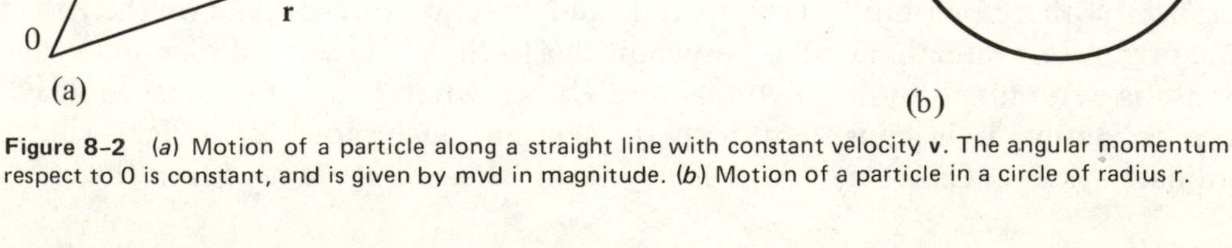

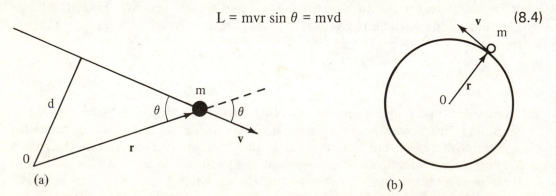

Figure 8-2 (*a*) Motion of a particle along a straight line with constant velocity **v**. The angular momentum with respect to 0 is constant, and is given by mvd in magnitude. (*b*) Motion of a particle in a circle of radius r.

Another common example of a system which has angular momentum is the uniform rotation of a particle in a circle, as in Figure 8-2(*b*). If the center of the circle is the origin, then the angular momentum is simply mvr in magnitude, since the velocity vector is always perpendicular to **r**. In other words,

$$L = mvr \qquad (\text{when } r \perp v) \tag{8.5}$$

The direction of **L** in this case is perpendicular to the plane of the circle, as determined by the right-hand rule. If the sense of rotation is counterclockwise, as in Figure 8-2(*b*), **L** points out of the paper. Likewise, if the sense of rotation of m is clockwise, **L** points into the paper. Typical examples of this type of motion include satellites in circular orbits, a ball swinging in a circle on the end of a string and the motion of charged particles moving perpendicular to a constant magnetic field. As a numerical exercise, the student should show that the angular momentum of a proton moving in a circular orbit 2 m in radius, with a speed of 4×10^5 m/sec, is given by 1.34×10^{-21} kg m^2/sec.

8.2 RELATIONSHIP BETWEEN CHANGE OF ANGULAR MOMENTUM AND TORQUE

An important relation can be derived which relates the torque acting on a particle and the time rate of change of its angular momentum. First, recall that the torque is defined by the expression

$$\boldsymbol{\tau} = \mathbf{r} \times \mathbf{F}$$

But from Newton's second law, $\mathbf{F} = d\mathbf{p}/dt$; therefore, $\boldsymbol{\tau}$ can be written as

$$\boldsymbol{\tau} = \mathbf{r} \times \frac{d\mathbf{p}}{dt} \tag{8.6}$$

Now let us differentiate Equation (8.2) with respect to time:

$$\frac{d\mathbf{L}}{dt} = \frac{d}{dt}(\mathbf{r} \times \mathbf{p}) = \mathbf{r} \times \frac{d\mathbf{p}}{dt} + \frac{d\mathbf{r}}{dt} \times \mathbf{p} \tag{8.7}$$

It should be noted that the *last* term in this expression is *zero,* since it can be written as $\mathbf{v} \times m\mathbf{v}$ and $\mathbf{v} \times \mathbf{v} = 0$ by definition of the cross product. Therefore, comparing Equation (8.6) with Equation (8.7), we see that

$$\boldsymbol{\tau} = \frac{d\mathbf{L}}{dt} \tag{8.8}$$

where the origins of $\boldsymbol{\tau}$ and **L** are the same. This expression is the rotational analog of $\mathbf{F} = d\mathbf{p}/dt$. It says that *the net torque acting on a particle is equal to the time rate of change of its angular momentum.* It is the basic equation used in treating rotating particles and is also applicable to rotating rigid bodies having finite dimensions.

Example 8.1

An airplane of mass M flies parallel to the earth's surface at an altitude h, with a constant acceleration of $2\mathbf{i}$ m/sec^2 (Figure 8-3). An observer on the ground notes that the airplane has a velocity of $200\mathbf{i}$ m/sec when it is directly above him. (a) What is the torque acting on the airplane relative to the observer when the airplane is located at the vector position $\mathbf{r} = (x\mathbf{i} + h\mathbf{j})$ m?

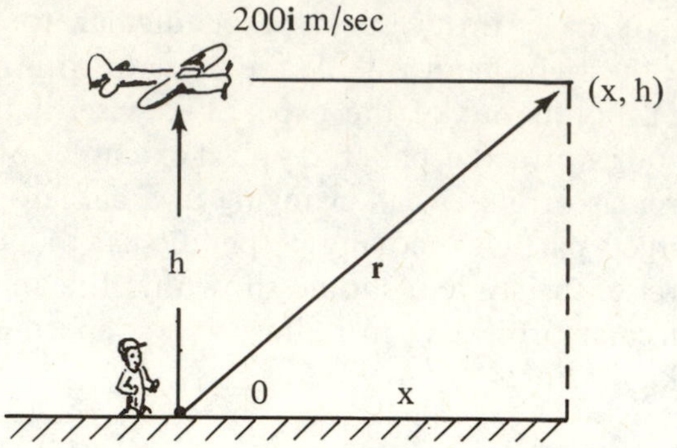

Figure 8-3

$$\boldsymbol{\tau} = \mathbf{r} \times \mathbf{F} = \mathbf{r} \times M\mathbf{a} = M\,[x\mathbf{i} + h\mathbf{j}] \times 2\mathbf{i}\ \text{kg}\,\frac{\text{m}^2}{\text{sec}^2}$$

$$\boldsymbol{\tau} = -2\,Mh\,\mathbf{k}\ \text{Nm}$$

That is, $\boldsymbol{\tau}$ points *into* the paper and is *constant* in magnitude.

(b) What is the angular momentum of the airplane relative to the observer at any time t?

$$\mathbf{L} = \mathbf{r} \times \mathbf{p} = \mathbf{r} \times M\mathbf{v} = M\mathbf{r} \times (\mathbf{v}_0 + \mathbf{a}t)$$

$$\mathbf{L} = M\,(x\mathbf{i} + h\mathbf{j}) \times (200\mathbf{i} + 2t\mathbf{i})\ \ \text{kg m}^2/\text{sec}^2$$

$$\mathbf{L} = -Mh\,(200 + 2t)\,\mathbf{k}\ \ \text{kg m}^2/\text{sec}^2$$

therefore, $\mathbf{L}$ and $\boldsymbol{\tau}$ are parallel vectors. Note, however, that the magnitude of $\mathbf{L}$ changes in time.

†(c) Show that $\boldsymbol{\tau} = d\mathbf{L}/dt$, that is, verify Equation (8.8).
From (b), we have

$$\boldsymbol{\tau} = \frac{d}{dt}\,[-Mh\,(200 + 2t)\,\mathbf{k}] = -2\,Mh\,\mathbf{k}$$

This agrees with the result of (a), so the relation is satisfied.

8.3 ANGULAR MOMENTUM OF A SYSTEM OF PARTICLES

The angular momentum of a system of particles is obtained by taking the vector sum of the individual angular momenta about some given point in an inertial reference frame. If a system has n particles, then

$$\mathbf{L} = \mathbf{L}_1 + \mathbf{L}_2 + \ldots + \mathbf{L}_n = \sum_{i=1}^{n} \mathbf{L}_i \tag{8.9}$$

Since the individual momenta may change in time, the total angular momentum may change in time. However, the angular momentum of the system can only change due to external torques, that is, torques arising from external forces. The net torque due to internal forces is *zero,* since by Newton's third law the internal forces occur in action-reaction pairs; hence, the torque arising from these pairs is zero. Therefore, for a system of particles, we can write

$$\boldsymbol{\tau}_{ext} = \frac{d\mathbf{L}}{dt} \tag{8.10}$$

That is, the net torque due to external forces acting on a system of particles equals the time rate of change of the total angular momentum.

8.4 RELATIONSHIP BETWEEN TORQUE AND ANGULAR ACCELERATION

Let us consider the circular motion of a particle of mass m about a point 0 as in Figure 8–4. In Chapter 4 we found that such motion is conveniently described by polar coordinates and the unit vectors $\hat{r}$ and $\hat{\theta}$, pointing along **r** and perpendicular to **r**, respectively.

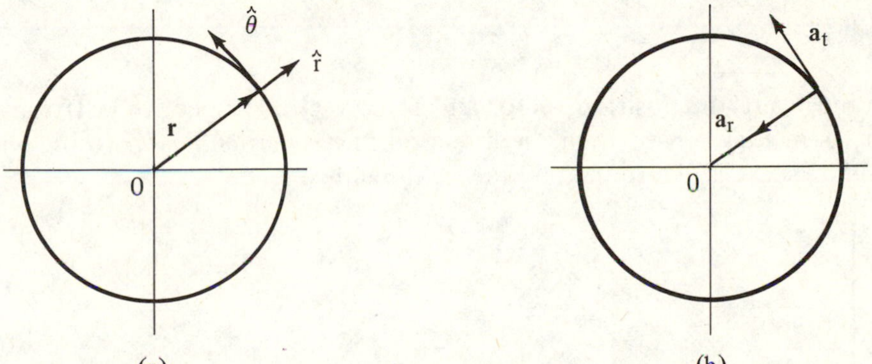

Figure 8–4 (*a*) Definition of unit vectors in polar coordinates. (*b*) Radial and tangential components of acceleration for circular motion.

We found that the total acceleration of a particle moving in a circle can be written as the vector sum of a radial component and a tangential component:

$$\mathbf{a} = \mathbf{a}_t + \mathbf{a}_r = r\alpha\hat{\theta} - \frac{v^2}{r}\hat{r}$$

If we apply Newton's second law to this motion, we have

$$\Sigma\mathbf{F} = m\mathbf{a} = mr\alpha\hat{\theta} - \frac{mv^2}{r}\hat{r} \tag{8.11}$$

Alternatively, we can write Equation (8.11) as the vector sum of two components, one tangent to the circle, $\mathbf{F}_t$, and one pointing radially inwards, $\mathbf{F}_r$. In other words,

$$\Sigma \mathbf{F} = \mathbf{F}_t + \mathbf{F}_r$$

where

$$F_t = mr\alpha$$

$$F_r = -\frac{mv^2}{r} \qquad (8.12)$$

The tangential component is of interest since it relates the applied force to the angular acceleration. That is, there is no angular acceleration when $F_t = 0$. Multiplying F_t by r, we see that

$$rF_t = mr^2\alpha$$

But rF_t is the *definition* of the *torque* acting on m about 0; therefore, we have

$$\tau = mr^2\alpha \qquad (8.13)$$

The vector property of α can be described with reference to the definition of torque. The torque vector τ is defined by the expression

$$\boldsymbol{\tau} = \mathbf{r} \times \mathbf{F}$$

where **r** and **F** are the position vector and force vector, respectively. If **r** and **F** are coplanar, as in Figure 8-5, then the direction of $\boldsymbol{\tau}$ is *perpendicular* to the plane of **r** and **F** and its sense is determined by the right-hand rule.

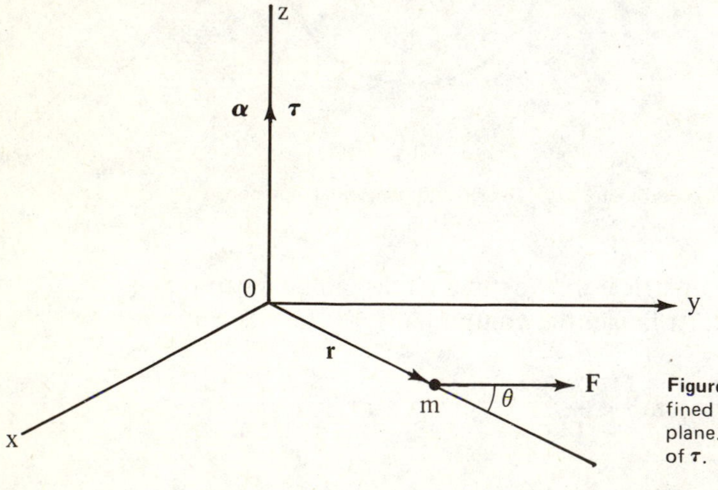

Figure 8-5 The direction of the torque vector $\boldsymbol{\tau}$ as defined by the relation $\boldsymbol{\tau} = \mathbf{r} \times \mathbf{F}$, when **r** and **F** are in the xy plane. The angular acceleration $\boldsymbol{\alpha}$ is also in the direction of $\boldsymbol{\tau}$.

Therefore, we can write the vector relation between $\boldsymbol{\tau}$ and $\boldsymbol{\alpha}$ as

$$\boldsymbol{\tau} = mr^2\boldsymbol{\alpha} \qquad (8.14)$$

where $\boldsymbol{\alpha}$ is in the direction of $\boldsymbol{\tau}$ as in Figure 8–4. Since the vector $\boldsymbol{\tau}$ has a magnitude

$$\boldsymbol{\tau} = \mathrm{rF} \sin \theta$$

we see that the component of **F** perpendicular to **r** gives rise to a torque about 0. Hence, if the "sense" of rotation of m is counterclockwise, looking down the z axis in Figure 8–4, then $\boldsymbol{\tau}$ and $\boldsymbol{\alpha}$ point in the +z direction. If the "sense" of rotation is clockwise looking down the z axis, $\boldsymbol{\tau}$ and $\boldsymbol{\alpha}$ point in the –z direction. In either case, the proper direction is governed by the right-hand rule.

8.5 ROTATIONAL KINETIC ENERGY AND MOMENTA OF INERTIA

Let us consider a rigid body as a collection of small particles and assume that the body rotates about a fixed axis, say, the z axis. Suppose the rigid body rotates about z with a constant angular velocity ω, as in Figure 8–6. Then each particle of the body has some kinetic energy governed by its velocity and mass. If we call the mass of the ith particle Δm_i, and its speed is v_i, then this particle has a kinetic energy

$$K_i = \frac{1}{2} \Delta m_i v_i^2$$

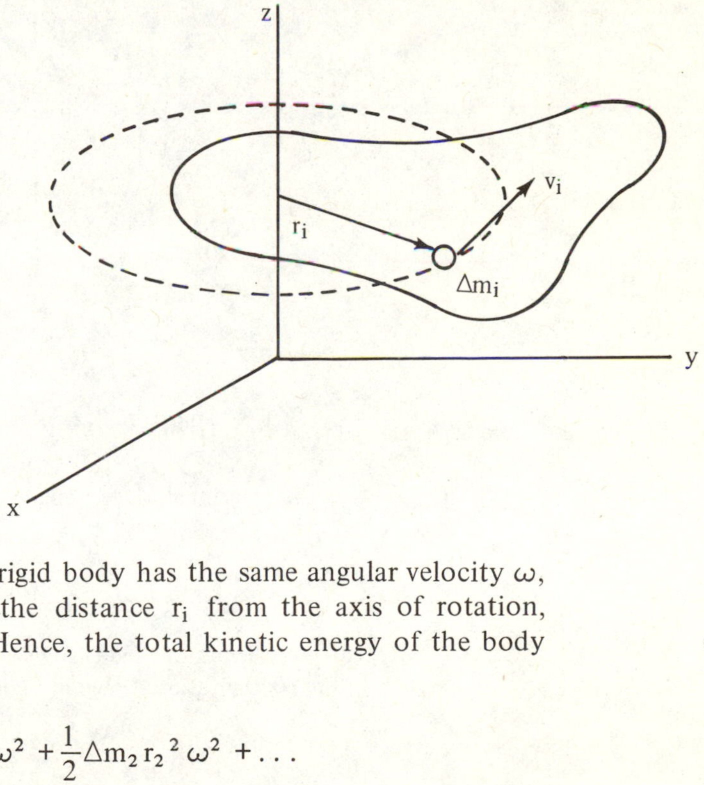

Figure 8–6 The rotation of a rigid body about the z axis. Each element Δm_i moves in a circle about z as shown by the dotted path.

However, whereas every particle in the rigid body has the same angular velocity ω, individual linear velocities depend on the distance r_i from the axis of rotation, according to the expression $v_i = r_i\omega$. Hence, the total kinetic energy of the body becomes

$$K = \Sigma K_i = \frac{1}{2} \Delta m_1 r_1{}^2 \omega^2 + \frac{1}{2} \Delta m_2 r_2{}^2 \omega^2 + \ldots$$

or,

$$K = \frac{1}{2} (\sum_i \Delta m_i r_i{}^2) \omega^2 \tag{8.15}$$

The constant in the parenthesis is called the *moment of inertia,* denoted by I.

$$I = \sum_i \Delta m_i r_i^2 \tag{8.16}$$

Therefore, we can write the kinetic energy of the *rotating* body as

$$K = \frac{1}{2} I\omega^2 \tag{8.17}$$

Note that I has units of ML^2 (kg m^2 in SI units and g cm^2 in cgs). In addition, our result should not be thought of as a new kind of energy. It is still ordinary kinetic energy since it was derived from a sum over individual kinetic energies of the particles making up the body. However, the form given by Equation (8.17) is a convenient one for a rotating rigid body. We sometimes refer to this as the *rotational kinetic energy*. I is analogous to M in linear motion, and ω is analogous to v; therefore, $\frac{1}{2}I\omega^2$ is analogous to $\frac{1}{2}Mv^2$.

Example 8.2

Three masses of 2 kg, 3 kg, and 4 kg are connected by light, rigid rods of equal length 2 m as shown in Figure 8-7. (a) Determine the moment of inertia of the system if the rotation occurs about an axis through 0, perpendicular to the plane of the figure.

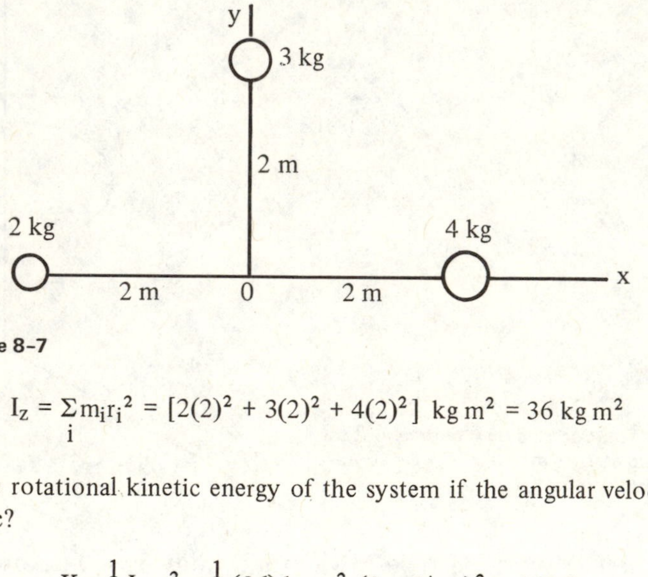

Figure 8-7

$$I_z = \sum_i m_i r_i^2 = [2(2)^2 + 3(2)^2 + 4(2)^2] \text{ kg m}^2 = 36 \text{ kg m}^2$$

(b) What is the rotational kinetic energy of the system if the angular velocity about this axis is 3 rad/sec?

$$K = \frac{1}{2} I_z \omega^2 = \frac{1}{2}(36) \text{ kg m}^2 \text{ (3 rad/sec)}^2 = 162 \text{ J}$$

(c) Repeat the calculations in (a) and (b) if the rotation occurs about the y axis, with the system in the configuration shown in Figure 8-7.

$$I_y = [2\,(2)^2 + 3\,(0)^2 + 4\,(2)^2] \text{ kg m}^2 = 24 \text{ kg m}^2$$

$$K = \frac{1}{2} I_y \omega^2 = \frac{1}{2}(24) \text{ kg m}^2 \text{ (3 rad/sec)}^2 = 108 \text{ J}$$

Therefore, we see that the moment of inertia and the rotational kinetic energy depend on the axis of rotation.

In order to calculate the moment of inertia of a *rigid body*, we can use Equation (8.16) and take the limit of this expression as Δm_i goes to zero. In this limit, we replace the summation by an integral, and Δm_i by dm. Therefore, I becomes

$$I = \int r^2 \, dm \tag{8.18}$$

Here again, the value of I depends on the axis of rotation of the rigid body. For example, the moment of inertia of a uniform *hoop* of mass M, radius R, about an axis perpendicular to the hoop through its center is MR^2, since $r = R =$ constant in Equation (8.18).

†Example 8.3

(a) Let us calculate the moment of inertia of a homogeneous rigid rod of length l about an axis perpendicular to the rod through one end [see Figure 8-8(*a*)].

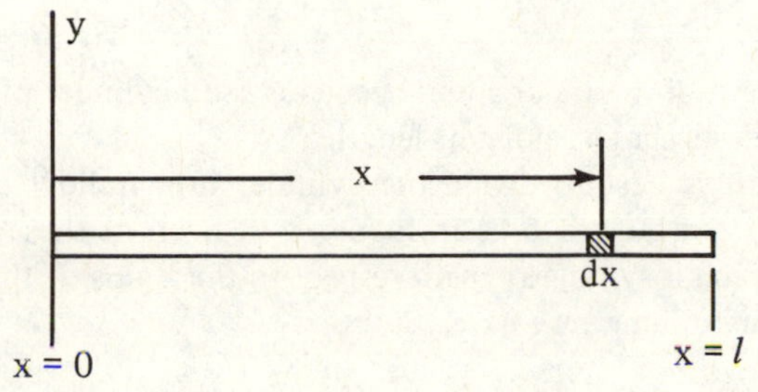

Figure 8-8(*a*)

If the rod has a mass per unit length λ, then we can say that $dm = \lambda dx$. Therefore, the integral given by Equation (8.18) reduces to a one-dimensional problem, with $r = x$. The moment of inertia about the y axis is

$$I_y = \int x^2 \, dm = \lambda \int_0^l x^2 \, dx = \frac{\lambda x^3}{3}\Big]_0^l = \frac{\lambda l^3}{3} \tag{8.19}$$

However, the *total* mass of the rod is $M = \lambda l$, so I_y reduces to

$$I_y = \frac{1}{3} M l^2 \tag{8.20}$$

(b) Let us repeat the calculation if the axis of rotation is perpendicular to the rod, but passes through the center of the rod, as in Figure 8-8(*b*).

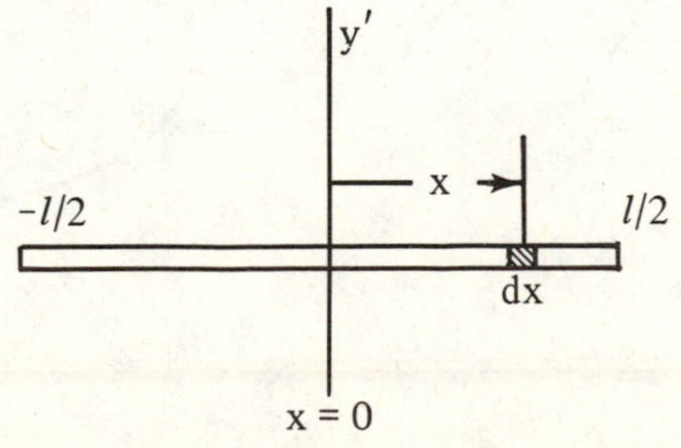

Figure 8-8(*b*)

This calculation is the same as in part (a), except our limits of integration range from $-l/2$ to $l/2$.

$$\therefore\ I_{y'} = \frac{\lambda x^3}{3}\Big]_{-l/2}^{l/2} = \frac{\lambda l^3}{12}$$

Again, since $M = \lambda l$, this reduces to

$$I_{y'} = \frac{1}{12} M l^2 \tag{8.21}$$

Comment: Since $I_{y'} < I_y$, it follows that it requires less energy to rotate the rod about y′ than about y. Also, the "easiest" axis of rotation is x, that is, the axis that coincides with the rod.

†Example 8.4

What is the moment of inertia of a homogeneous solid cylinder of radiums R, mass M, about the axis through its center, along its length?

In this problem, it is best to divide the cylinder into hollow cylinderical shells of thickness dr, radius r and length l, as in Figure 8-9. Each of these shells has a volume $dV = (2\pi r\, dr)\, l$ and each is symmetric with respect to the z axis. If the density is denoted by ρ (the mass per unit volume), we have

$$M = \rho V$$

$$dm = \rho\, dV = \rho l 2\pi r\, dr$$

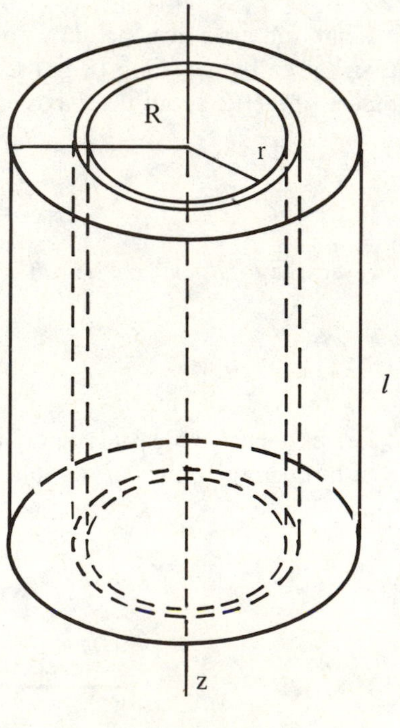

Figure 8-9

Therefore, we can write I as

$$I = \int r^2\,dm = 2\pi\rho l \int_0^R r^3\,dr$$

$$I = 2\,\pi\rho l\;R^4/4 = \pi\rho l\;R^4/2$$

But since $M = \rho V = \rho\pi R^2 l$, I reduces to

$$I = \frac{1}{2} MR^2 \qquad (8.22)$$

As we have seen, the moments of inertia for bodies with simple geometry and high symmetry are relatively easy to calculate, as long as the axis of rotation coincides with the axis of symmetry. The calculation of I for an arbitrary axis of rotation can be somewhat cumbersome, even for a body having a high degree of symmetry. However, there is a useful theorem which can be used to obtain the moment of inertia of a body about any axis parallel to an axis through the center of mass. If the moment of inertia of the body about an axis through the center of mass, I_{cm}, is known, then the moment of inertia about any axis parallel to this, a distance d away, is given by

$$I = I_{cm} + Md^2 \qquad (8.23)$$

This is known as the *parallel axis theorem,* and its proof is left as an exercise. The result can be used to obtain the moment of inertia of various bodies already discussed, as well as objects such as spheres, discs and hollow cylinders. A compilation of some useful data is given in the accompanying chart. Note that the results on the right column of this chart follow from the parallel axis theorem.

8.6 PURE ROTATIONAL MOTION OF A RIGID BODY

Consider a rigid body which is constrained to rotate about a fixed axis. If a single force **F** acts on that body, as in Figure 8–10(*a*), and assuming that the force **F** lies in the xy plane, then the body will rotate about the z axis, and the torque acting on the body is given by $\boldsymbol{\tau} = \mathbf{r} \times \mathbf{F}$.

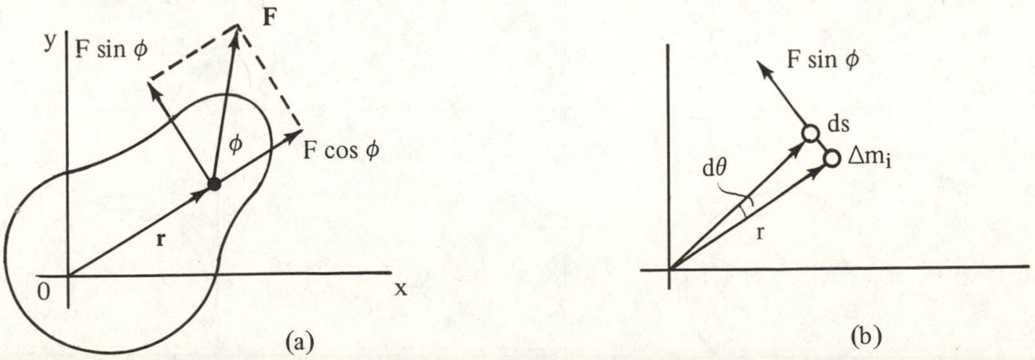

Figure 8–10 (*a*) A single force F acting on a rigid body, pivoted at 0. (*b*) Rotation of a mass element Δm_i through an angle $d\theta$.

MOMENTS OF INERTIA FOR SEVERAL COMMON RIGID BODIES

Hoop
$I = MR^2$
axis
R

Hoop
$I = 2MR^2$
axis
R

Solid cylinder or disc
axis
R
$I = \frac{1}{2}MR^2$

Solid cylinder or disc
axis
R
$I = \frac{3}{2}MR^2$

Long thin rod
$I = \frac{1}{12}Ml^2$
axis
l

Long thin rod
$I = \frac{1}{3}Ml^2$
axis
l

Solid sphere
axis
R
$I = \frac{2}{5}MR^2$

Solid sphere
axis
R
$I = \frac{7}{5}MR^2$

Thin spherical *shell*
axis
R
$I = \frac{2}{3}MR^2$

Thin spherical *shell*
axis
R
$I = \frac{5}{3}MR^2$

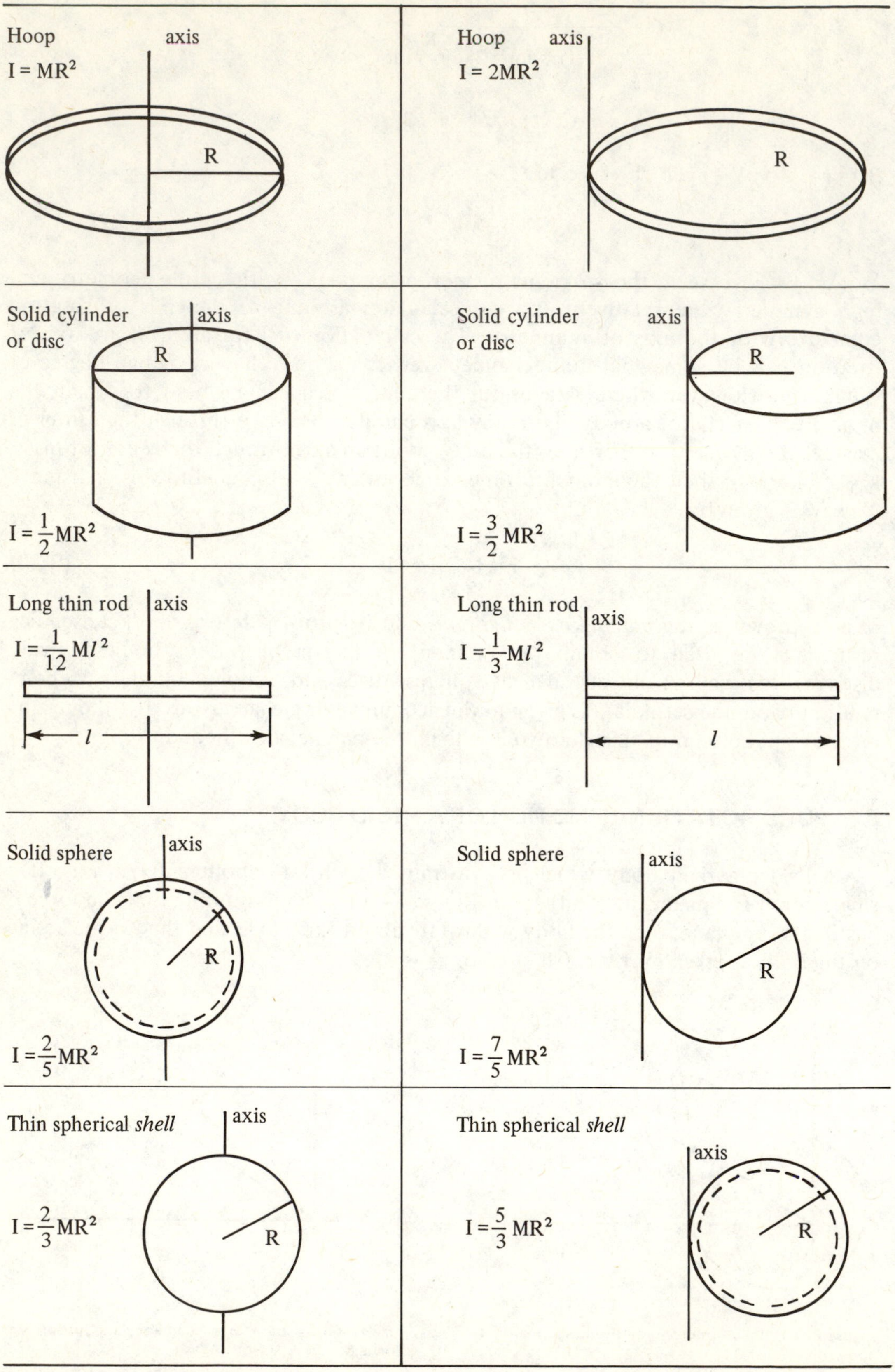

Consider one element Δm_i located at the vector position r. This element will rotate through an infinitesimal angle $d\theta$ in the time dt, as in Figure 8-10(*b*). Note that the component of force parallel to the displacement ds is F sin ϕ. Therefore, the work done by this component on this mass element for this displacement is

$$dW = (F \sin \phi)\, ds = (F \sin \phi)\, r d\theta$$

However, by definition, the magnitude of $\boldsymbol{\tau}$ is rF sin ϕ; therefore, we can write dW as

$$dW = \tau d\theta$$

This is analogous to dW = Fdx for translational motion.

The rate at which work is done is defined as power. Therefore, in rotational motion we see that the instantaneous power P becomes

$$P = \frac{dW}{dt} = \tau \frac{d\theta}{dt} = \tau\omega \qquad (8.24)$$

This is analogous to P = Fv for translational motion.

As in the case of particles discussed in previous sections, we expect that the torque will be proportional to the angular acceleration of the body. We can show that τ is proportional to α for a rigid body by considering the results of Section 8.4, namely Equation (8.13). This states that the torque on a particle about some origin is given in magnitude by

$$\tau = mr^2 \alpha$$

If we consider the rigid body as a collection of particles rotating about the z axis, then the torque on the rigid body can be written as

$$\tau = (\Sigma \Delta m_i r_i^2)\alpha = I\alpha \qquad (8.25)$$

In general, more than one force can act on the rigid body to give a net torque τ. We can rewrite the last equation as

$$\tau = I\alpha = I\frac{d\omega}{dt} = I\frac{d\omega}{d\theta}\frac{d\theta}{dt}$$

or,

$$\tau = I\omega \frac{d\omega}{d\theta}$$

Therefore,

$$dW = \tau d\theta = I\omega d\omega$$

and the total work done in the angular displacement θ_0 to θ is given by

$$W = \int_{\theta_0}^{\theta} \tau d\theta = \int_{\omega_0}^{\omega} I\omega d\omega$$

Since I is a constant, this reduces to

$$W = \frac{1}{2} I\omega^2 - \frac{1}{2} I\omega_0{}^2 \tag{8.26}$$

In other words, the work done by external forces acting on a rigid body equals the change in *rotational kinetic energy*. Of course, we are assuming the body is constrained to rotate about a fixed axis. Equation (8.24) is analogous to $W = \frac{1}{2} mv^2 - \frac{1}{2} mv_0{}^2$ in translational motion. We can think of I as a measure of the resistance of a body to change its rotational motion under the influence of external forces.

Finally, the *angular momentum* of a rigid body rotating about a fixed axis can be shown to be $L = I\omega$. We can write Equation (8.24) as

$$\tau = I\alpha = I \frac{d\omega}{dt}$$

Since I is a constant, this is equivalent to

$$\tau = \frac{d}{dt}(I\omega)$$

But the general relation between τ and the angular momentum of a body is $\tau = dL/dt$; therefore, by comparison, we see that

$$L = I\omega \tag{8.27}$$

This expression is analogous to $P = mv$ for translational motion.

Example 8.5

Determination of the moment of inertia of a circular disc. A common experiment performed in the laboratory is the determination of the moment of inertia of a uniform solid disc of radius R, mass M. The disc is mounted on a frictionless axle and a light string is wrapped around its rim, as in Figure 8-11(*a*). A weight is hung on the string to provide a constant tension T. (a) Find the angular acceleration of the disc.

From the free body diagram of the disc [Figure 8-11(*c*)], we see that the torque on the disc about an axis perpendicular to the disc through its center is given by

$$\tau = TR$$

But $\tau = I\alpha$; therefore,

$$I\alpha = TR$$

or,

$$\alpha = \frac{TR}{I} \tag{1}$$

(b) Find an expression for the linear acceleration of the suspended mass m, and relate this to α. What is the tension in the string?

Newton's second law applied to the suspended mass follows directly from the free body diagram in Figure 8-11(*b*), where $mg > T$.

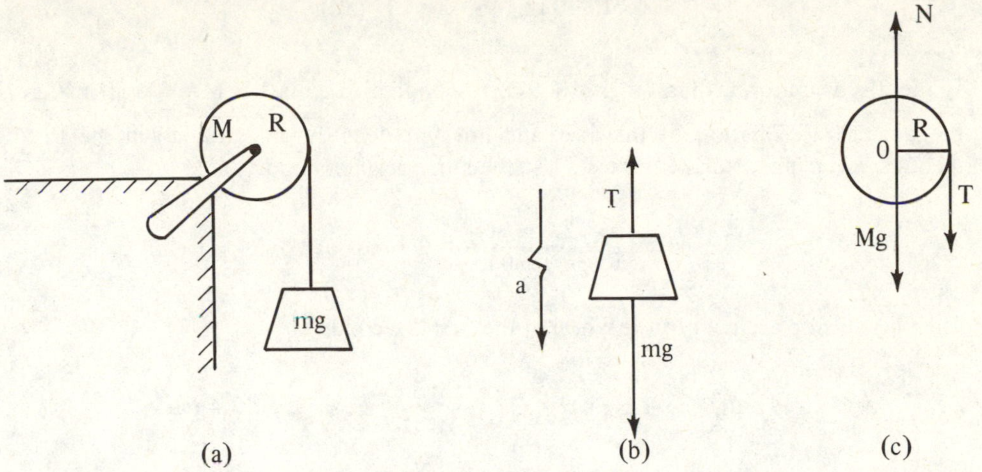

Figure 8-11 (*a*) Experimental arrangement for determining the moment of inertia of a solid disc of radius R, mass M. (*b*) Free body diagram of the suspended mass m. (*c*) Free body diagram for the disc.

$$\Sigma F_y = T - mg = - ma$$

$$\therefore \; a = \frac{mg - T}{m} \qquad (2)$$

This linear acceleration is related to the angular acceleration of the disc through the relation $a = R\alpha$. That is, any point on the rim of the wheel has a *tangential* acceleration equal to a. Therefore, from (1) and (2) we get

$$a = R\alpha = \frac{TR^2}{I} = \frac{mg - T}{m}$$

or,

$$T = \frac{mg}{\frac{mR^2}{I} + 1} \qquad (3)$$

Likewise, solving for a and α gives

$$a = \frac{g}{1 + I/mR^2} \qquad (4)$$

$$\alpha = \frac{a}{R} = \frac{g}{R + I/mR} \qquad (5)$$

(c) In a specific experiment, it is found that $a = 2.0 \text{ m/sec}^2$, $R = 30$ cm and $m = 0.50$ kg. What is the moment of inertia of the disc? What is its mass?

Solving for I in Equation (4), we find that

$$I = mR^2 \left[\frac{g}{a} - 1\right]$$

Therefore,

$$I = 0.50 \text{ kg } (0.30 \text{ m})^2 \left[\frac{9.8}{2.0} - 1\right] = 0.18 \text{ kg m}^2$$

If the disc is uniform, then $I = \frac{1}{2}MR^2$, corresponding to a mass of M = 4 kg. (d) What is the angular acceleration of the disc, and how far does the mass m fall in the first 2 seconds, assuming it starts from rest? Use the numerical data in part (c).

$$\alpha = \frac{a}{R} = \frac{2.0 \text{ m/sec}^2}{0.30 \text{ m}} = 6.7 \text{ rad/sec}^2$$

The angular displacement of the wheel in the first 2 seconds is

$$\theta - \theta_0 = \omega_0 t + \frac{1}{2}\alpha t^2 = \frac{1}{2}\left(6.7\frac{\text{rad}}{\text{sec}^2}\right)(2 \text{ sec})^2 = 13.4 \text{ rad}$$

Therefore, the distance m falls in the first 2 seconds is the distance a point on the rim moves in the first 2 seconds.

$$s = R(\theta - \theta_0) = 0.30 \text{ m } (13.4 \text{ rad}) \cong 4.0 \text{ m}$$

(e) What is the angular momentum of the disc at t = 2 sec?
Its angular velocity at t = 2 sec is

$$\omega = \not{\omega_0} + \alpha t = \left(6.7\frac{\text{rad}}{\text{sec}^2}\right)(2 \text{ sec}) = 13.4\frac{\text{rad}}{\text{sec}}$$

Therefore, its angular momentum, L, is given by

$$L = I\omega = 0.18 \text{ kg m}^2 \left(13.4\frac{\text{rad}}{\text{sec}}\right) = 2.4 \text{ kg m}^2/\text{sec}$$

Example 8.6

A uniform solid disc of radius **R**, mass **M**, is mounted on a frictionless axle as in Figure 8-12. A second smaller disc of radius r, mass m, is fastened to the large disc and a rope is wrapped around the smaller disc. A constant tension T is maintained on the rope. (a) What is the angular acceleration of the system?

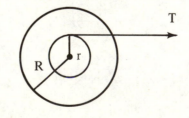

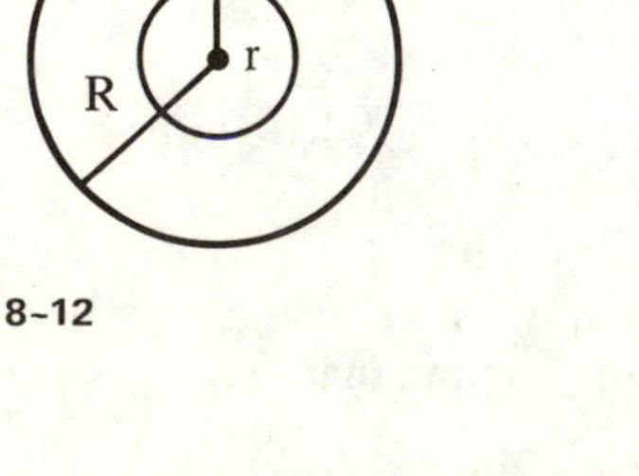

Figure 8-12

Solution

The torque on the system is Tr, but since $\tau = I\alpha$, we have

$$\tau = Tr = I\alpha$$

$$\alpha = \frac{Tr}{I}$$

But the moment of inertia is the sum of the moments of inertia for both discs, that is, $I = MR^2/2 + mr^2/2$. Therefore,

$$\alpha = \frac{Tr}{\frac{MR^2}{2} + \frac{mr^2}{2}} \tag{1}$$

(b) Find a numerical value for I and α if T = 20 N, r = 10 cm, R = 50 cm, M = 4 kg and m = 1 kg.

$$I = \frac{1}{2}(MR^2 + mr^2) = 2\text{ kg }(0.5\text{ m})^2 + 0.5\text{ kg }(0.10\text{ m})^2 \cong 0.50\text{ kg m}^2 \tag{2}$$

$$\alpha \cong \frac{20\text{ N }(0.10\text{ m})}{0.50\text{ kg m}^2} = 4\text{ rad/sec}^2 \tag{3}$$

(c) If the system starts from rest, what is the angular velocity and angular momentum after 5 sec?

$$\omega = \cancel{\omega_0} + \alpha t = 4\text{ rad/sec}^2\ (5\text{ sec}) = 20\text{ rad/sec} \tag{4}$$

$$L = I\omega = \frac{1}{2}(MR^2 + mr^2)\,\omega$$

$$L \cong 0.50\text{ kg m}^2\ (20\text{ rad/sec}) = 10\text{ kg m}^2/\text{sec} \tag{5}$$

(d) Show that the expression $\tau = dL/dt$ is satisfied in this case, using the results of (b). Since T remains constant, τ is also constant, and is given by

$$\tau = I\alpha \cong 0.50\text{ kg m}^2 \left(\frac{4\text{ rad}}{\text{sec}^2}\right) = 2\text{ kg}\frac{\text{m}^2}{\text{sec}^2} \tag{6}$$

For constant torque, we can write the change in angular momentum as

$$\Delta L = \tau \Delta t$$

In this example, Δt = 5 sec; therefore,

$$\Delta L = 2\text{ kg m}^2/\text{sec}^2\ (5\text{ sec}) = 10\text{ kg m}^2/\text{sec}$$

This agrees with (5), since $L_0 = 0$ (the system starts from rest); consequently, $\Delta L = L$, and our relation is satisfied.

(e) What is the angular displacement of the wheel in the first 5 sec?

$$\theta - \theta_0 = \omega_0 t + \frac{1}{2}\alpha t^2 = \frac{1}{2}\left(4\frac{\text{rad}}{\text{sec}^2}\right)(5\text{ sec})^2 = 50\text{ rad} \tag{7}$$

(f) Show that the work done by the force T in the first 5 seconds equals the change in rotational kinetic energy. That is, verify the work-energy theorem, Equation (8-26).

The work done by T is Ts, where s is the total distance a point on the rim of the *smaller* disc moves in 5 seconds. Making use of (7), we find that

$$s = r(\theta - \theta_0) = 0.10 \text{ m } (50 \text{ rad}) = 5 \text{ m}$$

Hence,

$$W = Ts = 20 \text{ N } (5 \text{ m}) = 100 \text{ J} \quad (8)$$

The initial rotational kinetic energy is zero, and the final rotational kinetic energy is $\frac{1}{2}I\omega^2$, where $\omega = 20$ rad/sec. Therefore,

$$\Delta K = \frac{1}{2}I\omega^2 = \frac{1}{2}(0.50 \text{ kg m}^2)(20 \text{ rad/sec})^2 = 100 \text{ J} \quad (9)$$

Comparing (8) and (9), we see that the work-energy theorem is verified.

8.7 CONSERVATION OF ANGULAR MOMENTUM

In section 8.3 we found that the resultant external torque acting on a system is equal to the change in angular momentum, that is,

$$\tau_{ext} = d\mathbf{L}/dt$$

Therefore, we conclude that *if* $\tau_{ext} = 0$, $\mathbf{L}$ = constant, or the *angular momentum of the system is conserved.* This applies to both a system of particles *or* a rigid body. For a system of particles, we can write

$$\mathbf{L} = \mathbf{L}_1 + \mathbf{L}_2 + \mathbf{L}_3 + \ldots = \Sigma \mathbf{L}_i = \text{constant} \quad (\text{if } \tau_{ext} = 0) \quad (8.28)$$

Likewise, for a rigid body rotating about a fixed axis, we can write its angular momentum as

$$L_0 = I_0 \omega_0 \quad (8.29)$$

If the moment of inertia of the body changes by a *redistribution of its mass,* but $\tau_{ext} = 0$, then the angular velocity of the body must change according to the *conservation of angular momentum* expression

$$I\omega = I_0 \omega_0 = \text{constant} \quad (\text{if } \tau_{ext} = 0) \quad (8.30)$$

There are many examples of conservation of angular momentum in everyday life. One common example is the increase in the angular speed of figure skaters upon pulling their hands and feet close to their bodies in a spin movement. There are no external torques on the skater and the increase in angular velocity is due solely to the decrease in I as a result of the redistribution of mass. Other examples which make use of the principle of conservation of angular momentum include the variation of rotational motion of acrobats and divers.

Example 8.7

A horizontal platform in the shape of a cylindrical disc rotates in a horizontal plane about a frictionless axle as in Figure 8-13. The platform has a mass of 100 kg and a radius of 2 m. A man, whose mass is 60 kg, walks slowly from the rim of the platform towards the center. (a) If the angular velocity of the platform and man is 2 rad/sec when the man is at the rim, what is the angular velocity when the man has reached a point 0.5 m from the center?

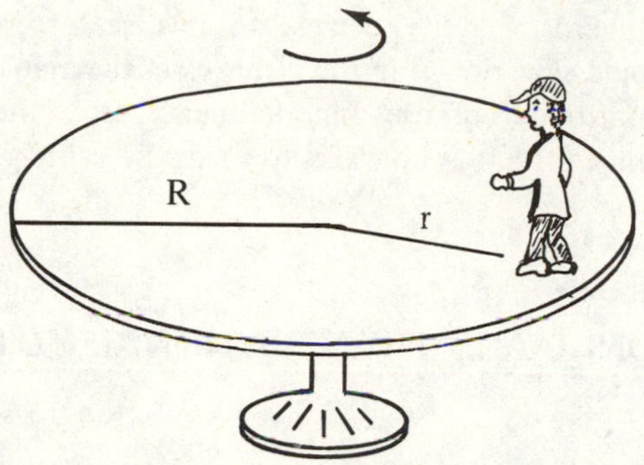

Figure 8-13

The moment of inertia of the platform about the axis of rotation is

$$I_p = \frac{1}{2}MR^2 = \frac{1}{2}100 \text{ kg } (2 \text{ m})^2 = 200 \text{ kg m}^2$$

The initial moment of inertia of the man about the axis of rotation, when he is at the rim is

$$I_m^0 = mR^2 = 60 \text{ kg } (2 \text{ m})^2 = 240 \text{ kg m}^2$$

Therefore, the total moment of inertia of the *system* when the man is at the rim is

$$I_0 = I_p + I_m^0 = 440 \text{ kg m}^2$$

When the man is at $r = 0.5$ m his moment of inertia about the axis of rotation changes to

$$I_m = mr^2 = 60 \text{ kg } (0.5 \text{ m})^2 = 15 \text{ kg m}^2$$

Hence, the *final* moment of inertia of the *system* is

$$I = I_p + I_m = 215 \text{ kg m}^2$$

Since angular momentum is conserved, and $\omega_0 = 2$ rad/sec, we have

$$I\omega = I_0\omega_0$$

$$215\ \omega = 440\ (2) \quad \text{or} \quad \omega = 4.1 \text{ rad/sec}$$

(b) Calculate the initial and final kinetic energy of the system. How do you account for the difference, if any?

$$K_0 = \frac{1}{2} I_0 \omega_0{}^2 = \frac{1}{2}(440 \text{ kg m}^2)\left(2 \frac{\text{rad}}{\text{sec}}\right)^2 = 880 \text{ J}$$

$$K = \frac{1}{2} I\omega^2 = \frac{1}{2}(215 \text{ kg m}^2)\left(4.1 \frac{\text{rad}}{\text{sec}}\right)^2 \cong 1810 \text{ J}$$

This result may seem surprising. The kinetic energy of the system *increases!* This increase in kinetic energy is accounted for by the fact that the man does work in walking towards the center of the platform. We might say that the man does positive work on the system, which is transferred into an increase in kinetic energy. The platform, in effect, is tending to "push" the man outwards, and if he were to reverse his path and walk radially outwards, the system would slow down. In the latter case, the man does negative work on the system and the system therefore loses kinetic energy. Does the force on the man in the radial direction remain constant as he walks towards the center?

8.8 COMBINED ROTATIONAL AND TRANSLATIONAL MOTION OF A RIGID BODY

Thus far in this chapter we have only discussed the dynamics of a rigid body constrained to rotate about a fixed axis. Now, we will examine the dynamics of a rigid body which is both rotating and translating. Consider a body such as a sphere which is rolling on an inclined plane. The sphere rotates about its center of mass, hence it has a rotational kinetic energy given by $\frac{1}{2} I\omega^2$, where I is the moment of inertia about a parallel axis through the center of mass. In addition, the center of mass has a translational motion; therefore, the sphere also has kinetic energy given by $\frac{1}{2} Mv^2$, where v is the velocity of the center of mass. The total kinetic energy is given by

$$K = \frac{1}{2} I\omega^2 + \frac{1}{2} Mv^2 \tag{8.31}$$

If the body is rolling *without slipping,* then we have a relation between v and ω based on geometrical considerations, namely, $s = R\theta$; therefore, $v = R\omega$. Therefore, *for pure rolling motion,* we can write Equation (8.31) as

$$K = \frac{1}{2}\left(\frac{1}{R^2} + M\right) v^2 \tag{8.32}$$

To illustrate the utility of this result we shall make use of the principle of conservation of energy in treating the motion of a rigid body rolling down an inclined plane. Although there is a force of friction present (and necessarily so to cause rotational motion), there is no loss of mechanical energy, since the point of contact of the body has *zero* instantaneous velocity, that is, there is no relative motion between this contact point and the surface. Of course, if the rigid body is allowed to slide, mechanical energy would not be conserved. We shall also show that the results of the energy method approach are consistent with the methods of dynamics.

Example 8.8

A solid homogeneous sphere of mass M, radius R, rolls down an inclined plane without slipping. The height of the incline is h, the angle of inclination is θ, and the sphere starts from rest at the top (Figure 8-14). (a) What is the speed of the center of mass of the sphere when it reaches the bottom?

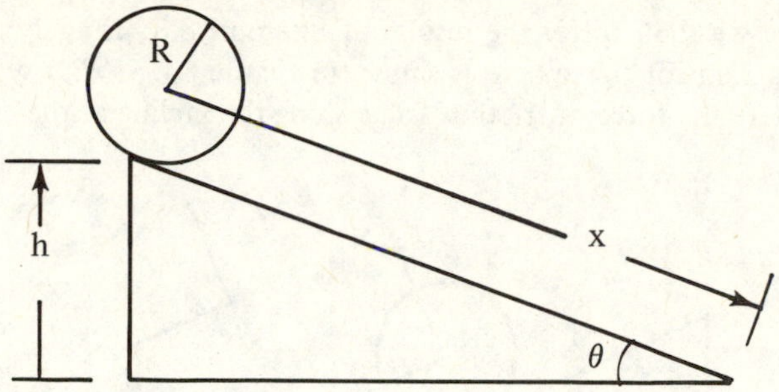

Figure 8-14

The kinetic energy of the sphere at the top is zero, and its kinetic energy at the bottom is given by Equation (8.32), with $I = \frac{2}{5}MR^2$. Since the decrease in gravitational potential energy is Mgh, we have

$$Mgh = \frac{1}{2}\left(\frac{1}{R^2} + M\right) v^2$$

$$Mgh = \frac{1}{2}\left(\frac{2}{5}M + M\right) v^2$$

or,

$$v = \sqrt{\frac{10}{7}gh}$$

(b) What is the acceleration of the center of mass?

Since $h = x \sin\theta$, we can write the energy expression as

$$\frac{7}{10}v^2 = gx \sin\theta$$

If we differentiate both sides with respect to time, we get

$$\frac{7}{10}\left(2v\frac{dv}{dt}\right) = g \sin\theta \frac{dx}{dt}$$

But by definition, $v = dx/dt$ and $a = dv/dt$, so this reduces to

$$a = \frac{5}{7}g \sin\theta$$

It is interesting to note that this result is independent of the mass and radius of the sphere! That is, *all homogeneous solid spheres would experience this same acceleration.* If we repeated this calculation for a solid cylinder, a hoop or a hollow sphere, we would find that the result is essentially the same, with a different numerical factor appearing before g sin θ. These calculations are left as exercises.

Example 8.9

In this exercise we shall verify the results of Example 8.8, using dynamical methods. The free body diagram for the sphere is shown in Figure 8-15. The symbols have their usual meaning, where the force of friction f acts along the incline at the contact point.

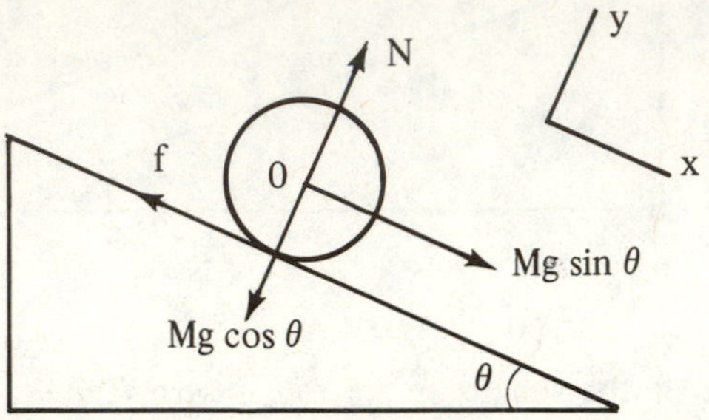

Figure 8-15

Newton's second law applied to the center of mass motion gives

$$\Sigma F_x = Mg \sin\theta - f = Ma \quad (1)$$

$$\Sigma F_y = N - Mg\cos\theta = 0 \quad (2)$$

Now let us write an expression for the torque acting on the sphere. A *convenient* origin to choose is through 0, the center of the sphere, since N and Mg go through this point and have zero moment arms. Therefore, only the force of friction enters the torque equation and its moment arm is R about 0.

$$\tau = fR = I\alpha$$

Since $I = \frac{2}{5}MR^2$ and $\alpha = a/R$, this becomes

$$f = \frac{I\alpha}{R} = \frac{\left(\frac{2}{5}MR^2\right)\frac{a}{R}}{R} = \frac{2}{5}Ma \quad (3)$$

Substituting (3) into (1) gives

$$a = \frac{5}{7}g \sin\theta \quad (4)$$

This agrees with the result of Example 8-8.

Since the acceleration of the center of mass, a, is constant, and the initial velocity of the sphere is zero, we can use the relation $v^2 = 2ax$ to obtain the speed of the sphere at the bottom. Since $x = h/\sin\theta$, we have

$$v^2 = 2ax = 2\left(\frac{5}{7} g \sin\theta\right)\frac{h}{\sin\theta}$$

or,

$$v = \sqrt{\frac{10}{7} gh}$$

Again, this is in agreement with the more direct energy method.

The *minimum* coefficient of friction necessary to maintain *pure rolling motion* can be obtained by using the relation $f = \mu N$. Using this together with (2), (3) and (4) gives

$$f = \mu N = Mg \cos\theta = \frac{2}{5} Ma$$

$$\mu = \frac{2}{5}\frac{a}{g\cos\theta} = \frac{2}{5}\left(\frac{\frac{5}{7} g \sin\theta}{g\cos\theta}\right)$$

or,

$$\mu = \frac{2}{7}\tan\theta$$

If μ is smaller than this, the motion of the sphere will be a combination of rolling and sliding.

8.9 PROGRAMMED EXERCISES

1.A

A particle of mass m has a velocity **v**. At some instant, its vector position is **r**. What is the angular momentum of the particle?

By definition,

$$\mathbf{L} = m\,\mathbf{r} \times \mathbf{v}$$

1.B

Is the value of the angular momentum unique? Explain.

L is unique only if the origin is specified and fixed.

1.C

Can the angular momentum of a particle be zero? Explain.

Yes. **L** can be zero in three ways: (1) when $\mathbf{v} = 0$; (2) when $\mathbf{r} = 0$; and finally, (3) when $\mathbf{r} \| \mathbf{v}$.

1.D

What are the units of **L**?

$[\mathbf{L}] = ML^2/T$

1.E

At $t = 0$, a particle whose mass is 3 kg has a velocity $\mathbf{v}_0 = (3\mathbf{i} + 2\mathbf{j})$ m/sec, and its vector position is $\mathbf{r}_0 = 2\mathbf{i}$ m. What is its angular momentum relative to the origin?

$$\mathbf{L}_0 = m\,\mathbf{r}_0 \times \mathbf{v}_0$$

$$\mathbf{L}_0 = 3 \text{ kg } (2\mathbf{i}) \times (3\mathbf{i} + 2\mathbf{j}) \text{ m}^2/\text{sec}$$

$$\mathbf{L}_0 = 12\,\mathbf{k} \text{ kg m}^2/\text{sec}$$

Note that $\mathbf{L}_0$ points in the +z direction.

1.F

At some later time t, the same particle has a velocity of $\mathbf{v} = (3\mathbf{i} - 2\mathbf{j})$ m/sec, and its vector position is $\mathbf{r} = (3\mathbf{i} - \mathbf{j})$ m. What is its angular momentum relative to the origin?

$$\mathbf{L} = m\,\mathbf{r} \times \mathbf{v}$$

$$\mathbf{L} = 3 \text{ kg } (3\mathbf{i} - \mathbf{j}) \times (3\mathbf{i} - 2\mathbf{j}) \text{ m}^2/\text{sec}$$

$$\mathbf{L} = 3\,[-6\mathbf{k} + 3\mathbf{k}] \text{ kg m}^2/\text{sec}$$

$$\mathbf{L} = -9\mathbf{k} \text{ kg m}^2/\text{sec}$$

1.G

What is the change in angular momentum of the particle in this time interval?

$$\Delta\mathbf{L} = \mathbf{L} - \mathbf{L}_0$$

$$\Delta\mathbf{L} = -9\mathbf{k} - 12\mathbf{k}$$

$$\Delta\mathbf{L} = -21\mathbf{k} \text{ kg m}^2/\text{sec}$$

2 Consider two particles of equal mass M, rotating about a point midway between the particles. The particles are separated by a distance 2 R, and each rotates with constant speed v.

2.A

What is the magnitude of the angular momentum for each particle with respect to 0?

Note:
$v_c = 0$

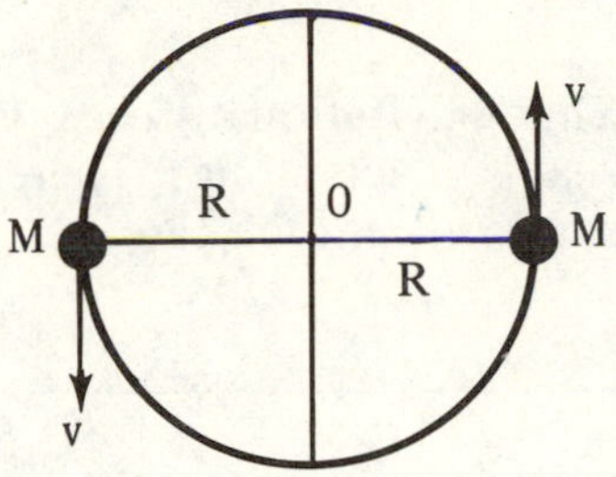

Since $\mathbf{v} \perp \mathbf{r}$ for *each* particle, each has an angular momentum given by

$$L = MvR \qquad (1)$$

2.B

What is the magnitude of the *total* angular momentum with respect to 0? What is its direction?

The total angular momentum is simply twice that given by (1).

$$L_{total} = 2MvR \qquad (2)$$

Its direction, as determined by the right hand rule, is *out* of the paper if the rotation is ccw.

2.C

Determine the total angular momentum with respect to a point located at one of the particles.

In this case, one of the particles has $L = 0$, since $r = 0$. The other has

$$L = Mv\,(2\,R) = 2\,MvR \qquad (3)$$

2.D

Note that the results (2) and (3) are identical. As a matter of fact, $L_{total} = 2\,MvR$ *regardless* of the origin chosen in this case. What general theorem does this result suggest?

When the velocity of the center of mass of a system of particles is *zero*, that is, when $v_c = 0$, L_{total} is *independent* of the origin chosen as the reference.

2.E

Write an expression for the total moment of inertia of the system about 0.

$$I = MR^2 + MR^2$$

$$I = 2MR^2 \qquad (4)$$

2.F

Show that the expression $L = I\omega$, where $\omega = v/R$, gives the same results as (2).

Using (4), we see that

$$L = I\omega = (2MR^2)\,\frac{V}{R} = 2MvR$$

2.G

Does the total angular momentum change in time? What can you say about the torque acting on the system?

Since v = constant, L_{total} is a constant. In fact, L for each mass is constant in this example. Since L is constant, then

$$\tau = dL/dt = 0 \qquad (5)$$

2.H

If these were two gravitational masses rotating about their common center, they exert a gravitational force on each other. How can the torque then be zero?

The gravitational force is a *central* force, that is, it acts along **r**. Therefore, $\boldsymbol{\tau} = \mathbf{r} \times \mathbf{F} = 0$, since $\mathbf{r} \| \mathrm{F}$ in each case.

2.I

Since $\boldsymbol{\tau} = 0$, and internal forces do not contribute to $\boldsymbol{\tau}$, what can you conclude about the external forces on the system?

There are two possibilities: 1. There are no external forces. 2. The external forces act in such a way that $\tau_{ext} = 0$.

2.J

What is the *total* kinetic energy of the system? Use the ordinary translational kinetic energy expression, $K = \frac{1}{2}mv^2$.

$$K_{tot} = \frac{1}{2}Mv^2 + \frac{1}{2}Mv^2$$

$$K_{tot} = Mv^2 = MR^2\omega^2 \qquad (6)$$

2.K

Show that the expression for K, that is, (6), is consistent with the idea that the kinetic energy can also be written as $\frac{1}{2}I\omega^2$.

Using (4) for I, we have

$$K = \frac{1}{2}I\omega^2 = \frac{1}{2}(2\,MR^2)\,\omega^2$$

$$K = MR^2\omega^2$$

3 A cylinder of radius R = 0.5 m has a moment of inertia $I = 2\ \mathrm{kg\ m^2}$ about an axis of rotation. A rope is tied around the perimeter of the cylinder and a constant tension of 10 N is maintained. For parts 3.A through 3.E, assume the axle is frictionless.

3.A

What is the magnitude of the torque due to the 10 N force about the axis of rotation?

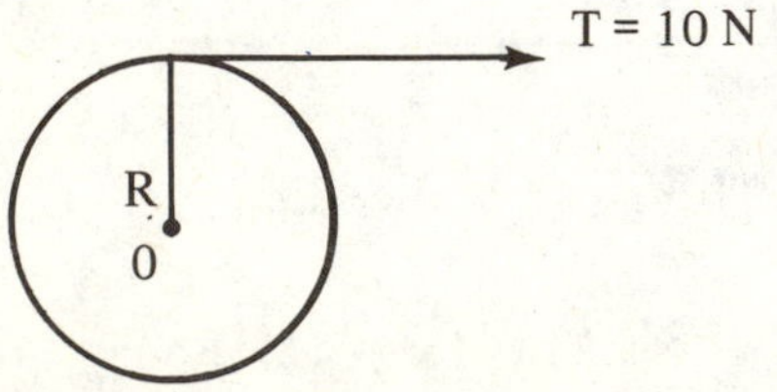

Cross-section of cylinder

The axis of rotation is through 0, coinciding with the symmetry axis.

$$\tau = TR = 10\ \mathrm{N} \times 0.5\ \mathrm{m},$$

$$\tau = 5\ \mathrm{N\ m} \qquad (1)$$

3.B

What is the *change* in angular momentum of the cylinder in a time interval of 4 sec?

Since $\tau = dL/dt$, and τ is constant, we can write the change in L as

$\Delta L = \tau \Delta t$, or

$\Delta L = 5 \text{ Nm } (4 \text{ sec})$

$\Delta L = 20 \text{ kg m}^2/\text{sec}$ (2)

3.C

If the cylinder starts from rest, what is its angular speed after 4 sec? Use the fact that $L = I\omega$.

Since $L = I\omega$, and $\Delta L = L$ (that is, $L_0 = 0$), then

$$\omega = \frac{L}{I} = \frac{20 \text{ kg m}^2/\text{sec}}{2 \text{ kg m}^2}$$

or,

$$\omega = 10 \text{ rad/sec} \qquad (3)$$

3.D

What is the angular acceleration of the wheel? Recall that $\tau = I\alpha$.

$$\alpha = \frac{\tau}{I} = \frac{5 \text{ Nm}}{2 \text{ kg m}^2}$$

$\alpha = 2.5 \text{ rad/sec}^2$ (4)

Note that we can also use $\omega = \alpha t$ to get (3), with $t = 4$ sec.

3.E

Through what angle (in rad) does the wheel rotate in the first 4 sec?

Setting $\omega_0 = 0$ and $\theta_0 = 0$ gives

$$\theta = \frac{1}{2}\alpha t^2$$

$$\theta = \frac{1}{2}(2.5)(4)^2 = 20 \text{ rad}$$

3.F

Suppose the axle of rotation is *not* frictionless and a constant tension of 10 N is maintained. Assuming the radius of the axle is 0.1 m, and the force of friction is constant and equal to 20 N, show a diagram of the forces giving rise to a torque about the axle.

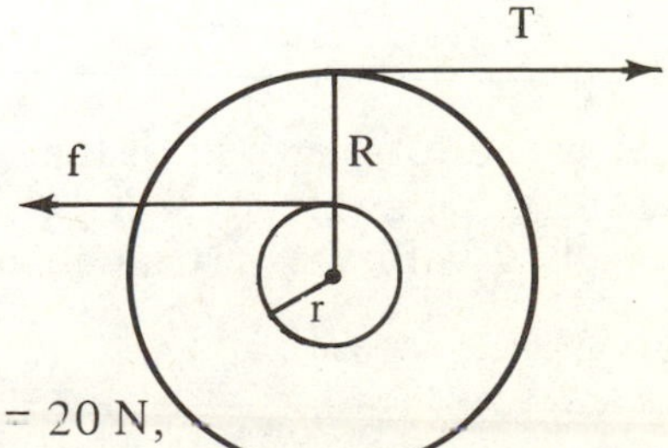

$T = 10$ N, $f = 20$ N,

$R = 0.5$ m, $r = 0.1$ m.

3.G

What is the magnitude and direction of the *net torque* acting about the axis? Note that the torque due to T is different in magnitude and direction compared to the torque due to f.

The torque due to T is *into* the paper and has a magnitude of 5 Nm. The torque due to f is *out* of the paper and has a magnitude fr = 2 Nm. Hence, the *net* torque is *into* the paper and has a magnitude of 3 Nm.

3.H

In this case, what is L, ω, α and θ for the wheel after 4 sec of motion (that is, repeat parts 3.B, 3.C, 3.D and 3.E, but use the fact that τ_{net} = Nm)?

$L = \tau_{net}\,\Delta t = 3\,(4) = 12 \text{ kg m}^2/\text{sec}$

$\omega = L/I = 12/2 = 6 \text{ rad/sec}$

$\alpha = \frac{\tau}{I} = \frac{3}{2} = 1.5 \text{ rad/sec}^2$

$\theta = \frac{1}{2}\alpha t^2 = \frac{1}{2}(1.5)\,(4)^2 = 12 \text{ rad}$

4 A rigid, light rod of length $2l$ rotates in a *vertical* plane about a frictionless axis through its center. A mass m_1 is attached at one end, and a mass m_2 is at the opposite end, where $m_1 > m_2$. The system is released from rest as shown in Figure (*a*) below, and at some later time the rod makes an angle θ with the vertical, as in Figure (*b*).

4.A

What is the total mechanical energy initially? Take the reference point at the center of the rod.

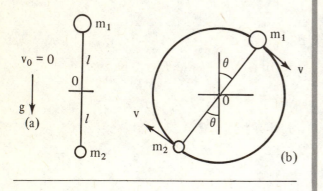

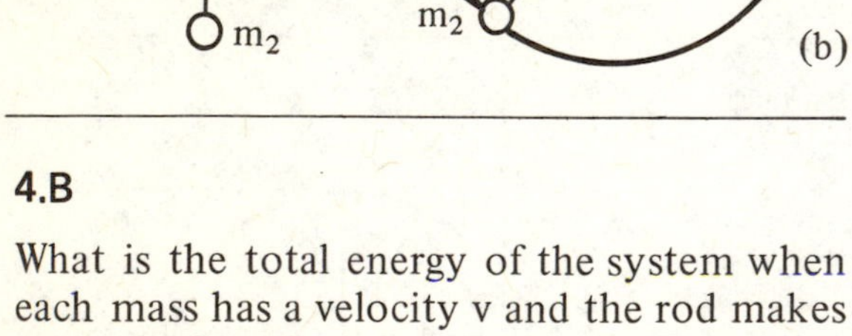

Since $v_0 = 0$, $K_0 = 0$, so

$E_0 = m_1 gl - m_2 gl$

$E_0 = (m_1 - m_2)\,gl$ (1)

4.B

What is the total energy of the system when each mass has a velocity v and the rod makes an angle θ with the vertical as in Figure (b) above?

$K = \frac{1}{2} m_1 v^2 + \frac{1}{2} m_2 v^2$

$U_g = m_1 gl \cos\theta - m_2 gl \cos\theta$

$E = K + U_g$

$E = \frac{1}{2}(m_1 + m_2)\,v^2 + (m_1 - m_2)\,gl\cos\theta$ (2)

4.C

Find an expression for the velocity in terms of m_1, m_2, g, l and θ.

Since energy is conserved, $E = E_0$, so from (1) and (2), we get

$$\frac{1}{2}(m_1 + m_2)v^2 = (m_1 - m_2)gl[1 - \cos\theta]$$

$$v = \left\{\frac{2(m_1 - m_2)}{m_1 + m_2} gl[1 - \cos\theta]\right\}^{1/2} \quad (3)$$

4.D

What is the angular velocity of the wheel at the angle θ?

Since both masses move in a circle of radius l, then $\omega = v/l$, where v is given by (3).

4.E

Calculate the moment of inertia of the system about 0.

$$I = m_1 l^2 + m_2 l^2$$

$$I = (m_1 + m_2)l^2 \quad (4)$$

4.F

What is the angular momentum of the system at the angle θ?

$$L = I\omega = \left(\frac{I}{l}\right) v$$

$$L = (m_1 + m_2)\, lv \quad (5)$$

where v is given by (3).

4.G

What is the *net torque* on the system at the angle θ?

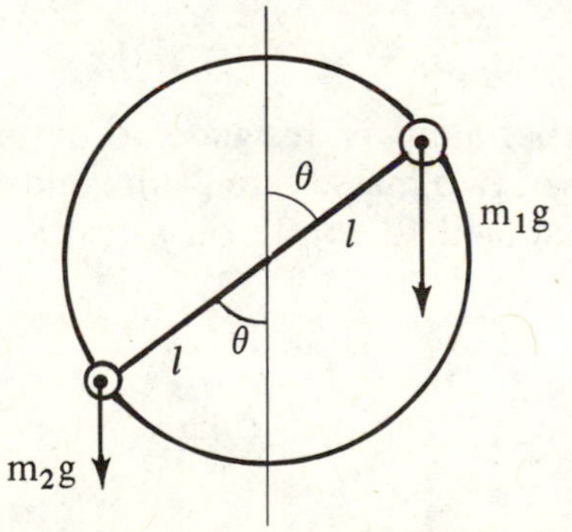

$\tau_1 = m_1 gl \sin\theta$ *into* the paper.

$\tau_2 = m_2 gl \sin\theta$ *out* of the paper.

$$\tau_{net} = \tau_1 - \tau_2 = (m_1 - m_2)\, gl \sin\theta \quad (6)$$

where τ_{net} is *into* the paper.

4.H

Find the angular acceleration of the system at the angle θ.

Since $\tau = I\alpha$, we can use (4) and (6) together to get

$$\alpha = \frac{\tau}{I} = \left(\frac{m_1 - m_2}{m_1 + m_2}\right)\frac{g}{l}\sin\theta \quad (7)$$

4.I

For what angles is α zero? When is the absolute value of α a maximum?

From (6) we see that $\alpha = 0$ when $\theta = 0$ or π rad, that is, when the rod is vertical. Likewise, $|\alpha|$ is a maximum for $\theta = \pi/2$ or $3\pi/2$ rad, that is, when the rod is horizontal.

4.J

When $\theta = \pi/2$, the weight vectors have no components along the rod. At this angle, apply Newton's second law to the *radial* direction, and find the tensions T_1 and T_2 in the two halves of the rod. Note that $T_1 \neq T_2$. What does this result suggest?

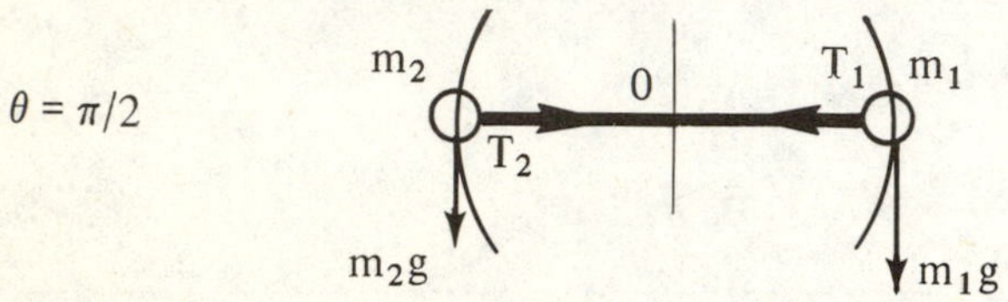

$$\Sigma F_r \text{ on } m_1 = T_1 = \frac{m_1 v^2}{l}$$

$$\Sigma F_r \text{ on } m_2 = T_2 = \frac{m_2 v^2}{l}$$

where v at $\pi/2$ can be obtained from (3), with $\cos\theta = 0$. Since $m_1 > m_2$, $T_1 > T_2$. This means there is a *net* force on the axle due to the rod's rotation.

4.K

When $\theta = \pi$, we found that $\alpha = 0$, so $a_t = 0$. At this angle, apply Newton's second law to the radial direction to find T_1 and T_2. Note that the results differ from those of 4.J.

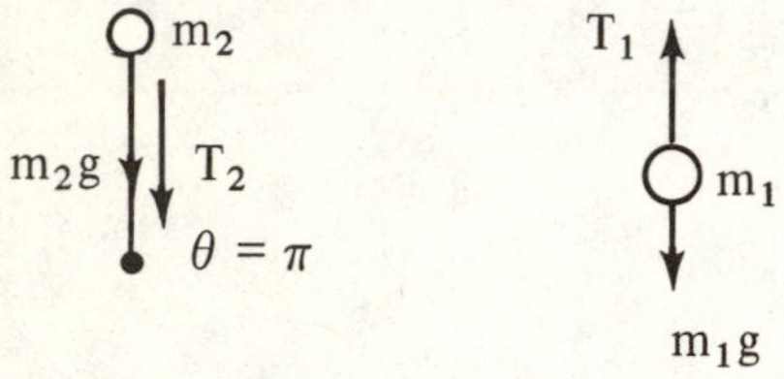

Note that at this angle, $\cos(\pi) = -1$, so from (3) we see that v is a max.

$$\Sigma F_r \text{ on } m_1 = T_1 - m_1 g = \frac{m_1 v^2}{l}$$

$$\Sigma F_r \text{ on } m_2 = T_2 + m_2 g = \frac{m_2 v^2}{l}$$

Again, $T_1 > T_2$.

5 This exercise is similar to Example 8.5, where an experiment is designed to determine the moment of inertia of a disc of radius R, mass M. In this case, two masses are at the ends of a cord which passes over the disc as shown below. The disc is assumed to rotate on a frictionless axle, and $m_1 > m_2$. (The rope doesn't slip.)

5.A

Draw free body diagrams for m_1 and m_2.

Note: $m_1 > m_2$, and $T_1 \neq T_2$

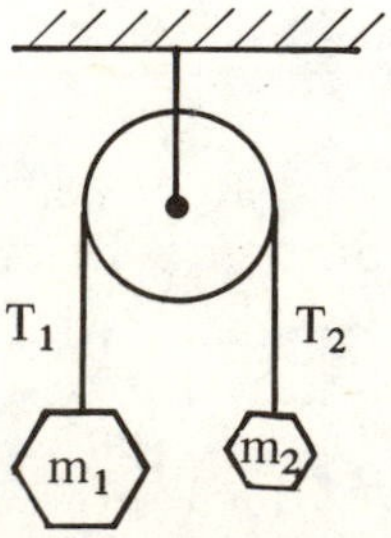

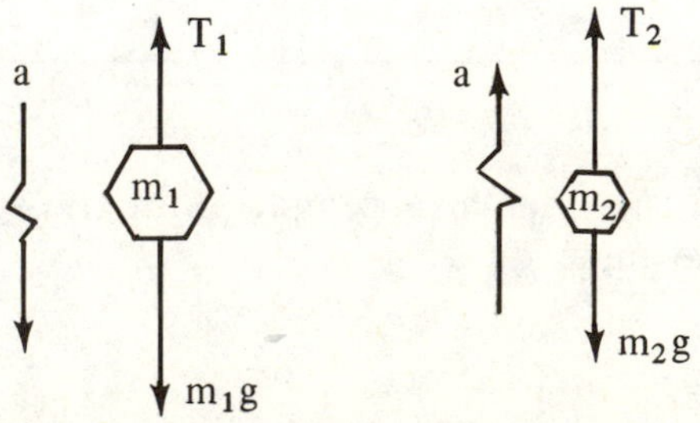

5.B

In 5.A it was stated that $T_1 \neq T_2$. Why must this be true?

Since the rope doesn't slip on the disc, the disc will only rotate if there is a *net torque* acting on it. This will only occur when $T_1 \neq T_2$. In fact, $T_1 > T_2$ if $m_1 > m_2$.

5.C

Write equations of motion for m_1 and m_2. Let the acceleration be a.

$$\Sigma F_y \text{ on } m_1 = T_1 - m_1 g = -m_1 a \quad (1)$$

$$\Sigma F_y \text{ on } m_2 = T_2 - m_2 g = m_2 a \quad (2)$$

5.D

Draw a free body diagram for the disc.

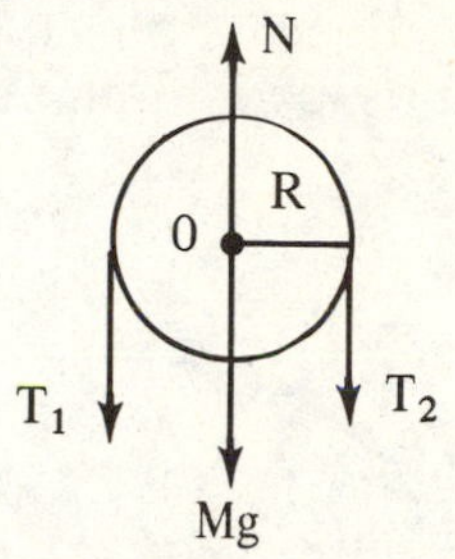

5.E

Write an equation for the *net torque* on the disc about the axis of rotation. Note that N and Mg pass through 0, so they have *zero* torque about this point.

The torque due to T_1 is $T_1 R$ and is *out* of the paper. The torque due to T_2 is $T_2 R$ and is *into* the paper. Hence, $\tau = T_1 R - T_2 R$

$$\tau = (T_1 - T_2)R \quad (3)$$

where τ is *out* of the paper, since $T_1 > T_2$.

5.F

Use the expression $\tau = I\alpha$ and (3) to obtain relations for α and the acceleration a in terms of T_1, T_2 and R.

$$\tau = I\alpha = (T_1 - T_2)R$$

$$\alpha = (T_1 - T_2)\frac{R}{I} \quad (4)$$

Since $a = R\alpha$, we have

$$a = (T_1 - T_2)\frac{R^2}{I} \quad (5)$$

5.G

Obtain an expression for a in terms of m_1, m_2 and I by solving (1), and (2) and (5), simultaneously.

From (1) and (2) we have

$$T_1 - T_2 = (m_1 - m_2)g - (m_1 + m_2)a \quad (6)$$

Substituting this into (5), and simplifying gives

$$a = \frac{(m_1 - m_2)g}{(m_1 + m_2) + I/R^2} \quad (7)$$

5.H

Substitute (7) into (6) and show that $T_1 > T_2$.

$$T_1 - T_2 = (m_1 - m_2)\,g - \frac{(m_1 - m_2)\,g}{1 + I/(m_1 + m_2)\,R^2}$$

$$T_1 - T_2 = (m_1 - m_2)\,g \left\{1 - \frac{(m_1 + m_2)\,R^2}{(m_1 + m_2)\,R^2 + I}\right\}$$

$\therefore T_1 > T_2$, since $m_1 > m_2$

6 A uniform rod of length l, mass M, is free to rotate about one end on a frictionless pivot. A point mass m is fixed to the lower end of the rod. The center of mass of the rod is raised a distance h above its lowest point as shown below.

6.A

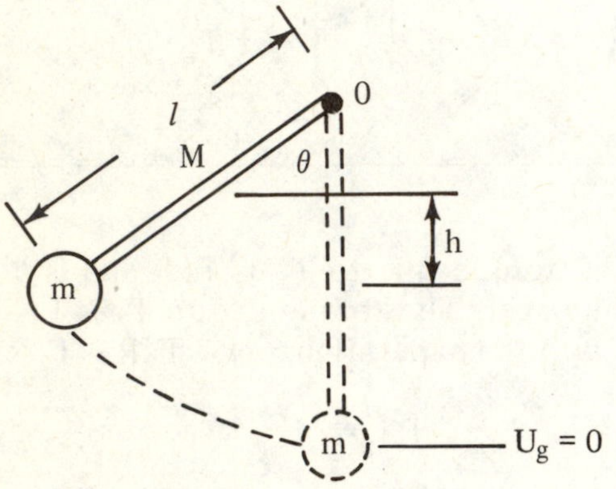

What is the moment of inertia of the system about an axis through 0, perpendicular to the plane of rotation?

$$I = I_{rod} + I_m$$

From the chart on page 168, we see that

$$I_{rod} = \frac{1}{3} M l^2$$

$\therefore$

$$I = \frac{1}{3} M l^2 + m l^2$$

$$I = \left(\frac{1}{3} M + m\right) l^2 \qquad (1)$$

6.B

What is the change in potential energy of the rod as it moves from its highest position to its lowest? Call the lowest point of the rod the reference of $U_g = 0$.

$$\Delta U_g(\text{rod}) = Mg\frac{l}{2} - Mg\left(h + \frac{l}{2}\right)$$

$$\Delta U_g(\text{rod}) = -\,Mgh \qquad (2)$$

6.C

What is the corresponding change in potential energy of the mass m?

Note that m is lowered by a distance 2 h.

$$\Delta U_g(m) = 0 - mg(2\,h)$$

$$\Delta U_g(m) = -2\,mgh \qquad (3)$$

6.D

What is the total change in potential energy of the system?

$$\Delta U_g = -\ Mgh - 2mgh$$

$$\Delta U_g = -(M + 2m)\ gh \qquad (4)$$

6.E

What is the initial kinetic energy of the system at the instant the rod is released?

Obviously, $K_0 = 0$.

6.F

What is the final kinetic energy of the system at the lowest point? Let ω be the angular velocity at this point.

$$K = \frac{1}{2} I\omega^2$$

where I is given by (1)

$$K = \frac{1}{2}\left(\frac{1}{3}M + m\right)\omega^2 \qquad (5)$$

6.G

Write an expression for the work-energy theorem applied to this situation.

$$W = \Delta K + \Delta U_g = 0$$

$$\frac{1}{2}\left(\frac{1}{3}M + m\right)\omega^2 - (M + 2m)\ gh = 0 \qquad (6)$$

6.H

Calculate the final angular velocity at the lowest point from (6).

$$\omega = \left\{\frac{2\,(M + 2m)\ gh}{\frac{1}{3}M + m}\right\}^{1/2} \qquad (7)$$

6.I

What is the angular momentum of the system at the lowest point relative to 0?

$$L = I\omega$$

where I is given by (1) and ω is given by (7).

6.J

Is the angular momentum constant? Explain.

No. L is a *maximum* at the lowest point, *zero* at the start. It is not constant because $\tau_{ext} \neq 0$.

6.K

What is the magnitude of the external torque, τ_{ext}, on the system when the rod makes an angle θ with the vertical?

$$\tau_{ext} = Mg\frac{l}{2}\sin\theta + mgl\sin\theta$$

$$\tau_{ext} = \left(\frac{M}{2} + m\right) gl\sin\theta \qquad (8)$$

6.L

What equation of motion would one have to solve to obtain L at any angle θ?

$$\tau_{ext} = I\alpha = I\frac{d^2\theta}{dt^2}$$

$$-\left(\frac{M}{2} + m\right) gl \sin\theta = I\frac{d^2\theta}{dt^2} \quad (9)$$

6.M

Obviously, the solution to (9) is not readily obtained. A simple solution is obtained when θ is assumed to be small. However, the energy method is shown to be a powerful technique in this situation, with no assumptions made on θ.

No answer.

7 A large, solid cylinder of mass M, radius R rotates about its axis on a frictionless bearing with an angular velocity ω_0. A small piece of putty, of mass m and initial velocity v_0, collides and sticks to the edge of the cylinder as shown.

7.A

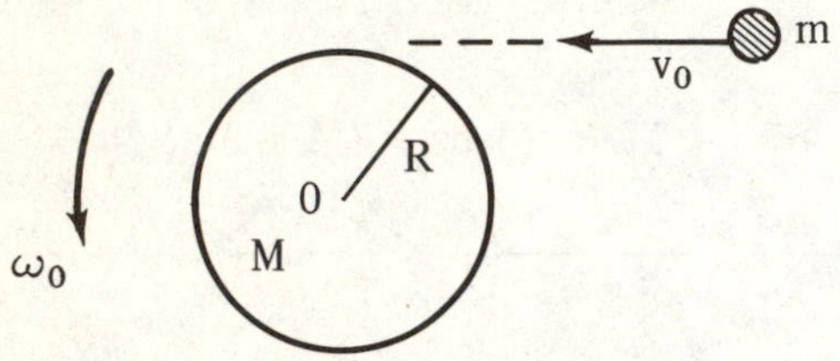

What is the angular momentum of the cylinder about 0 *before* the collision?

$$L_c = I_0 \omega_0$$

$$L_c = \frac{1}{2} MR^2 \omega_0 \quad (1)$$

The direction of $\mathbf{L_c}$ is *out* of the paper (for ccw rotation).

7.B

What is the angular momentum of the piece of putty relative to 0 *before* the collision?

$$L_m = mv_0 R \quad (2)$$

where the direction of $\mathbf{L_m}$ is also *out* of the paper.

7.C

What is the *total* angular momentum of the system *before* the collision?

$$L = L_c + L_m = \frac{1}{2} MR^2 \omega_0 + mv_0 R \quad (3)$$

7.D

If the putty were moving to the *right* before the collision, what would be the total angular momentum?

In this case, L_m would be $mv_0 R$ in magnitude, but its direction would be *into* the paper relative to 0.

$$\therefore L = \frac{1}{2} MR^2 \omega_0 - mv_0 R$$

7.E

After m collides and sticks to the cylinder, the angular velocity changes to ω. Now what is the angular momentum of the system?

$L = I_{system}\omega$

$$L = \left(\frac{1}{2}MR^2 + mR^2\right)\omega \quad (4)$$

7.F

Is angular momentum conserved in this collision? That is, can we equate (3) to (4)? Explain.

Yes. There are no external forces on the system, so $\tau_{ext} = \frac{dL}{dt} = 0$, $\therefore L$ = constant.

7.G

Use the fact that L is conserved to find the final angular velocity of the system.

Setting (3) equal to (4) gives

$$\omega = \frac{\frac{1}{2}MR^2\omega_0 + mv_0R}{\frac{1}{2}MR^2 + mR^2} \quad (5)$$

7.H

Suppose M = 2 kg, R = 20 cm, ω_0 = 3 rad/sec, m = 0.5 kg and v_0 = 5 m/sec. Find a numerical value for ω.

Substitute these values into (5). The result gives

$$\omega = 10.3 \text{ rad/sec} \quad (6)$$

7.I

What is the initial kinetic energy of the system?

$$K_0 = \frac{1}{2}mv_0^2 + \frac{1}{2}I_c\omega_0^2$$

$$K_0 = \frac{1}{2}mv_0^2 + \frac{1}{4}MR^2\omega_0^2 \quad (7)$$

K_0 = 6.43 J

7.J

What is the kinetic energy of the system right after the collision?

$$K = \frac{1}{2}I_{system}\omega^2$$

$$K = \frac{1}{2}\left(\frac{1}{2}MR^2 + mR^2\right)\omega^2 \quad (8)$$

Substituting (6) into (8), and the numerical values for m, M and R gives

$$K = 3.18 \text{ J}$$

7.K

What is the loss in kinetic energy due to this collision? Explain the result.

The loss in $K = (6.43 - 3.18)$ J

$$\text{loss} = 3.25 \text{ J}$$

This is a completely inelastic collision; hence, it is not surprising that $K < K_0$.

8 A solid cylinder of mass M, radius R, is pulled by a horizontal force F along a rough surface. The cylinder rolls without slipping, and F acts at a distance r from the center.

8.A

Draw a free body diagram for the cylinder.

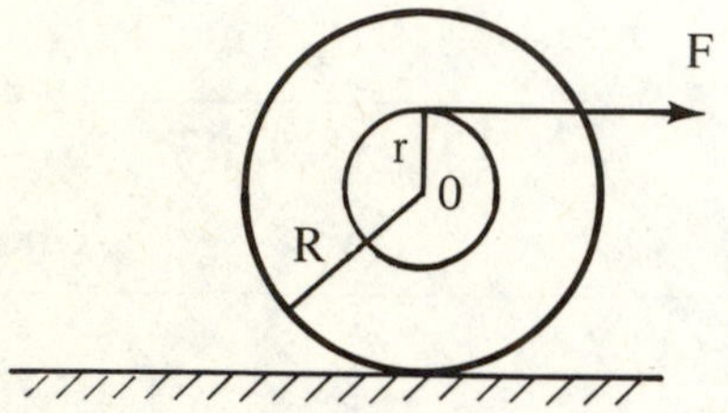

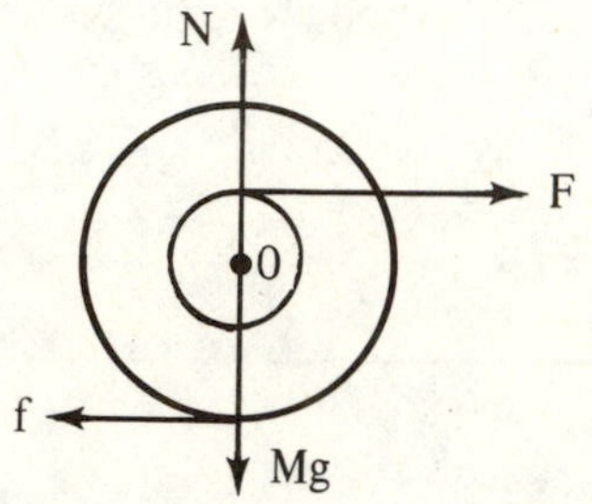

8.B

Write equations of motion corresponding to $\Sigma F = ma$ for the x and y directions.

$$\Sigma F_x = F - f = Ma \quad (1)$$

$$\Sigma F_y = N - Mg = 0 \quad (2)$$

8.C

Write an expression for the torque about 0, and relate this to the angular acceleration through the realtion $\tau = I\alpha$.

$$\tau_0 = fR + Fr = I_0 \alpha$$

but

$$I_0 = \frac{1}{2} MR^2 \text{, so}$$

$$fR + Fr = \left(\frac{1}{2} MR^2\right) \alpha \quad (3)$$

8.D

What is the condition such that the cylinder rolls without slipping?

If the cylinder rolls without slipping, then the linear displacement x is related to the angular displacement θ by $x = R\theta$. Differentiating this with respect to time gives $v = R\omega$. Differentiating a second time gives $a = R\alpha$.

8.E

Eliminate f from (1) and (3) to obtain a relation between acceleration and the force F.

Substitute $f = F - Ma$ from (1) into (3). This gives

$$a = \frac{2(R + r)}{3MR} F \tag{4}$$

8.F

What is the *minimum* value of μ such that the cylinder will roll without slipping?

We set $f = \mu N = \mu Mg$, and substitute this into (1). Making use of (4) gives

$$\mu_{min} = \frac{F(R - 2r)}{3MgR} \tag{5}$$

8.G

What is the allowed range of values for μ_{min}? Note that negative values are not physically meaningful.

From (5) we see that the largest value of μ_{min} corresponds to $r = 0$. The smallest value is *zero*, corresponding to $r = R/2$. Thus,

$$0 \leqslant \mu_{min} \leqslant \frac{F}{3Mg}$$

8.10 SUMMARY

Angular momentum of a particle:	$\mathbf{L} = \mathbf{r} \times \mathbf{p} = m\mathbf{r} \times \mathbf{v}$
Torque acting on a particle:	$\boldsymbol{\tau} = \mathbf{r} \times \mathbf{F} = \dfrac{d\mathbf{L}}{dt}$
Moment of inertia for a system of particles:	$I = \Sigma m_i r_i^2$
Moment of inertia of a rigid body:	$I = \int r^2\,dm$
Work-energy theorem for a rotating rigid body:	$W = \int \tau d\theta = \frac{1}{2} I\omega^2 - \frac{1}{2} I\omega_0{}^2$
Conservation of angular momentum:	$I\omega = I_0\,\omega_0$ (when $\tau_{ext} = 0$)

A tabulation of other useful expressions in rotational motion, together with the analogous expressions for translational motion, is presented in Table 8-1.

TABLE 8-1 COMPARISON OF USEFUL EXPRESSIONS IN ROTATIONAL MOTION WITH THOSE OF TRANSLATIONAL MOTION

Translational Motion		Rotational Motion	
Displacement	$x - x_0$	Angular displacement	$\theta - \theta_0$
Velocity	$v = dx/dt$	Angular velocity	$\omega = d\theta/dt$
Acceleration	$a = dv/dt$	Angular acceleration	$\alpha = d\omega/dt$
Resultant force	$\mathbf{F} = M\mathbf{a}$	Resultant torque	$\tau = I\alpha$
Mass	M	Inertia	I
a = constant	$v = v_0 + at$ $x - x_0 = v_0 t + \frac{1}{2} at^2$ $v^2 = v_0{}^2 + 2a(x - x_0)$	α = constant	$\omega = \omega_0 + \alpha t$ $\theta - \theta_0 = \omega_0 t + \frac{1}{2}\alpha t^2$ $\omega^2 = \omega_0{}^2 + 2\alpha(\theta - \theta_0)$
Work	$W = \int F dx$	Work	$W = \int \tau d\theta$
Kinetic energy	$K = \frac{1}{2} Mv^2$	Kinetic energy	$K = \frac{1}{2} I\omega^2$
Power	$P = \mathbf{F} \cdot \mathbf{v}$	Power	$P = \boldsymbol{\tau} \cdot \boldsymbol{\omega}$
Linear momentum	$\mathbf{p} = m\mathbf{v}$	Angular momentum	$\mathbf{L} = I\omega$
Resultant force	$\mathbf{F} = \dfrac{d\mathbf{p}}{dt}$	Resultant torque	$\boldsymbol{\tau} = \dfrac{d\mathbf{L}}{dt}$

8.11 PROBLEMS

1. Two particles having masses of m and 3m, respectively, move in opposite directions along a straight line as in Figure 8-16. Each particle has a speed v, where the particle of mass m moves to the right and the particle of mass 3m moves to the left. What is the total angular momentum of the system (a) with respect to point A? (b) with respect to point B? (c) with respect to point C? (Find both magnitudes and directions.)

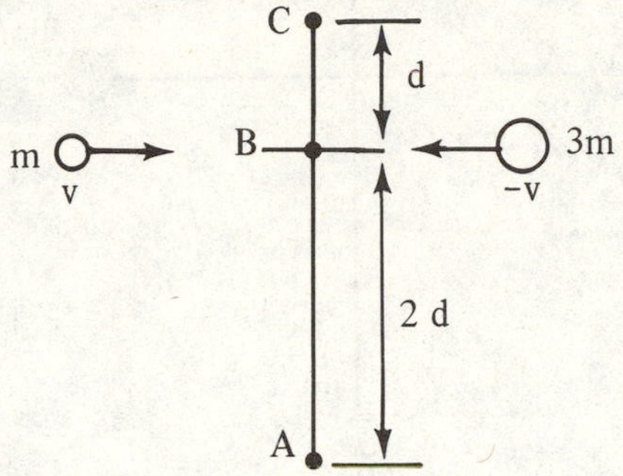

Figure 8-16

2. Assume that the earth rotates about the sun in a circular orbit, with the sun fixed at the center. (a) Calculate the angular momentum of the earth with respect to the sun, treating the earth as a point mass. (b) What is the translational kinetic energy of the earth? (c) What is the rotational kinetic energy of the earth due to its spin about its own axis, assuming it is a uniformly dense sphere? (d) To appreciate the scale of the translational kinetic energy, determine how long this energy would last if it could be used to supply one billion watts to a city (recall that 1 W = 1 J/sec). Use the fact that the mass of the earth is 6.0×10^{24} kg, its mean orbital speed is 3.0×10^{4} m/sec, and the earth-sun separation is 1.5×10^{11} m.

3. A particle of mass 1 kg moves along a straight line with a constant acceleration of 4**j** m/sec^2. At t = 0 the particle is located at the point (2,0) m, and its velocity is $\mathbf{v}_0$ = 3**j** m/sec. (a) What is the angular momentum of the particle about the origin at t = 0? (b) What is the vector position of the particle at t = 1 sec? (c) What is the velocity of the particle at t = 1 sec? (d) Determine the angular momentum of the particle about the origin at t = 1 sec. (e) Calculate the change in angular momentum of the particle in the interval t = 0 to t = 1 sec. (f) What is the torque acting on the particle?

4. A particle of mass m, located at the vector position **r**, has a momentum **p**. (a) If **r** and **p** both have three components, that is, $\mathbf{r} = x\mathbf{i} + y\mathbf{j} + z\mathbf{k}$ and $\mathbf{p} = p_x\mathbf{i} + p_y\mathbf{j} + p_z\mathbf{k}$, show that the angular momentum of the particle relative to the origin also has three components, given by $L_x = yp_z - zp_y$, $L_y = zp_x - xp_z$ and $L_z = xp_y - yp_x$. (b) If the particle moves only in the xy plane, prove that $L_x = L_y = 0$, but $L_z \neq 0$.

5. Four particles are located at the corners of a polygon as shown in Figure 8-17. Assuming that the particles are connected by light, rigid rods, (a) calculate the moment of inertia of the system about 0. (b) If the system rotates about the z axis with an angular velocity ω, calculate the kinetic energy of the system. (c) Repeat (b), assuming the rotation occurs about the y axis. (d) Determine the angular momentum of the system relative to 0 if the rotation occurs about the z axis.

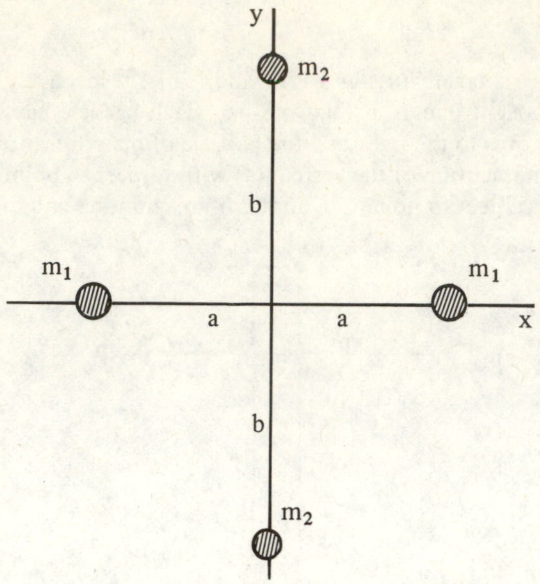

Figure 8-17

6. (a) A uniform solid disc of radius R, mass M, is free to rotate on a frictionless pin located at a point on the rim, as in Figure 8-18. If the disc is released from rest in the position shown, what is the velocity of the center of mass when it reaches the dotted position? (b) Repeat this calculation if the object is a uniform hoop. (See chart, page 168, for moments of inertia.)

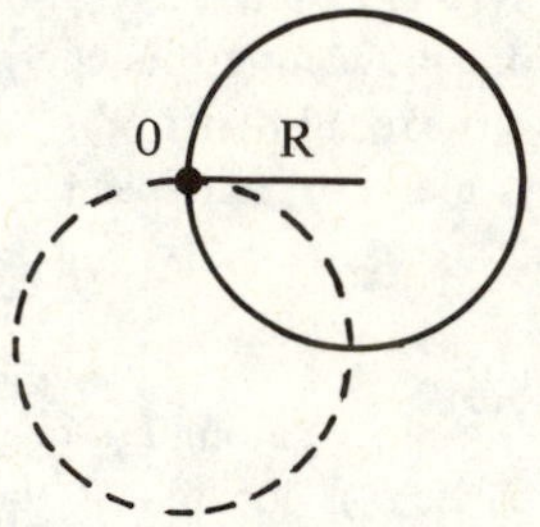

Figure 8-18

†7. (a) A uniform disc of radius R, mass M, rotates about an axis in the *plane* of the disc, passing through its center as in Figure 8-19. Show that the moment of inertia about this axis is $I_y = \frac{1}{4}MR^2$. *Hint:* The mass per unit area, σ, can be written as $\sigma = M/\pi R^2$, and $dm = 2\ \sigma y\ dr$, where the variables are defined in Figure 8-19. (b) Use the parallel axis theorem to prove that the moment of inertia about the axis labeled y′ is $I_y{}' = \frac{5}{4}MR^2$.

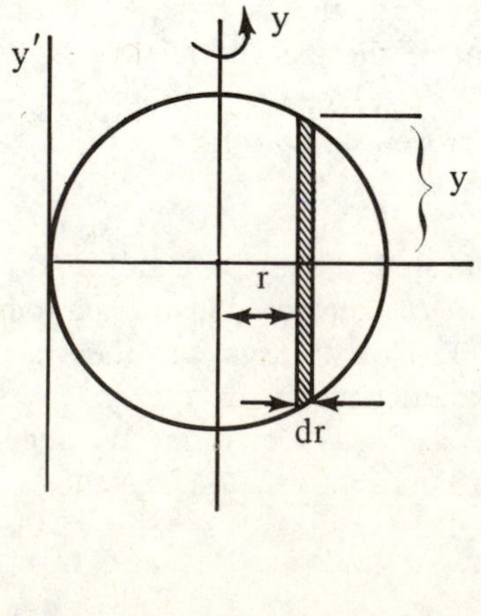

Figure 8-19

8. Three uniform rods, each of mass M, length l, are connected as shown in Figure 8-20. What is the moment of inertia for rotation about (a) axis y? (b) axis y′? (c) axis x? (d) axis x′?

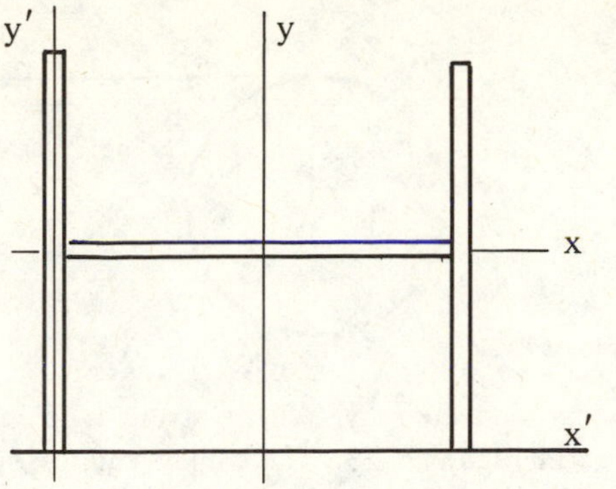

Figure 8-20

9. A particle of mass m is given a velocity $-v_0\mathbf{j}$ at the point (−d, 0), and proceeds to accelerate in the earth's field as in Figure 8-21. (a) Find an expression for the angular momentum as a function of time with respect to 0. (b) Calculate the torque $\boldsymbol{\tau}$ acting on m at any time relative to 0. (c) Use your results to (a) and (b) to verify the relation $\boldsymbol{\tau} = dL/dt$.

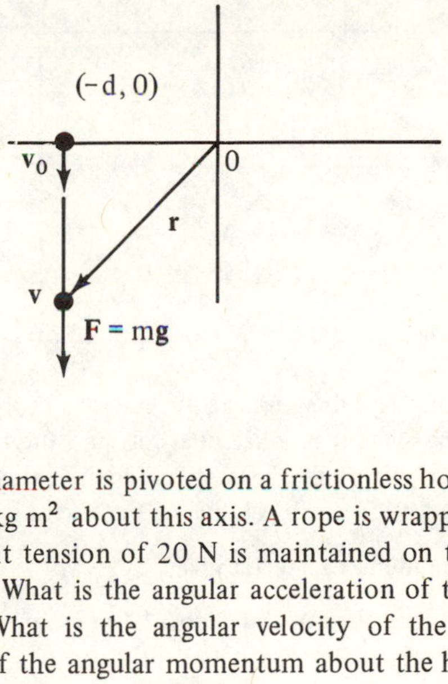

Figure 8-21

10. A flywheel 1 m in diameter is pivoted on a frictionless horizontal axis and has a moment of inertia equal to 5 kg m^2 about this axis. A rope is wrapped around the outside of the flywheel and a constant tension of 20 N is maintained on the rope. The flywheel starts from rest at t = 0. (a) What is the angular acceleration of the flywheel while the rope is under tension? (b) What is the angular velocity of the wheel at t = 3 sec? (c) Calculate the magnitude of the angular momentum about the horizontal axis at t = 3 sec? (d) What is the kinetic energy of the flywheel at t = 3 sec? (e) How much rope is unwound in the first 3 seconds?

11. A uniform rod of mass 100 g and 50 cm in length rotates in a horizontal plane about a fixed, vertical, frictionless pin through its center. Two small beads, each of mass 30 g, are mounted on the rod such that they are able to slide without friction along the rod. The beads are initially held by catches at positions 10 cm on each side of center of the rod, at which time the system rotates at an angular velocity of 20 rad/sec. Suddenly, the catches are released and the small beads slide outwards along the rod. (a) Find the angular velocity of the system at the instant the beads reach the ends of the rod. (b) Find the angular velocity of the rod after the beads fly off the ends.

12. A small particle of mass m = 10 g and speed v_0 = 5 m/sec collides and sticks to the edge of a uniform solid sphere of mass M = 1 kg and radius R = 20 cm, as in Figure 8-22. (a) If the sphere is initially at rest, find the angular speed of the system after the collision. (b) How much energy is lost in this collision?

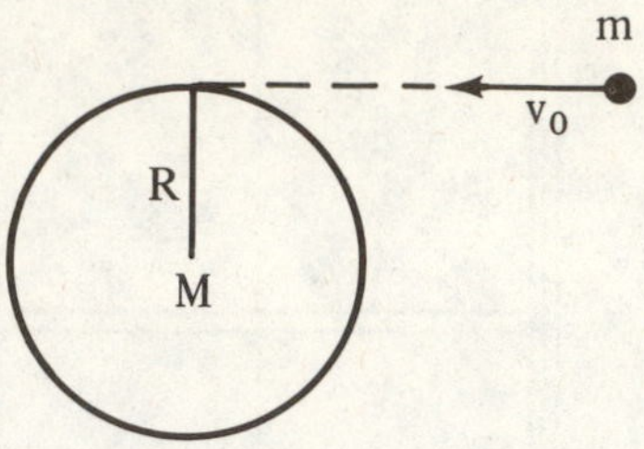

Figure 8-22

13. A mass m_1 is connected by a light string to a mass m_2, which slides on a frictionless surface, as in Figure 8-23. The pulley rotates on a frictionless axle, has a radius R, and a moment of inertia I. Assume the string does not slip on the pulley. (a) Find the acceleration of the system. (b) Find the tensions T_1 and T_2 in the two strings. (c) Obtain numerical values for T_1, T_2 and a if m_1 = 5 kg, m_2 = 10 kg, I = 1 kg m², and R = 20 cm.

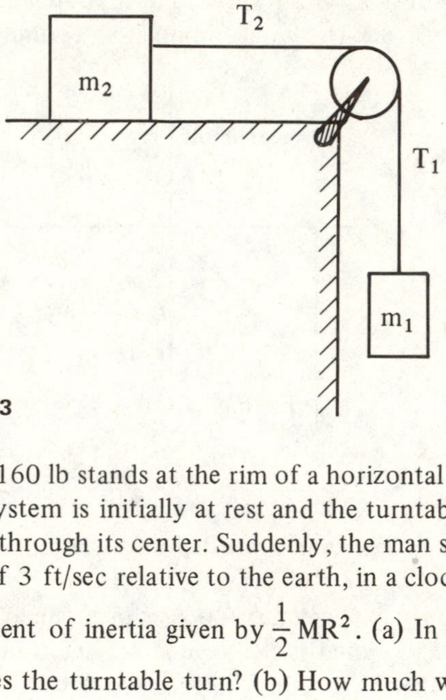

Figure 8-23

14. A man weighing 160 lb stands at the rim of a horizontal turntable of radius 10 ft and weight 320 lb. The system is initially at rest and the turntable is free to rotate about a frictionless vertical axle through its center. Suddenly, the man starts walking around the rim at a constant speed of 3 ft/sec relative to the earth, in a clockwise direction. Assume the turntable has a moment of inertia given by $\frac{1}{2}MR^2$. (a) In what direction and with what angular velocity does the turntable turn? (b) How much work does the man do to set this system in motion?

15. The combination of an external force and a friction force produces a constant total torque of 24 Nm on a pivoted wheel. The external force is applied for 5 sec, during which time the angular velocity of the wheel increases from zero to 10 rad/sec. The external force is then removed and the wheel is brought to rest by friction in 50 sec. Find (a) the moment of inertia of the wheel, (b) the friction torque and (c) the total number of revolutions of the wheel.

16. A solid uniform cylinder of mass M, radius R, is supported with frictionless bearings on a horizontal axle as in Figure 8-24. Two weights, each of mass m, hang from cords wrapped around the cylinder. (a) Find the tension in each cord. (b) Determine the acceleration of each mass. Assume that $I = \frac{1}{2}MR^2$ for the cylinder.

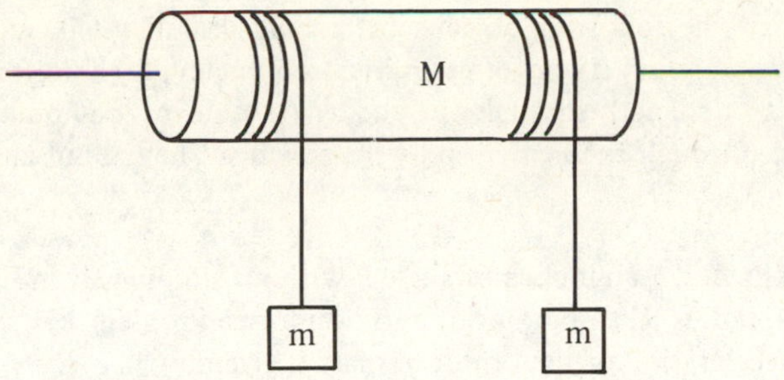

Figure 8-24

17. A block of mass M is attached to a cord passing through a hole in a frictionless, horizontal surface as in Figure 8-25. The block is originally rotating in a circle of radius r_0, with an angular velocity ω_0. The cord is then pulled from below, shortening the radius of the circle to r. (a) What is the angular velocity of M at the radius r? (b) How much work is done in moving M from r_0 to r? (c) Obtain numerical values for ω and W if M = 50 g, r_0 = 30 cm, ω_0 = 5 rad/sec and r = 10 cm.

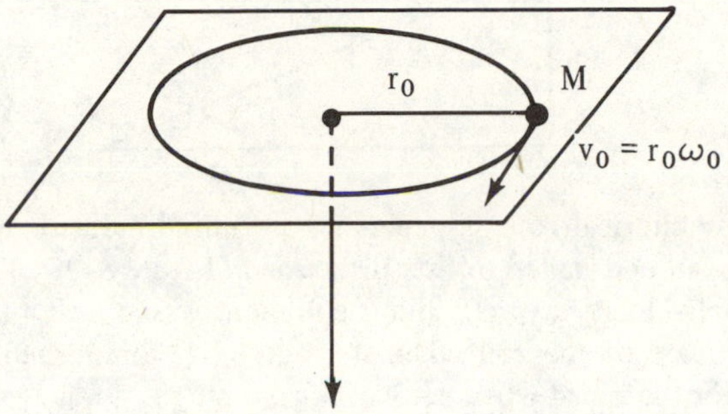

Figure 8-25

***18.** A long, uniform rod of length *l*, mass M, is pivoted about a horizontal, frictionless axis through one end. The rod is released from rest in a vertical position, with the center of mass above the horizontal as in Figure 8-26. *When the rod is horizontal,* find (a) the angular velocity of the rod, (b) the angular acceleration of the rod, (c) the x and y components of acceleration of the center of mass and (d) the horizontal and vertical components of the force on the rod by the pivot.

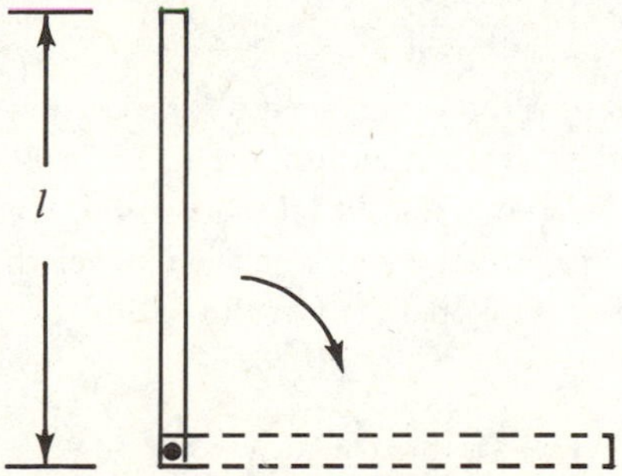

Figure 8-26

19. The *radius of gyration,* k, of a rigid body of mass M with respect to some axis is defined to be the distance of a point mass M from that axis, such that the moment of inertia with respect to that axis can be written as $I = Mk^2$. Therefore, $k = \sqrt{I/M}$. Find the radius of gyration of the following objects: (a) a solid disc about its axis of cylindrical symmetry, (b) a hoop about its axis of cylindrical symmetry, (c) a uniform rod about an axis perpendicular to its length through one end, (d) a uniform rod about an axis perpendicular to its length through the center and (e) a spherical shell about an axis through its center.

20. A body rolls down an incline of angle θ without slipping, as in Figure 8-27. The body has a mass M, radius R, and radius of gyration k (see Problem 19), where $k = \sqrt{I/M}$. (a) Show that the acceleration of the center of mass is given by

$$a = \frac{g \sin \theta}{1 + k^2/R^2}$$

(b) Determine explicit values for a if the object is (1) a solid disc, (2) a hoop, (3) a hollow sphere. (c) Assuming the disc, hoop and hollow sphere start from the top at rest, determine their speeds at the bottom. (d) What are the *minimum* coefficients of friction necessary to maintain pure rolling motion for the disc, the hoop and the hollow sphere?

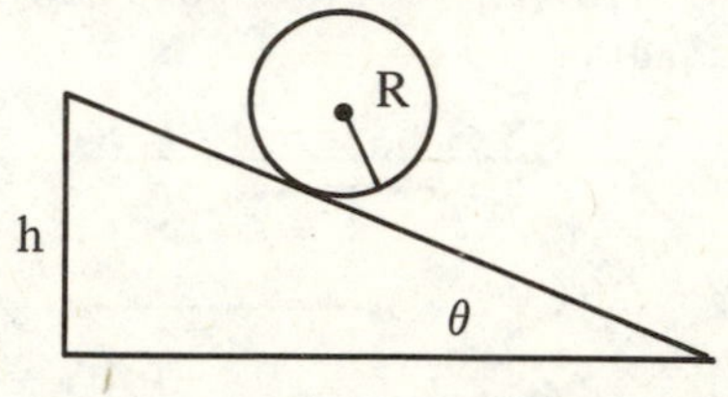

Figure 8-27

***21.** A large cylindrical roll of paper, whose initial radius is R, lies on a horizontal surface with its open end nailed to the surface as in Figure 8-28. The roll is given a slight shove (so its initial velocity $v_0 \approx 0$) and it commences to unroll. (a) Determine the speed of the center of mass of the roll when its radius has diminished to r. (b) Calculate a numerical value for the speed when r = 1 mm, assuming R = 6 m. (c) What happens to the energy of the system when the paper is completely unrolled? *Hint:* Assume the roll has a uniform mass density and apply energy principles.

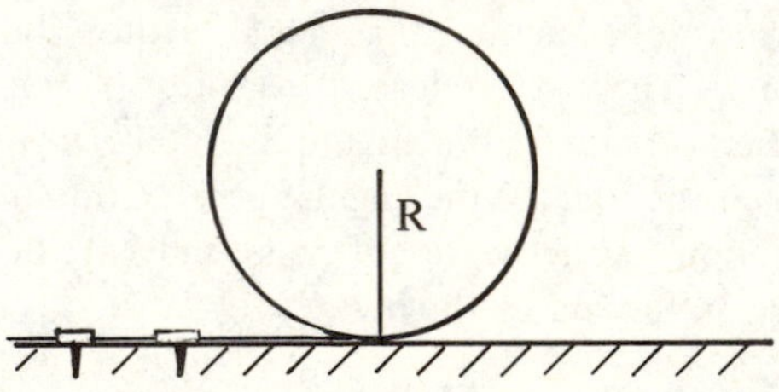

Figure 8-28

***22.** A uniform roll of ribbon is allowed to unroll from rest from the top of an incline of angle θ (Figure 8-29). The open end of the roll is nailed to the surface and the length of the incline is l. (a) How long does it take the roll to reach the bottom? (b) What is the speed of the center of mass of the roll when it reaches the bottom? *Hint:* The acceleration of the center of mass is independent of the radius.

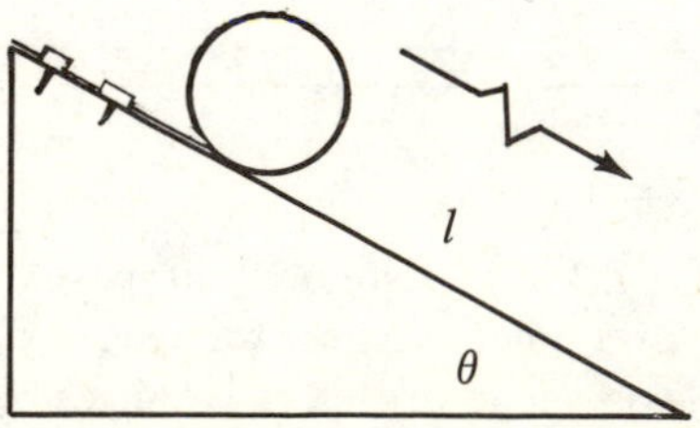

Figure 8-29

23. The system shown in Figure 8-30 consists of a pulley whose moment of inertia is I, connected to a mass M on a frictionless incline, which, in turn, is connected to a spring whose force constant is k. The pulley is wound counterclockwise so as to stretch the spring a distance d from its unstretched position. (a) If the system is then released from rest from this position, determine the angular velocity of the pulley when the spring is unstretched. (b) Obtain a numerical value for the angular velocity at this point if $I = 1\ \text{kg m}^2$, $k = 50$ N/m, $x = 0.2$ m, $M = 0.5$ kg, $R = 0.3$ m and $\theta = 37°$.

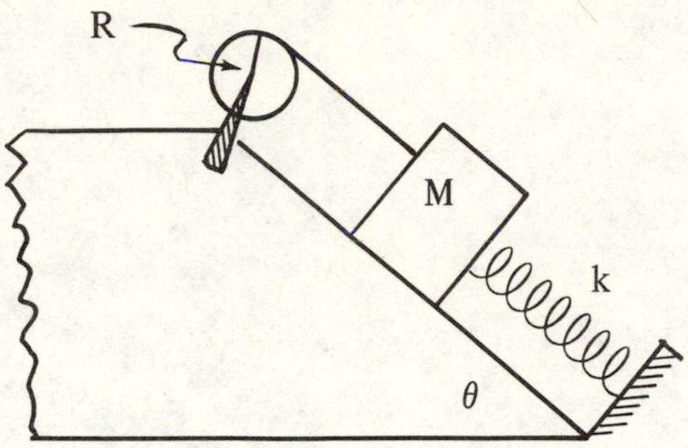

Figure 8-30

24. A uniform solid sphere of radius R is placed on the inside surface of a hemispherical bowl of radius a. The sphere is released from rest at an angle θ with the vertical and rolls *without slipping* as in Figure 8-31. Determine the angular speed of the sphere when it reaches the bottom of the bowl.

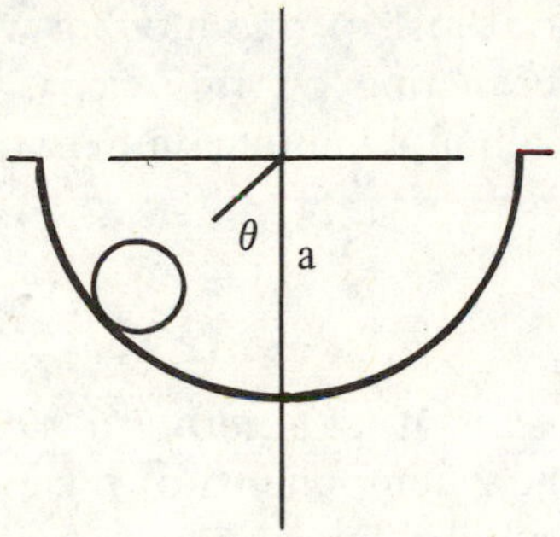

Figure 8-31

†25. An object with a circular cross section has a radius R and moment of inertia $I = Mk^2$, where k is its radius of gyration (see Problem 19). The object is given an initial velocity v_0 on a rough horizontal plane (Figure 8-32). Assuming the object starts its motion with *no* rotation, and the coefficient of friction is μ, find (a) the time it takes for *pure* rolling motion to occur, (b) the linear velocity of the center of mass when pure rolling occurs and (c) the distance the object moves at the instant pure rolling sets in. *Hint:* Note that the frictional force provides both the linear acceleration and angular acceleration, and $v_c = R\omega$ when pure rolling commences. Therefore, from Newton's second law, the acceleration of the center of mass is $a_c = -\mu g$, and from $\tau = I\alpha$ we obtain $\alpha = \mu g R/k^2$.

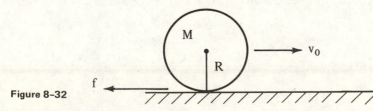

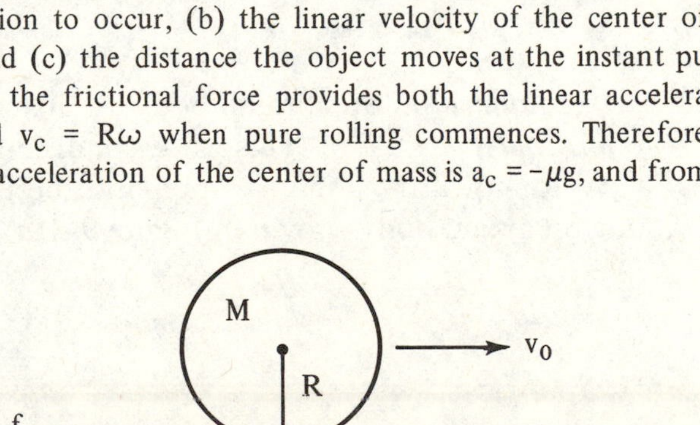

Figure 8-32

9

BODIES IN EQUILIBRIUM

9.1 BASIC CONCEPTS AND DEFINITIONS

In this chapter we will deal with special problems involving rigid bodies which are in equilibrium and which remain in equilibrium even under the action of external forces. First, it must be understood that the word "equilibrium" does not necessarily imply that the body is at rest. An object at rest is a special case of equilibrium. In general, an object at rest or one moving with constant velocity will only do so if the resultant force acting on it is zero. This is, in fact, the *first condition of equilibrium.* That is, a body will remain in equilibrium only if

$$\sum_n \mathbf{F}_n = 0 \tag{9.1}$$

Since Equation (9.1) is a *vector sum* of all *external forces,* this necessarily implies that the vector sum of the x, y and z components of the external forces is also zero. Therefore, the first condition of equilibrium can also be written as

$$\begin{aligned} \Sigma \mathbf{F}_{xn} &= 0 \\ \Sigma \mathbf{F}_{yn} &= 0 \\ \Sigma \mathbf{F}_{zn} &= 0 \end{aligned} \tag{9.2}$$

The *second condition of equilibrium* of a rigid body is that *the sum of all torques relative to any point must be zero.* This is a condition of rotational equilibrium. Torque is a vector quantity and is, perhaps, best described by considering one force **F** acting on a rigid body shown in Figure 9-1(*a*).

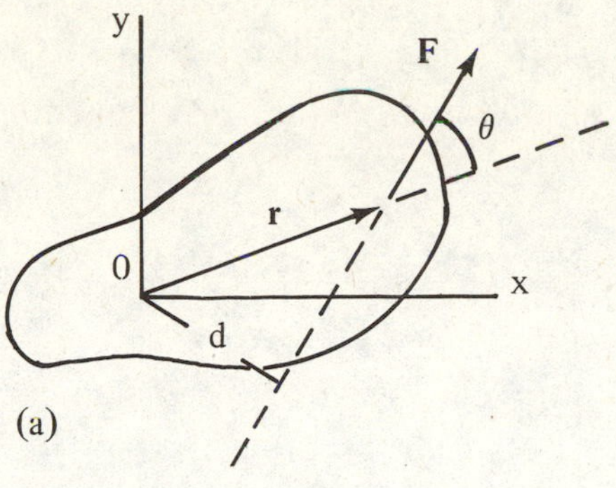

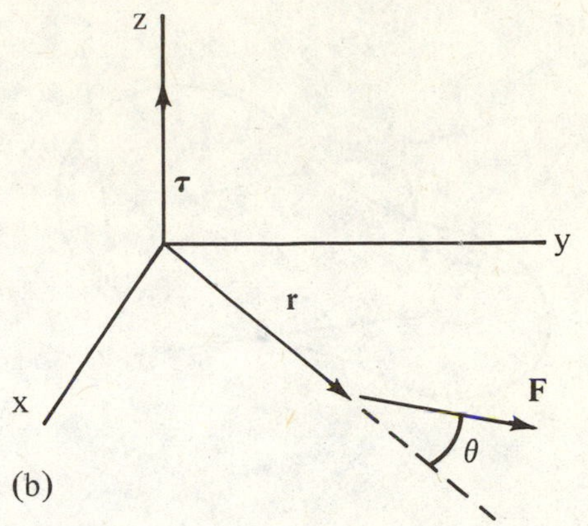

Figure 9–1 (*a*) Torque of the force **F** about 0 is given by $\boldsymbol{\tau} = \mathbf{r} \times \mathbf{F}$. (*b*) Vector relation between $\boldsymbol{\tau}$, **r**, and **F**.

The torque of the force **F** about 0 is given by the vector product

$$\boldsymbol{\tau} = \mathbf{r} \times \mathbf{F} \tag{9.3}$$

If **r** and **F** are both in the xy plane, the direction of $\boldsymbol{\tau}$ as defined by the right hand rule is in the z direction as shown in Figure 9–1(*b*). This result can be obtained by writing $\mathbf{r} = x\mathbf{i} + y\mathbf{j}$, $\mathbf{F} = F_x\mathbf{i} + F_y\mathbf{j}$, giving for the vector product

$$\boldsymbol{\tau} = (x\mathbf{i} + y\mathbf{j}) \times (F_x\mathbf{i} + F_y\mathbf{j}) = (xF_y - yF_x)\mathbf{k} \tag{9.4}$$

whose magnitude is

$$\tau = xF_y - yF_x \tag{9.5}$$

Obviously, from Equation (9.5) we see that τ can be positive or negative, depending on the location of the arbitrary point 0. In practice, however, it is simple to write the torque τ as the *product of F* times the *perpendicular* to the *line of action* of **F** which is labeled d in Figure 9–1(*a*). Therefore, the torque of F about 0 is

$$\tau = Fd \tag{9.6}$$

where d is called the *moment arm*.

Now consider two forces $\mathbf{F}_1$ and $\mathbf{F}_2$ acting on a rigid body as shown in Figure 9–1(*c*). If both forces are in the xy plane and have moment arms d_1 and d_2, respectively, the torques of $\mathbf{F}_1$ and $\mathbf{F}_2$ are given by

$$\boldsymbol{\tau}_1 = F_1 d_1 \mathbf{k}$$

$$\boldsymbol{\tau}_2 = -F_2 d_2 \mathbf{k}$$

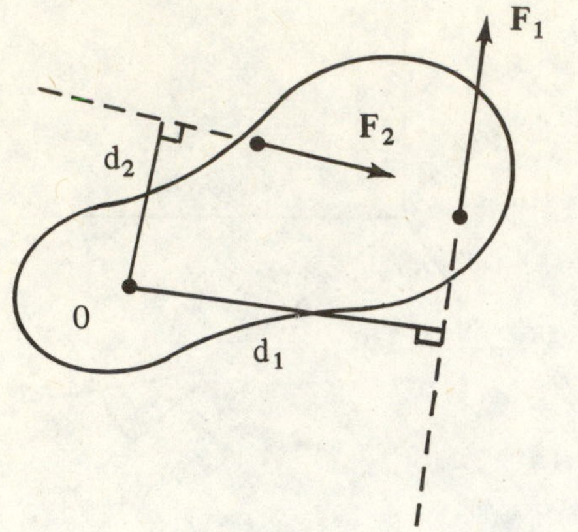

Figure 9-1(*c*). The moment arm of F_1 is d_1
The moment arm of F_2 is d_2.

The signs here are very important. It is easy to remember the appropriate signs by assigning τ *positive for a counterclockwise sense of the rotation about 0* (namely that due to $\mathbf{F}_1$) and τ *negative for a clockwise sense of rotation about 0* (that due to $\mathbf{F}_2$). If the body is in rotational equilibrium, *the second condition of equilibrium requires that*

$$\sum_n \boldsymbol{\tau}_n = \sum_n \mathbf{r}_n \times \mathbf{F}_n = 0 \tag{9.7}$$

Thus, in the case of the two forces shown in Figure 9-1(*c*)

$$\boldsymbol{\tau}_1 + \boldsymbol{\tau}_2 = (F_1 d_1 - F_2 d_2)\,\mathbf{k} = 0$$

or

$$F_1 d_1 = F_2 d_2$$

Since we will be dealing with problems involving forces in the xy plane, that is, those having x and y components only, $\boldsymbol{\tau}$ will always be in the +z or −z direction. Therefore, the *two conditions of equilibrium* that must be met in order that a rigid body be in translational and rotational equilibrium (no linear or angular acceleration) can be written as

$$\sum_n F_{xn} = 0, \quad \sum_n F_{yn} = 0$$

and

$$\sum_n \boldsymbol{\tau}_n = 0 \tag{9.8}$$

Example 9.1

A man wishes to push a book at constant velocity along a table, as shown in Figure 9-2(*a*). The magnitude of the force F he must exert to move the book must exactly equal the force of friction f between the book and the table in order that the book move at constant velocity. [To get the book in motion initially, he must exert a force somewhat greater than this since the static frictional forces are generally greater than sliding frictional forces.] Hence, the book moving with constant velocity is in dynamic equilibrium since the man's force on the book is equal in magnitude and opposite in direction to the frictional force. This situation is essentially equivalent to the book at rest on the table, with **F** = 0, as in Figure 9-2(*b*).

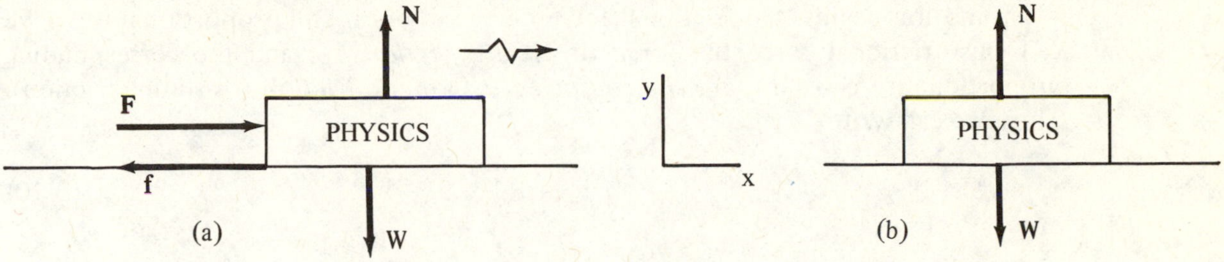

Figure 9-2 (*a*) A book in dynamic equilibrium. (*b*) A book in static equilibrium.

Note that in both Figures 9-2(*a*) and 9-2(*b*) there are two other external forces acting on the book. These are the gravitational force **W** (the weight of the book) and the "push" of the table on the book **N**, usually referred to as the normal force. The first condition of equilibrium can be written as

$$\Sigma \mathbf{F}_{xn} = \mathbf{F} + \mathbf{f} = 0$$

$$\Sigma \mathbf{F}_{yn} = \mathbf{N} + \mathbf{W} = 0$$

Of course, $\Sigma \mathbf{F}_{zn} = 0$, since there are no forces in the z direction. In unit vector notation, these relations become

$$F\mathbf{i} - f\mathbf{i} = 0$$

$$N\mathbf{j} - W\mathbf{j} = 0$$

or simply

$$F - f = 0$$

$$N - W = 0$$

Of course we could have written these immediately by inspection. However, it is good practice to write such conditions in vector form to remind ourselves that we are dealing with *vector* properties. It is especially useful in treating more complex problems.

When there is no relative motion between the book and the table, and there is an applied force **F**, there must be a frictional force between the book and the table to oppose **F** and keep the book at rest. We call this frictional force a *force of static friction* $\mathbf{f}_s$. This forcc has some maximum value corresponding to the value at which the book is on the verge of motion. When **F** exceeds this maximum force of static friction, the book will start to move. It is generally observed that this maximum force of static friction is approximately proportional to the magnitude of the normal force **N**. If we call the proportionality constant μ_s, *the coefficient of static friction,* we can write

$$f_s \leqslant \mu_s N \tag{9.9}$$

When motion begins, the frictional force decreases but is still proportional to N. We call this frictional force the *force of kinetic friction,* f_k, and the corresponding proportionality constant, μ_k, *the coefficient of kinetic friction* (or sliding friction). Thus, we can write

$$f_k = \mu_k N \tag{9.10}$$

Example 9.2

A common mistake that is made is to assume that the normal force is always equal in magnitude to the weight of the object. This is not always the case. The following example illustrates a situation where the normal force is less than W. Suppose a block is being pulled by a rope at constant speed along a rough horizontal surface with a force **F** as shown in Figure 9-3.

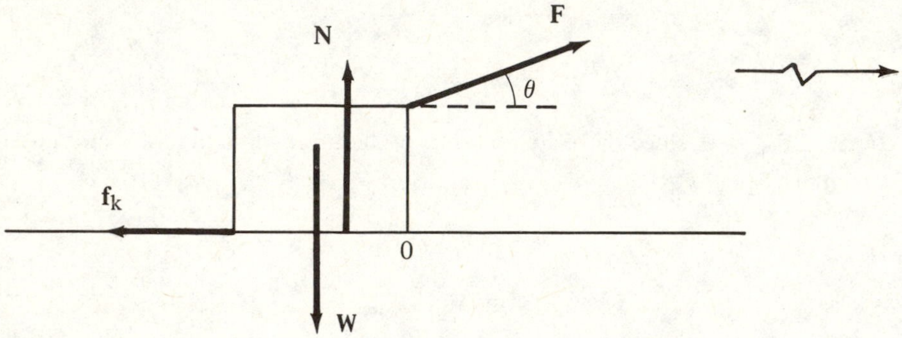

Figure 9-3 Diagram of external forces acting on a block of weight W moving on a rough surface with constant speed.

In order to analyze this problem, it is very important to draw a diagram, which will be referred to as a *free body diagram,* of all the external forces acting on the block. In this case, there are four forces, **N**, **W**, **F**, and $\mathbf{f_k}$. If the force **F** acts at an angle θ to the horizontal, the first condition of equilibrium gives

$$\mathbf{F} + \mathbf{f_k} + \mathbf{N} + \mathbf{W} = 0$$

We will not make use of the second condition of equilibrium in this example, since the block is sliding and must be in rotational equilibrium. However, it is generally more useful to write the first condition in component form, namely

$$\sum_n \mathbf{F}_{xn} = F \cos\theta \mathbf{i} - f_k \mathbf{i} = 0$$

$$\sum_n \mathbf{F}_{yn} = N\mathbf{j} + F \sin\theta \mathbf{j} - W\mathbf{j} = 0$$

Therefore, the magnitudes of the vector components are related by

$$F \cos \theta = f_k$$

$$N + F \sin \theta = W$$

But from Equation (9.10), $f_k = \mu_k N$, and the above equations can be solved simultaneously to give

$$N = \frac{W}{1 + \mu_k \tan \theta}$$

$$F = \frac{\mu_k N}{\cos \theta} = \frac{\mu_k W}{\cos \theta + \mu_k \sin \theta}$$

Suppose $W = 50$ lb, $\mu_k = 0.5$, and $\theta = 37^\circ$. The force F required to move this block at constant speed at the angle $\theta = 37^\circ$ is calculated to be

$$F = \frac{0.5\,(50)\text{ lb}}{0.8 + 0.5\,(0.6)} = 22.7\text{ lb}$$

and the frictional force $f_k = \mu_k N = F \cos \theta = 18.2$ lb. We see that this frictional force *decreases* as θ increases. This makes sense since as θ increases for a given value of F, N decreases (the force F tends to lift the block off the floor) and f_k decreases correspondingly.

Example 9.3

A uniform beam of weight 50 lb and length 12 ft is supported by a pivot at one end and a rope connected to a 200 lb weight through a frictionless pulley as shown in Figure 9-4. A man weighing 150 lb stands at the position x such that the beam is horizontal and the rope makes an angle of 37° with the horizontal. (a) Draw a free body diagram for the beam. (b) Find the position x at which the man must stand in order to keep the plank horizontal. Also, find the force exerted by the pivot on the beam.

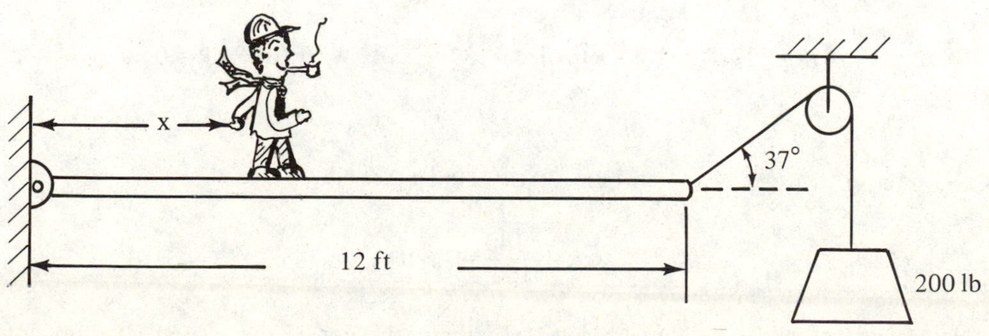

Figure 9-4

Solution

(a) T = 200 lb
w = 50 lb
W = 150 lb

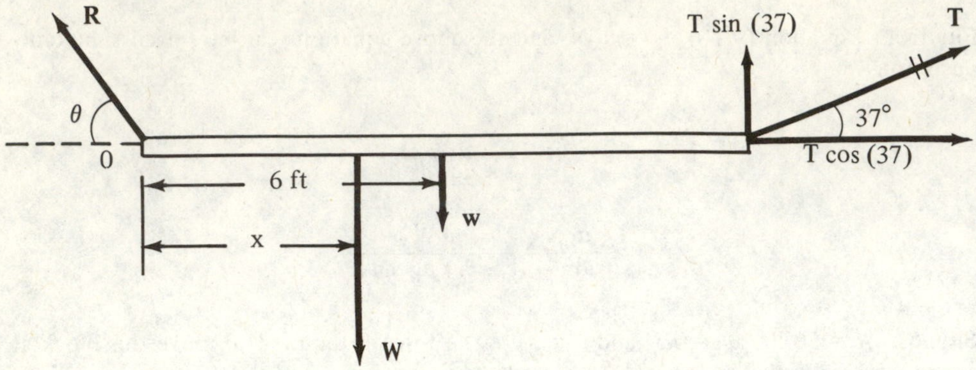

(b) First, it should be noted that the tension T in the rope is uniform and equal to 200 lb since, if we "isolate" the 200 lb weight, the rope must exert a force of 200 lb upward on it to keep the weight in equilibrium. Calling the weights of the man and beam **W** and **w**, respectively, and denoting the force exerted by the pivot by **R**, we have for the first condition of equilibrium, that is, $\mathbf{R} + \mathbf{W} + \mathbf{w} + \mathbf{T} = 0$. In component form, this becomes

$$R \sin \theta \mathbf{j} + T \sin (37) \mathbf{j} - W\mathbf{j} - w\mathbf{j} = 0$$

$$T \cos (37) \mathbf{i} - R \cos \theta \mathbf{i} = 0$$

Dropping the unit vector notation and substituting in W = 150, w = 50, T = 200 gives the equations

$$R \sin \theta - 80 = 0$$

$$160 - R \cos \theta = 0$$

Solving these two equations gives $\theta = 26.6^\circ$, R = 179 lb. The second condition of equilibrium can be written by first resolving **T** in components perpendicular and parallel to the beam. Note that the moment arm of the T cos (37) component is zero if we take the *left end of the beam* as the origin of coordinates. The moment arms of T sin (37), w, W and R are 12 ft, 6 ft, x and 0, respectively. Therefore,

$$\Sigma \tau_0 = [12\, T \sin (37) + (0)\, T \cos (37) - 6w - Wx + (0)\, R]\, \mathbf{k} = 0$$

or,

$$1440 - 300 - 150\, x = 0 \text{ ft lb}$$

$$x = \frac{1140}{150} = 7.6 \text{ ft}$$

Note that we could not have obtained x without the use of the second condition of equilibrium. Also, it is sometimes useful to construct a table of forces, moment arms and torques to "keep the books straight." In this example, with the origin at 0, we would have the following table.

Force Component	Moment Arm, ft	Torque Component, ft lb
T cos (37)	0	0
T sin (37)	12	+ 12 T sin (37) ccw
w	6	-6w cw
W	x	-Wx cw
R	0	0

The student should also find x using a *different* origin for the torques, say the right hand end of the beam. Of course, for this choice, the moment arms of the forces are different but the value for x *has* to be the same, that is, independent of the choice of the origin. This is a good method for checking your solution to a given problem. Sometimes the direction of R is not obvious from inspection. However, if the wrong direction of R_x or R_y (or both) is chosen, the solutions of the equations will yield negative values for those components, but the magnitudes will be correct.

9.2 PROGRAMMED EXERCISES

1.A

A particle has two external forces acting on it, given by **F** = (3i + 2j) N and $\mathbf{F}_2$ = (−2i + 3j) N. Is the particle in equilibrium? Explain your answer.

No. The first condition of equilibrium requires that ΣF = 0. In this example, ΣF ≠ 0.

$$\Sigma\mathbf{F} = \mathbf{F}_1 + \mathbf{F}_2 = (\mathbf{i} + 5\mathbf{j})\ \mathrm{N}$$

1.B

What are the x and y components of the resultant force acting on the particle?

Resultant force = (i + 5j) N. Therefore: F_x = 1 N; F_y = 5 N.

1.C

What angles does the resultant force make with the +x axis? Draw the resultant force in a vector diagram.

$$\tan\theta = \frac{F_y}{F_x} = 5$$

$$\theta = \arctan 5 \cong 79°$$

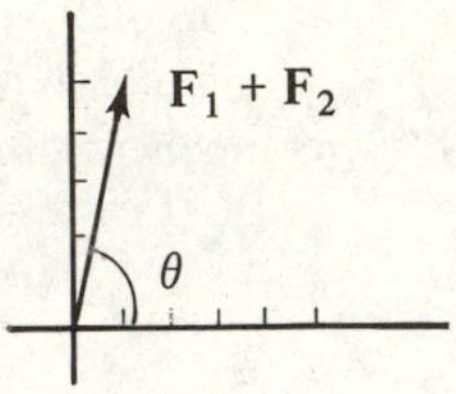

1.D

What third force $\mathbf{F}_3$ must be applied to the particle to cause an equilibrium condition?

$$\Sigma\mathbf{F} = \mathbf{F}_1 + \mathbf{F}_2 + \mathbf{F}_3 = 0$$

$$\mathbf{F}_3 = -(\mathbf{F}_1 + \mathbf{F}_2)$$

$$\mathbf{F}_3 = (-\mathbf{i} - 5\mathbf{j})\ \mathrm{N}$$

2.A

A long bar shown below has two forces acting on the ends, equal in magnitude, but opposite in direction. What does the first condition of equilibrium tell us in this situation?

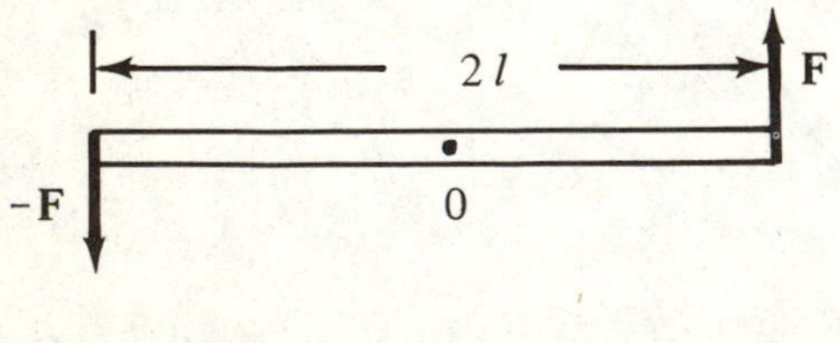

$$\Sigma\mathbf{F} = \mathbf{F} - \mathbf{F} = 0$$

The resultant force on the bar is *zero!*

2.B

Does the result of 2.A imply that the bar is in equilibrium?

No. The second condition is not satisfied, $\Sigma\tau \neq 0$.

2.C

Use the second condition of equilibrium to show that the bar is *not* in equilibrium.

Taking the torque equation about an axis through 0 gives

$$\Sigma \boldsymbol{\tau}_0 = Fl\mathbf{k} + Fl\mathbf{k} = 2\,Fl\mathbf{k} \neq 0$$

3.A

A force $\mathbf{F}_1 = (5\mathbf{i} - 2\mathbf{j})$ N acts on an arbitrarily shaped object at a position $\mathbf{r} = (3\mathbf{i} + \mathbf{j})$ m measured from a *fixed pivot* through 0. What is the torque $\boldsymbol{\tau}$ due to $\mathbf{F}_1$ about 0?

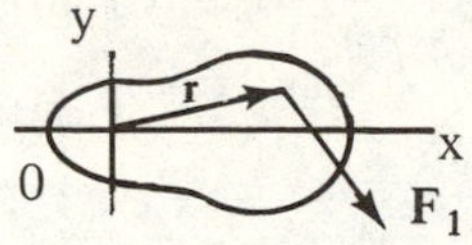

By definition, $\boldsymbol{\tau}_1 = \mathbf{r} \times \mathbf{F}_1$. However, **r** and $\mathbf{F}_1$ are both in the xy plane, so $\boldsymbol{\tau} = (xF_y - yF_x)\,\mathbf{k}$. In our example, x = 3, y = 1, $F_y = -2$, $F_x = 5$, so

$$\boldsymbol{\tau}_1 = (-6 - 5)\,\mathbf{k} = (-11\mathbf{k})\ \text{Nm}.$$

3.B

What is the meaning of the negative sign in your result for $\boldsymbol{\tau}$? What would a positive $\boldsymbol{\tau}$ indicate?

The negative sign simply indicates the *sense* of the rotation of the object about 0 is *clockwise,* corresponding to $\boldsymbol{\tau}$ in the *negative* z direction (−**k**). A positive sign would indicate a *counterclockwise* rotation. These follow from the right hand rule and the use of a right handed coordinate system.

3.C

If $\mathbf{F}_1$ is the only force acting on the object other than the force of the pivot (say a nail), is the object in equilibrium? Explain.

No. By the second condition of equilibrium, $\Sigma\boldsymbol{\tau} = 0$. In this case, $\Sigma\boldsymbol{\tau}_0 \neq 0$. The object will rotate in a clockwise sense about 0.

3.D

If the direction of **F** can be varied, is there any direction such that the object will be in equilibrium? Explain.

Yes. If **F** is along **r**, then its line of action passes through 0 and its moment arm is zero. Hence, $\boldsymbol{\tau} = 0$.

Since $\mathbf{r} \times \mathbf{r} = 0$, $\boldsymbol{\tau} = \mathbf{r} \times \mathbf{F} = \mathbf{r} \times F\mathbf{r} = 0$

3.E

A second force of 2 N is applied to the object in the +y direction. At what point on the body should the force be applied such that the body is in equilibrium?

$$F_2 = 2\mathbf{j}$$

From 3.A, $\boldsymbol{\tau}_1 = -11\mathbf{k}$; therefore, in order that $\Sigma\boldsymbol{\tau} = 0$, $\boldsymbol{\tau}_2 = -\boldsymbol{\tau}_1 = 11\mathbf{k}$. But $F_{2x} = 0$, $F_{2y} = 2$, so we have

$$\boldsymbol{\tau}_2 = 11\mathbf{k} = (xF_{2y} - \cancel{yF_{2x}})\,\mathbf{k}$$

$$\begin{cases} x = \dfrac{11}{2} = 5.5\ \text{m} \\ y = 0 \end{cases}$$

4.A

A uniform beam of weight W is supported at two points as shown below. Draw a free body diagram for the beam. Define all forces in your diagram.

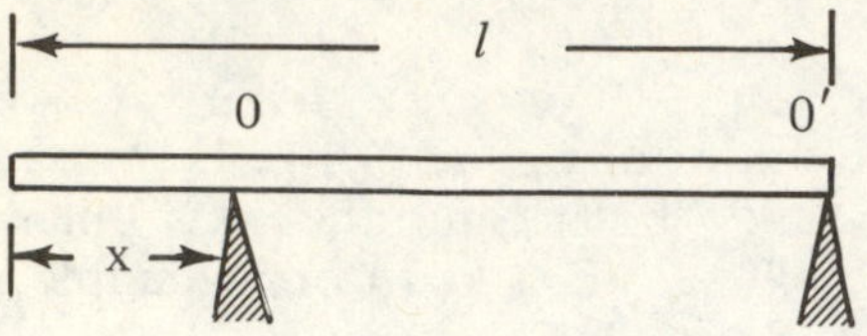

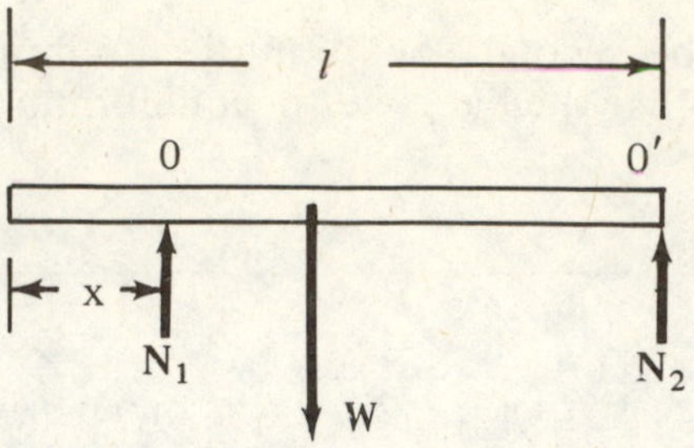

W = Force of gravity on the beam.

N_1 = Force of left support on the beam.

N_2 = Force of right support on the beam.

4.B

Write the equation for the first condition of equilibrium for the beam.

$\Sigma F = N_1 + N_2 + W = 0$

$N_1 + N_2 - W = 0$ (1)

Since N_1 and N_2 are up, W is down.

4.C

Using the results to 4.B, can you solve for the unknowns, assuming W is known? Explain.

No. From (1) we see that both N_1 and N_2 are unknown and cannot be obtained from this one condition alone. However, in the *special case* when $x = 0$, $N_1 = N_2 = W/2$.

4.D

Write the equation for the second condition of equilibrium for the beam. Take the origin of the torque equation about an axis through 0.

The moment arms for the forces N_1, W and N_2 are 0, $l/2 - x$ and $l - x$, respectively. Therefore, the second condition gives

$$\Sigma \tau_0 = N_1\ (0) - W\left(\frac{l}{2} - x\right) + N_2\ (l - x) = 0$$

$$N_2\ (l - x) - W\left(\frac{l}{2} - x\right) = 0 \qquad (2)$$

4.E

Can the forces of the supports on the beam be obtained using (2) alone? Explain. How would you find N_1 and N_2 if l, W and x are known?

No. Expression (2) only allows us to obtain N_2. However, we can then use (1) to find N_1, once N_2 is known. [That is, (1) and (2) constitute two equations with two unknowns.]

4.F

Write the second condition of equilibrium taking the origin about an axis through 0′. Explain what this gives. How will the final values for N_1 and N_2 differ from those obtained using (1) and (2)?

$$\Sigma\tau_{0'} = W\left(\frac{l}{2}\right) - N_1\,(l - x) = 0 \qquad (3)$$

This gives N_1, which can then be used in (1) to get N_2. The values for N_1 and N_2 *cannot* differ since the second condition is *independent* of the origin.

5.A

A beam of weight W and length $2l$ is supported by a light rope as shown. The floor is rough and the coefficient of static friction is μ. Draw a free body diagram for the beam.

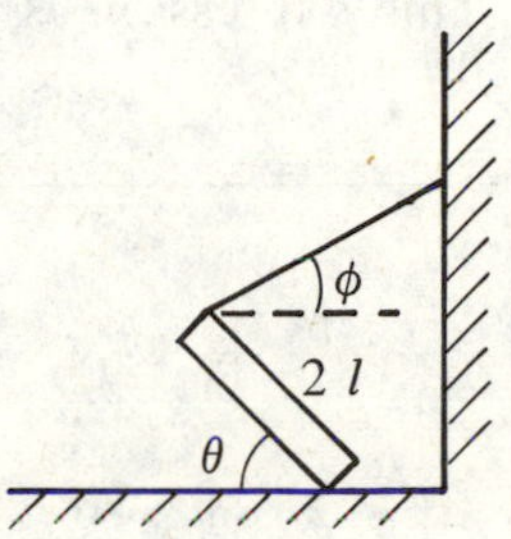

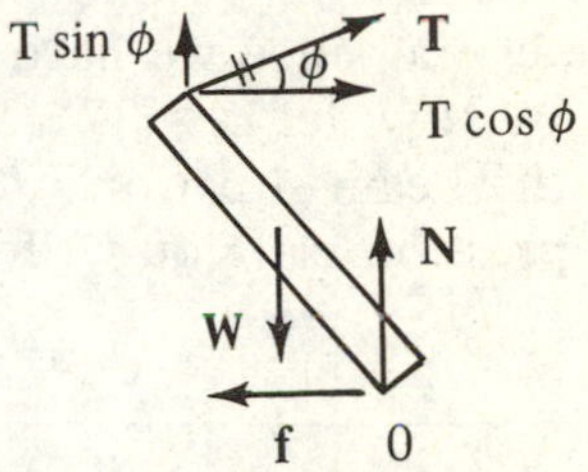

T = Tension
W = Force of gravity
N = Vertical component of the force of the floor on the beam
f = Frictional force of the floor on the beam

5.B

Write equations in component form for the first condition of equilibrium for the beam.

$$\Sigma F_x = T\cos\phi - f = 0 \qquad (1)$$

$$\Sigma F_y = T\sin\phi + N - W = 0 \qquad (2)$$

5.C

What are the moment arms of **T**, **W**, **N** and **f** about an axis through 0? Construct a table listing these moment arms and the torque about 0. Use the right hand rule for the sign of τ.

Since N and f go through 0, their moment arms and torques are *zero*.

Force	Moment Arm	Torque
$T\cos\phi$	$2\,l\sin\theta$	$-2\,Tl\cos\phi\sin\theta$
$T\sin\phi$	$2\,l\cos\theta$	$-2\,Tl\sin\phi\cos\theta$
W	$l\cos\theta$	$Wl\cos\theta$

5.D

Why are the torques due to the components of T negative, whereas the torque due to W is positive?

The sense of the "rotation" of $T\sin\phi$ and $T\cos\phi$ about 0 is clockwise corresponding to a vector $\boldsymbol{\tau}$ *into* the paper or in the direction $-\mathbf{k}$. The sense of W is counterclockwise, or a $\boldsymbol{\tau}$ *out* of the paper in the direction $+\mathbf{k}$.

5.E

Write the equation for the second condition of equilibrium using the axis through 0 as the origin of τ. Are the units of this equation the same as (1) and (2)?

From the Table in 5.C, we have

$\Sigma\tau_0 =$

$W l \cos\theta - 2T l (\sin\theta \cos\phi + \cos\theta \sin\phi = 0$ (3)

No, the units of (3) are ft lb, while (1) and (2) have units of lb.

5.F

The results given by (1), (2) and (3) are general since we have not specified θ, ϕ and W. Now let $\phi = 37°$, W = 50 lb, and suppose the beam slips for $\theta < 53°$. What further expression can you write for f_m, the maximum f?

When the beam is on the verge of slipping,

$$f_m = \mu N \quad (4)$$

This expression is valid at the *critical angle* $\theta = 53°$.

5.G

Rewrite (1), (2) and (3) using the numerical values given in 5.F above. Recall that

$$\cos(37) = \sin(53) = 0.8$$

$$\cos(53) = \sin(37) = 0.6$$

$$T(0.8) - f_m = 0 \quad (5)$$

$$T(0.6) + N - 50 = 0 \quad (6)$$

$$50(0.6) - 2T(0.64 + 0.36) = 0 \quad (7)$$

5.H

Find a value for T from (7).

$$T = \frac{30}{2} = 15 \text{ lb} \quad (8)$$

5.I

What is f_m equal to?

From (5) and (8) we have

$$f_m = 15(0.8) = 12 \text{ lb} \quad (9)$$

5.J

Find a value for N.

Using (6) and (8) gives

$$N = 50 - 15(0.6) = 41 \text{ lb} \quad (10)$$

5.K

Suppose the beam were at an angle $53° < \theta < 90°$. Would the results for f, T and N be less than, equal to, or greater than those values given by (8), (9) and (10)?

For values of θ greater than the critical angle 53° (and less than 90°), $f < f_m$; therefore, from (5), $T < 15$ lb and from (6), $N > 41$ lb.

5.L

What is the reaction force at the floor **R** when the beam is about to slip? Find both its magnitude and angle relative to the horizontal.

$$R = \sqrt{N^2 + f_m{}^2}$$

$$R = \sqrt{(41)^2 + (12)^2}$$

$$R \cong 42.6 \text{ lb}$$

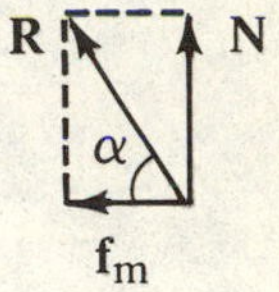

$$\alpha = \text{arc tan}\,\frac{N}{f_m} = \text{arc tan}\left(\frac{41}{12}\right)$$

$$\alpha \approx 74°.$$

Note: $\alpha \neq 53°$. That is, the reaction force does not lie along the beam.

9.3 PROBLEMS

1. A mass of 10 kg hangs from the ceiling by two ropes as shown in Figure 9-5. Find the tensions in the three ropes.

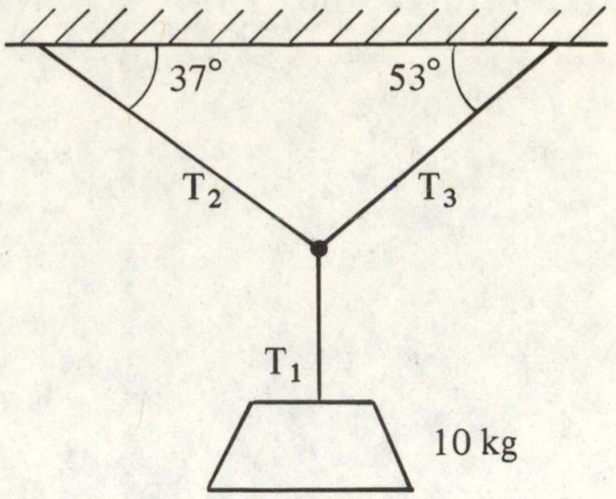

Figure 9-5

2. Four coplanar forces act on a body, their lines of action all passing through the origin. The four forces have magnitudes of 10 lb, 20 lb, 25 lb and 15 lb, and their directions are shown in Figure 9-6. Find (a) the components of the resultant force, (b) the magnitude and direction of the resultant force and (c) a fifth force which, when applied to this body, would make the resultant force zero.

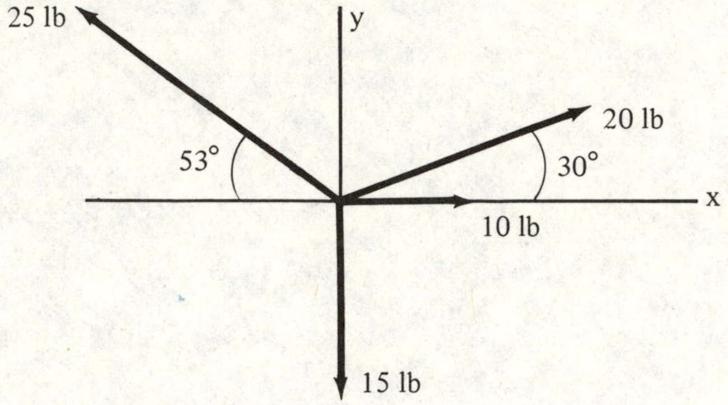

Figure 9-6

3. A block weighing 20 lb is placed on an inclined plane and connected to a 10 lb block by a light string passing over a frictionless pulley as shown in Figure 9-7. When the angle of inclination is 53°, the 20 lb block is on the verge of slipping *down* the plane. (a) Draw free body diagrams for both blocks. (b) Determine the equations of equilibrium for both blocks. (c) Find the coefficient of static friction μ_s and the frictional force.

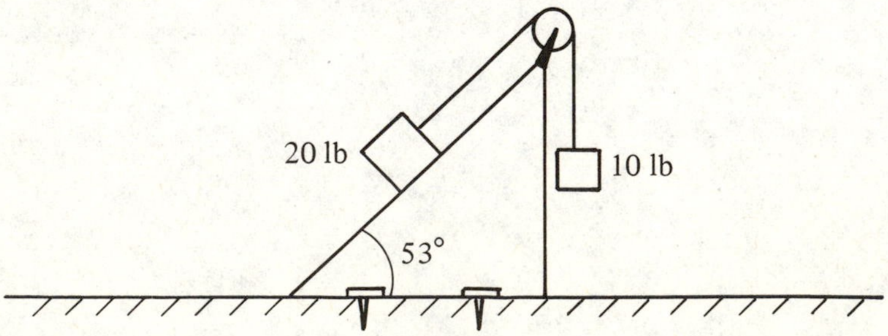

Figure 9-7

4. A meter stick pivoted at the 50 cm mark has weights of 300 g and 200 g hanging from it at the 10 cm and 60 cm marks, respectively. Determine the position at which one would hang a third weight of 400 g to keep the meter stick horizontal.

5. A standard technique for demonstrating that $\mu_s > \mu_k$ is to place an object such as a coin on a inclined plane. The angle of the incline θ_s for which the coin is on the verge of slipping is then measured and compared with the angle θ_k at which the coin moves down the inclined plane with constant speed. It is found that $\theta_s > \theta_k$. Show that $\mu_s = \tan\theta_s$ and $\mu_k = \tan\theta_k$.

6. A uniform beam of length 4 m and mass 10 kg supports a 20 kg mass as shown in Figure 9-8. (a) Draw a free body diagram for the beam. (b) Determine the tension in the supporting wire and the components of the reaction force at the pivot.

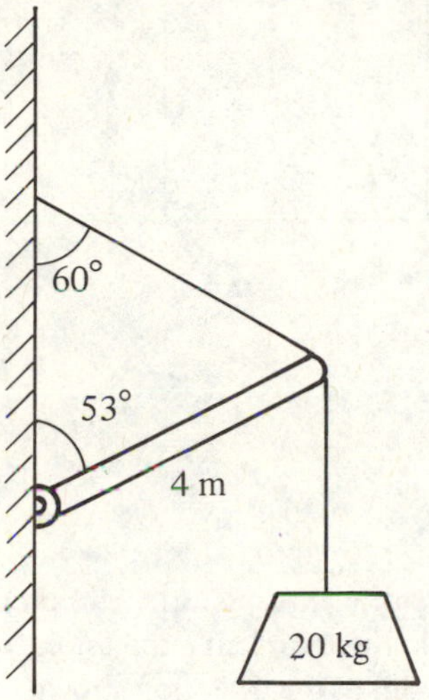

Figure 9-8

7. A monkey weighing 24 lb walks up a uniform ladder which weighs 30 lb as in Figure 9-9. The upper and lower ends of the ladder rest on frictionless rollers and the lower end of the ladder is fastened to the wall by a horizontal rope. The rope will support a maximum weight of 25 lb before breaking. (a) Draw a free body diagram for the ladder. (b) Find the tension in the rope when the monkey is one third the way up the ladder. (c) What is the maximum distance d the monkey can walk up the ladder before the rope breaks? Express your answer in terms of a fraction of the total length l.

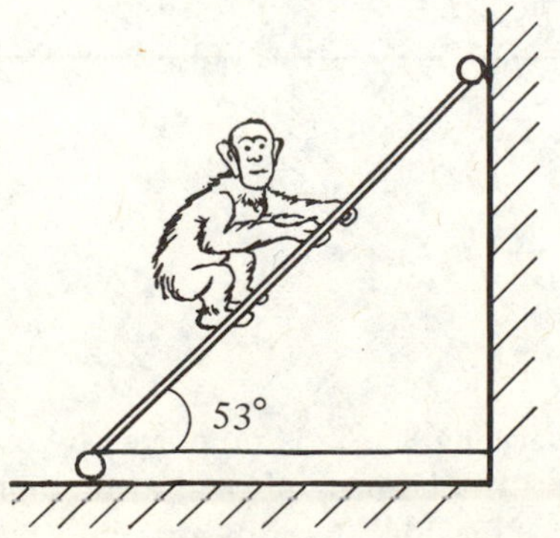

Figure 9-9

8. A rectangular block weighing 100 lb has a force F exerted on it at an angle of 37° with respect to the horizontal direction, as in Figure 9-10. (a) It is observed that when F = 50 lb and h = 1 ft, the block slides with constant speed. From this information, find the coefficient of sliding friction and the position of the normal force. (b) If F = 75 lb, find the value of h for which the block will just begin to tip from a stationary position.

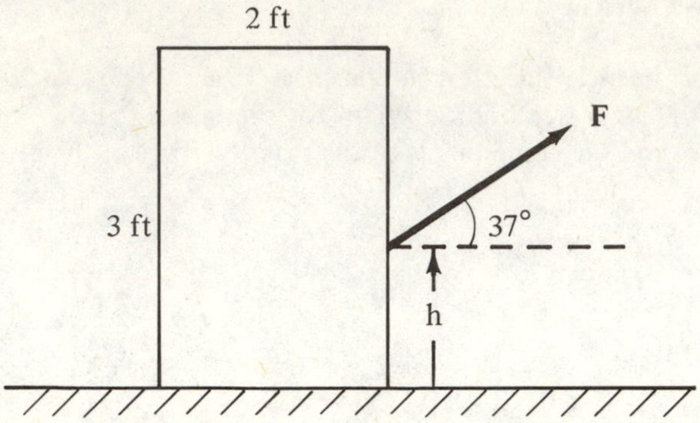

Figure 9-10

9. In Figure 9-11 the coefficients of static friction μ_s are the same for both blocks. Assuming W, θ, ϕ and μ_s are known, find an expression for W_x corresponding to the case where the blocks are *about* to slip in the direction shown.

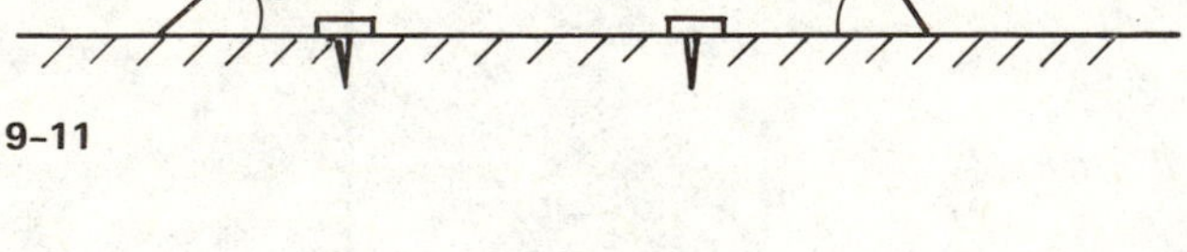

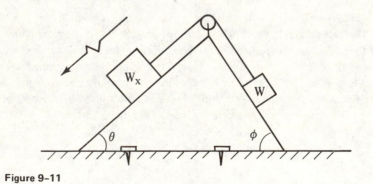

Figure 9-11

10. A uniform beam weighing 300 lb is supported by a cable connected at a point 3/4 the length of the beam. The beam is pivoted at the bottom, as in Figure 9-12. A weight of 500 lb hangs from the top of the beam. Find the tension in the supporting cable and the components of the reaction force on the beam at the hinge.

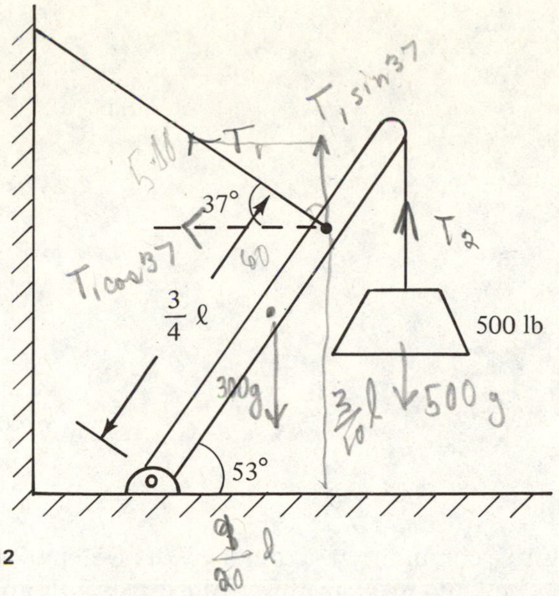

Figure 9-12

11. A uniform ladder weighing 50 lb is leaning against a wall. The ladder is observed to be in equilibrium for angles of $\pi/3 \leqslant \theta \leqslant \pi/2$, where θ is the angle between the ladder and the horizontal. Assuming the coefficients of static friction at the wall and floor are the same, obtain a value for μ_s.

***12.** A uniform beam of weight w is inclined at an angle θ to the horizontal with its upper end supported by a wire and its lower end resting on a rough floor, as in Figure 9-13. (a) If the coefficient of static friction is μ_s between the beam and floor, find an expression for the *maximum* weight W that can be suspended from the top before the beam slips. (b) What is the magnitude of the reaction force at the floor? (c) What is the magnitude of the force of the beam on the knot at the top?

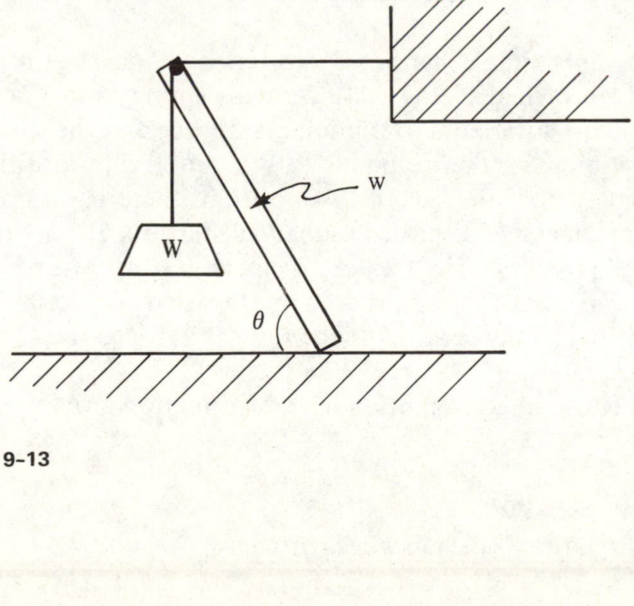

Figure 9-13

10

HARMONIC OSCILLATIONS

Oscillating motions are common to a number of systems, such as: a mass attached to a spring, the pendulum, atoms in a solid, stringed musical instruments and in some electrical circuits. Electromagnetic waves such as radar waves, radio waves and laser light are characterized by oscillating electric and magnetic field vectors. *Simple harmonic motions* correspond to the oscillation of an object between two spatial positions for an indefinite period of time, with no loss in mechanical energy. In real mechanical systems retarding (or frictional) forces are always present. Such forces will reduce the mechanical energy of the system in time, and the oscillations are said to be *damped.* Of course, an external driving force can be used to overcome this loss in energy, and the oscillations that result are called *forced oscillations.*

10.1 THE SIMPLE HARMONIC OSCILLATOR

An object can exhibit simple harmonic motion if a *linear restoring force* acts on it. Consider a mass attached to the end of a light spring as in Figure 10–1. The mass is assumed to be on a horizontal, frictionless surface and the spring has a force constant k. The position $x = 0$ corresponds to the equilibrium position where the mass would reside if left undisturbed. In this position, there is no horizontal force on m. However, if the mass is displaced a *small* distance x from equilibrium, the spring exerts a force given by $F = -kx$. This linear restoring force is sometimes referred to as Hooke's law. The minus sign means that F is to the *left* if the displacement x is positive, whereas F is to the *right* if x is negative. These two cases are described in Figure 10–1.

Newton's second law applied to motion in the x direction gives

$$F_x = -kx = ma_x \tag{10.1}$$

Since $a_x = d^2x/dt^2$, Equation (10.1) can be written as

$$\frac{d^2x}{dt^2} = -\omega^2 x \tag{10.2}$$

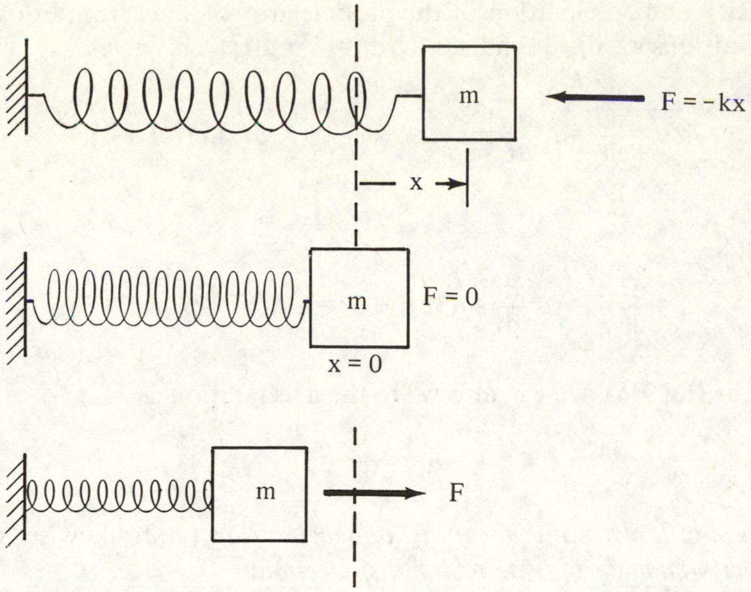

Figure 10-1 Oscillating motion of a mass on the end of a spring. Note the change in direction of the spring force as the mass goes through the equilibrium position.

where

$$\omega^2 = k/m \tag{10.3}$$

It is left as an exercise to show that a general solution of Equation (10.2) is

$$x = A \cos (\omega t + \delta) \tag{10.4}$$

where A is the amplitude of the vibration (that is, the maximum value of x), δ is a phase constant which is adjusted to meet the initial conditions of the motion and ω is the angular frequency. Since the function cos $(\omega t + \delta)$ repeats itself after a time $2\pi/\omega$, the *period* of the motion T (or the time for one complete oscillation) is given by

$$T = \frac{2\pi}{\omega} = 2\pi \sqrt{\frac{m}{k}} \tag{10.5}$$

and the frequency of motion is given by

$$f = \frac{1}{T} = \frac{1}{2\pi} \sqrt{\frac{k}{m}} \tag{10.6}$$

The unit of the angular frequency ω is rad/sec, while the frequency f has the unit cycles/sec, or $(\text{sec})^{-1}$. It has become common to call the unit one cycle per sec the Hertz (Hz).

The velocity and acceleration of the particle are obtained from Equation (10.4), using the definitions $v = dx/dt$ and $a = dv/dt = d^2x/dt^2$. This gives

$$v = \frac{dx}{dt} = A\frac{d}{dt}\cos(\omega t + \delta) = -A\,\omega \sin(\omega t + \delta) \tag{10.7}$$

and

$$a = \frac{dv}{dt} = -A\omega\frac{d}{dt}\sin(\omega t + \delta) = -A\omega^2\cos(\omega t + \delta) \tag{10.8}$$

Since $x = A\cos(\omega t + \delta)$, we can also write the acceleration as

$$a = -\omega^2 x \tag{10.9}$$

Therefore, *in simple harmonic motion the acceleration* (and, likewise, the force) *is always proportional and opposite to the displacement.*

In order to get a better "feel" for the meaning of the solution and the terms contained in it, let us consider two different initial conditions corresponding to two values of δ.

Case I

Suppose we release the mass from rest at $t = 0$ while it is in the stretched position $x = A$, as in Figure 10-2. We must then require that our solution for $x(t)$ be such that $x = A$ at $t = 0$. We sometimes refer to this as a *boundary condition.* This condition will be met by Equation (10.4) if we choose $\delta = 0$. That is, the solution $x = A\cos\omega t$ is the choice, since at $t = 0$, $\cos(0) = 1$ and $x = A$. Again, we get the velocity and acceleration as functions of time by taking the first and second derivatives of x with respect to time.

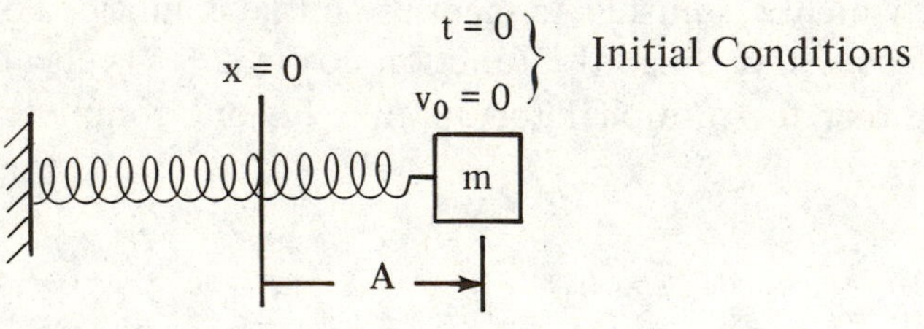

Figure 10-2 The appropriate solution to the equation of motion is $x = A\cos\omega t$.

$$x = A\cos\omega t \tag{10.10}$$

$$v = \frac{dx}{dt} = A\frac{d}{dt}\cos\omega t = -A\omega\sin\omega t \tag{10.11}$$

$$a = \frac{dv}{dt} = -A\omega\frac{d}{dt}\sin\omega t = -A\omega^2\cos\omega t \tag{10.12}$$

Note that Equation (10.11) makes physical sense, since we *also* require that the mass start off at rest at t = 0. In other words, $v_0 = 0$, and Equation (10.11) agrees with this condition, since at t = 0, sin(0) = 0, so $v_0 = 0$. Also, Equation (10.12) says that at t = 0, $a = -A\omega^2$, which is the maximum value. This makes sense, since at t = 0, x = A, the force is a maximum equal to −kA (to the left) and the acceleration is a maximum given by

$$|a_{max}| = \frac{kA}{m} = \omega^2 A \qquad (10.13)$$

Note also that at x = 0, the velocity of the mass is a *maximum* given by $|v_{max}| = A\omega$, while the acceleration is *zero* (since the spring is unstretched). At x = −A, the mass again comes to rest, so v = 0 and a = ωA to the *right*, since the spring force is now to the right. Diagrams of the displacement, velocity and acceleration of the mass *vs* time are shown in Figure 10-3 for Case I. Note that all three vary periodically in time.

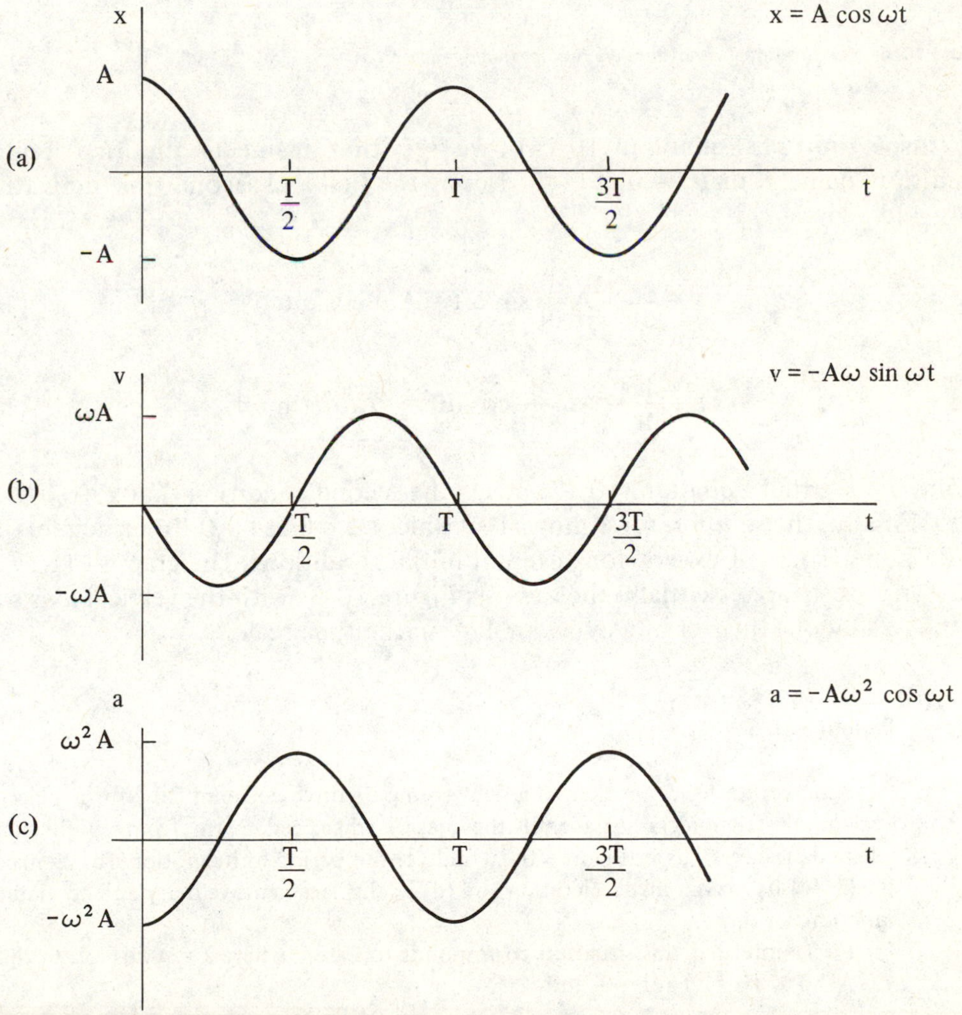

Figure 10-3 Representations of (*a*) the displacement, (*b*) the velocity, and (*c*) the acceleration as a function of time for an object moving in simple harmonic motion according to Case I.

Case II

Now suppose the mass is projected from the origin, $x = 0$, at $t = 0$ with an initial velocity $+v_0$, as in Figure 10-4. Our particular solution now must satisfy these conditions. The choice of δ which gives the required solution is $\delta = -\pi$, for in this case $\cos(\omega t - \pi) = \sin \omega t$, and we get

$$x = A \sin \omega t \tag{10.14}$$

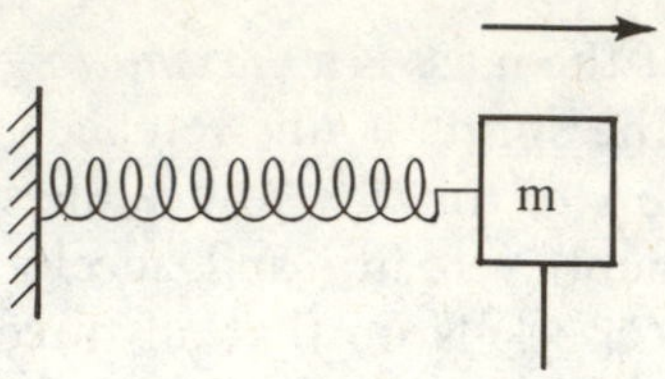

$v_0 = A\omega$ $\left.\begin{matrix} x = 0 \\ t = 0 \end{matrix}\right\}$ Initial Conditions

Figure 10-4 For this initial condition, we require that $x = A \sin \omega t$.

By inspection of Equation (10.14), we see that it meets the first boundary condition, namely, that $x = 0$ at $t = 0$. Taking the first and second time derivatives to get v and a gives

$$v = \frac{dx}{dt} = A\frac{d}{dt} \sin \omega t = A\omega \cos \omega t \tag{10.15}$$

$$a = \frac{dv}{dt} = A\omega\frac{d}{dt} \cos \omega t = -A\omega^2 \sin \omega t \tag{10.16}$$

Again, we see that Equation (10.15) meets the second condition that $v = v_0 = A\omega$ at $t = 0$. Finally, the solution for a shows that indeed $a = 0$ at $t = 0$. In principle, Case II is the same as Case I except for different initial conditions. The graphs of x, v and a *vs* t for Case II are essentially the *same* as Figure 10-3, with the vertical axis shifted to the *left* by one quarter of a cycle, or T/4 on the time scale.

Example 10.1

A mass of 50 g connected to a light spring of force constant 20 N/m is placed on a horizontal, frictionless surface. If the mass is displaced 5 cm from equilibrium and released from rest, as in Figure 10-2, find (a) the period of its motion, (b) its maximum speed, (c) its maximum acceleration and (d) its displacement, velocity and acceleration as functions of time.

First, note that this situation corresponds to Case I, where $x = A \cos \omega t$. In this case, $x = A = 5 \times 10^{-2}$ m at $t = 0$, and

$$\omega = \sqrt{\frac{k}{M}} = \sqrt{\frac{20 \text{ N/m}}{50 \times 10^{-3} \text{ kg}}} = 20 \frac{\text{rad}}{\text{sec}}$$

(a) $T = \frac{2\pi}{\omega} = \frac{2\pi}{20} = \frac{\pi}{10}$ sec

(b) $|v_{max}| = A\omega = 5 \times 10^{-2}$ m $\times$ $20 \frac{\text{rad}}{\text{sec}} = 1$ m/sec

(c) $|a_{max}| = A\omega^2 = 5 \times 10^{-2}$ m $\times$ $\left(20 \frac{\text{rad}}{\text{sec}}\right)^2 = 20$ m/sec^2

(d) Using Equations (10.7), (10.8) and (10.9) for Case I gives

$$x = A \cos \omega t = 5 \times 10^{-2} \cos 20t \text{ m}$$

$$v = -A\omega \sin \omega t = -\sin 20t \text{ m/sec}$$

$$a = -A\omega^2 \cos \omega t = -20 \cos 20t \text{ m/sec}^2$$

10.2 THE ENERGY OF THE SIMPLE HARMONIC OSCILLATOR

The kinetic energy of the oscillating mass can be obtained by using Equation (10.7).

$$K = \frac{1}{2} mv^2 = \frac{1}{2} m\omega^2 A^2 \sin^2 (\omega t + \delta) \quad (10.17)$$

The elastic potential energy stored in the spring for any displacement x is given by $\frac{1}{2}$ kx^2. Using Equation (10.4), we get

$$U = \frac{1}{2} kx^2 = \frac{1}{2} kA^2 \cos^2 (\omega t + \delta) \quad (10.18)$$

Since $\omega^2 = k/m$, we can write the *total energy* of the simple harmonic oscillator as

$$E = K + U = \frac{1}{2} kA^2 [\sin^2 (\omega t + \delta) + \cos^2 (\omega t + \delta)]$$

However, recall that $\sin^2 \theta + \cos^2 \theta = 1$, so

$$E = \frac{1}{2} kA^2 \quad (10.19)$$

In other words, the total energy is proportional to the square of the amplitude and is a constant of the motion. The latter is true, since there are no dissipative (or nonconservative) forces acting on the system. (The internal spring force is conservative.) Since the maximum speed is given by $A\omega$, we can also write the total energy as

$$E = \frac{1}{2} m{v_{max}}^2 = \frac{1}{2} mA^2 \omega^2 \quad (10.20)$$

Note that the energy in the system is *entirely* potential energy when $x = A$, where $v = 0$. Likewise, all of the energy is *kinetic* when $x = 0$, at which point $v = v_{max}$. Equating Equations (10.19) and (10.20) simply verifies that $\omega^2 = k/m$. For an arbitrary value of x, the energy is the sum of K and U.

Another useful form of the kinetic energy is obtained by using Equations (10.18) and (10.19), together with the fact that K = E − U. This gives

$$K = \frac{1}{2}kA^2 - \frac{1}{2}kx^2 = \frac{1}{2}k(A^2 - x^2) \tag{10.21}$$

Since $x \leqslant A$, we see that both K and U are *always* greater than or equal to zero.

Example 10.2

(a) Calculate the total energy of the system described in Example 10.1, using Equation (10.19). Repeat the calculation, using Equation (10.20).

$$E = \frac{1}{2}kA^2 = \frac{1}{2} \times 20\frac{N}{m} \times (5 \times 10^{-2}\,m)^2 = 2.5 \times 10^{-2}\,J$$

$$E = \frac{1}{2}mv_{max}{}^2 = \frac{1}{2}(50 \times 10^{-3}\,kg)\left(1\frac{m}{sec}\right)^2 = 2.5 \times 10^{-2}\,J$$

(b) Calculate the potential and kinetic energies of the system described in Example 10.1 for a displacement of 2 cm. Show that the sum equals E.

$$U = \frac{1}{2}kx^2 = \frac{1}{2} \times 20\frac{N}{m} \times (2 \times 10^{-2}\,m)^2 = 0.4 \times 10^{-2}\,J$$

$$K = \frac{1}{2}k(A^2 - x^2) = \frac{1}{2} \times 20\frac{N}{m}(25 - 4) \times 10^{-4} = 2.1 \times 10^{-2}\,J$$

Therefore,

$$K + U = 2.5 \times 10^{-2}\,J = E$$

10.3 THE SIMPLE PENDULUM

A *simple pendulum* consists of a mass m attached to a light string of length l, where the upper end of the string is fixed as in Figure 10-5. The motion of such a system is periodic and, for small angular displacements θ from the vertical, the equation of motion is that of a simple harmonic oscillator. To see this, consider the free body diagram for the mass m, as in Figure 10-5. The resultant force acting on m in the direction *tangent* to the circle of motion has a magnitude mg sin θ. Since this force is always directed towards $\theta = 0$ (that is, it is a restoring force), the equation of motion along the tangential direction becomes

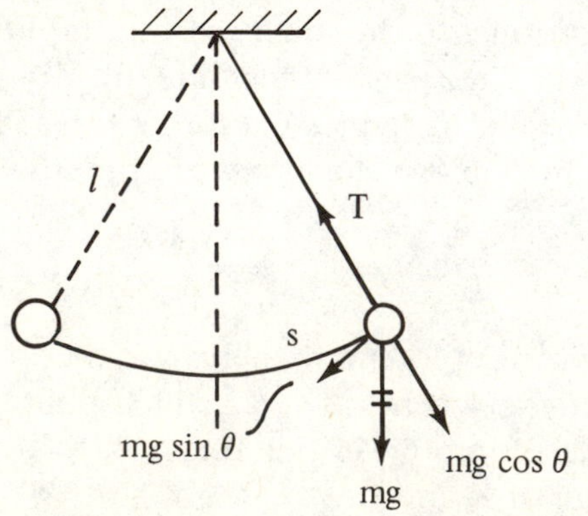

Figure 10-5 The simple pendulum.

$$\Sigma F_t = -\ mg \sin\theta = m\frac{d^2 s}{dt^2} \qquad (10.22)$$

where the tangential acceleration is given by $a_t = d^2 s/dt^2$, and $s = l\theta$. For *small* values of θ, recall that $\sin\theta \approx \theta$. Therefore, Equation (10.22) reduces to

$$-mg\,\theta \cong ml\frac{d^2\theta}{dt^2}$$

or,

$$\frac{d^2\theta}{dt^2} = -\frac{g}{l}\theta \qquad (10.23)$$

Note that Equation (10.23) is of *exactly* the same mathematical form as Equation (10.2), so we conclude that the solution for $\theta(t)$ is sinusoidal [that is, $\theta = \theta_0 \cos(\omega t + \delta)$] with an angular frequency given by

$$\omega = \sqrt{\frac{g}{l}} \qquad (10.24)$$

The *period* of the motion is given by

$$T = \frac{2\pi}{\omega} = \frac{1}{f} = 2\pi\sqrt{\frac{l}{g}} \qquad (10.25)$$

In other words, the period is only a function of the length of the string and the acceleration of gravity. The period does *not* depend on the mass, so we can conclude that all simple pendula of equal length oscillate with equal periods. It must be emphasized that Equations (10.24) and (10.25) are *only approximate* results which are applicable when θ is a small angle (typically $<5°$).

For those situations where somewhat larger angular displacements are involved, an approximate expression for the period of the pendulum is

$$T \cong 2\pi\sqrt{\frac{l}{g}}\left(1 + \frac{{\theta_0}^2}{16}\right) \qquad (10.26)$$

where θ_0 is the *maximum* angular displacement in radians. This expression is valid to a high degree of accuracy for angles as large as $\sim 25°$.

Example 10.3

(a) If you were to construct a simple pendulum such that its period would be 1.0 sec, what length should the string have?

From Equation (10.25) we get

$$l = g\frac{T^2}{4\pi^2} = 9.8\,\frac{m}{sec^2} \times \frac{(1\ sec)^2}{4\pi^2} \cong 0.25\ m$$

(b) Suppose the simple pendulum described in part (a) is taken to the moon, where the acceleration of gravity is ≅ 0.17 g. What would its period be on the moon?

$$T = 2\pi\sqrt{\frac{l}{g_m}} = 2\pi\sqrt{\frac{0.25\ \text{m}}{0.17 \times 9.8\ \text{m/sec}^2}} \cong 2.4\ \text{sec}$$

10.4 DAMPED OSCILLATIONS

In a realistic system, such as a mass oscillating at the end of a spring, other forces act which could retard the motion, such as frictional forces. These forces will cause the oscillation amplitudes to decrease in time, since mechanical energy is "removed" from the system by these nonconservative forces. If we *assume* that the mass at the end of the spring described in Figure 10–1 has an *additional resistive* force acting on it, proportional to its velocity, then Newton's second law gives

$$\Sigma F_x = -kx - \lambda v = ma \tag{10.27}$$

where λ is a constant. Since $v = \frac{dx}{dt}$ and $a = d^2x/dt^2$, the equation of motion can be written as

$$\frac{d^2x}{dt^2} = -\frac{\lambda}{m}\frac{dx}{dt} - \frac{k}{m}x \tag{10.28}$$

It is left as an exercise to show that a solution of Equation (10.28) is

$$x = Ae^{-\frac{\lambda}{2m}t}\cos(\omega t + \delta) \tag{10.29}$$

where

$$\omega = \sqrt{\frac{k}{m} - \left(\frac{\lambda}{2m}\right)^2} \tag{10.30}$$

That is, the displacement of the *damped* oscillator still varies sinusoidally in time, as in the case of the simple harmonic oscillator. However, the amplitude *decreases* exponentially in time, as shown by the exponential factor in Equation (10.29). In addition, the frequency of the damped oscillator is less than that of the simple harmonic oscillator (undamped). The displacement as a function of time for the damped oscillator is shown in Figure 10–6.

10.5 FORCED OSCILLATIONS

An oscillating body, such as the mass at the end of a spring or the bob of a pendulum, can be kept in motion by the force of an external agent. For example, a child on a swing can be kept in motion through periodic "pushes" of his playmate.

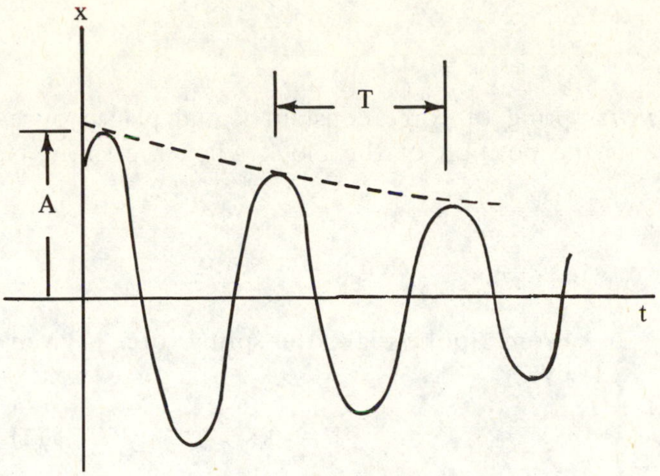

Figure 10-6 Displacement *vs* time for the damped oscillator.

This energy input must at least equal the energy losses due to friction if the amplitude of the swing is to remain constant. A maximum transfer of energy is obtained when the external agent pushes in the direction of motion of the oscillator.

A common example of a forced oscillator is a damped oscillator *driven* by an external force which varies as $F = F_0 \cos\omega t$, where F_0 is a constant. Adding this driving force to Equation (10.28) gives, for the equation of motion,

$$F_0 \cos\omega t - \lambda \frac{dx}{dt} - kx = m\frac{d^2x}{dt^2} \tag{10.31}$$

The general solution of Equation (10.31) will not be given, for it is very complicated. However, after a long period of time, when the energy input per cycle of oscillation equals the energy lost per cycle, a *steady-state* condition is reached. At this time, oscillations will commence with constant amplitude, and the *steady-state* solution of Equation (10.31) becomes

$$x = \frac{F_0/m}{\sqrt{(\omega^2 - \omega_0{}^2) + \left(\frac{\lambda\omega}{m}\right)^2}} \cos(\omega t + \delta) \tag{10.32}$$

where ω is the frequency of the driving force and $\omega_0 = \sqrt{k/m}$ is the frequency of the undamped oscillator.

From Equation (10.32) we see that the displacement of the mass will be a *maximum* when $\omega = \omega_0$, that is, when the frequency of the driving force *equals* the natural frequency of the oscillator. In addition, if the system is undamped ($\lambda = 0$), the amplitude of oscillation approaches ∞ as $\omega \to \omega_0$. The frequency ω_0 at which this occurs is called the *resonance frequency* of the system. The physical reason for the large amplitude oscillations is the fact that energy is always being added to the system when the driving force is in phase with the oscillations. Later, we shall see that certain circuits have such natural (or resonance) frequencies.

10.6 PROGRAMMED EXERCISES

1 A block of mass m is attached to a *light* spring of force constant k and placed on a horizontal, frictionless surface. $x = 0$ represents the position of the block when the spring is unstretched.

1.A

What is the magnitude and direction of the force on m when it is displaced a distance x from the origin?

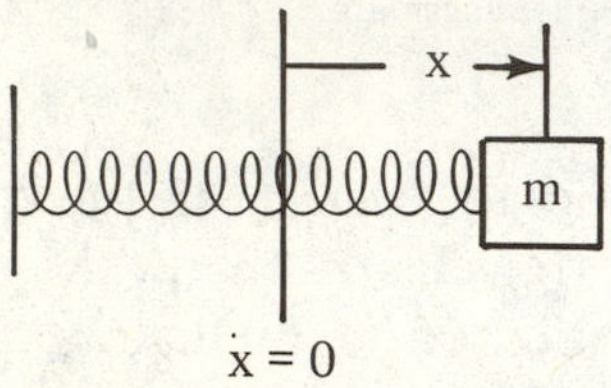

From Hooke's law, the spring force is given by

$$F = -kx \qquad (1)$$

This is consistent with the requirement that $F = 0$ at $x = 0$.

1.B

What is the physical meaning of the negative sign in (1)?

This implies that F is a *restoring* force; that is, F is to the *left* when x is positive, while F is to the *right* when x is negative.

1.C

Using Newton's second law, write the equation of motion in the x direction for m.

$$F = m\frac{d^2x}{dt^2} = -kx \qquad (2)$$

1.D

Equation (2) is a second order differential equation whose solution is well-known. What is a general form of this solution?

$$x = A\cos(\omega t + \delta) \qquad (3)$$

where A is the maximum displacement of the mass from $x = 0$.

---

†1.E

Use (3) and the definitions of velocity and acceleration to find v(t) and a(t) for the simple harmonic oscillator.

$$v = \frac{dx}{dt} = -A\omega\sin(\omega t + \delta) \qquad (4)$$

$$a = \frac{dv}{dt} = -A\omega^2\cos(\omega t + \delta) \qquad (5)$$

1.F

What are the *maximum* values of v and a?

From (4) and (5) we see that the maximum values are simply the coefficients of the sin and cos functions.

$$|v_{max}| = A\omega \qquad (6)$$

$$|a_{max}| = A\omega^2 \qquad (7)$$

1.G

What is the constant ω which appears in (3)?

ω is the natural angular frequency of oscillation, given by

$$\omega = \sqrt{\frac{k}{m}} \qquad (8)$$

1.H

What is the period of the oscillation (that is, the time for one complete cycle) and the frequency of oscillation.

$$T = \frac{2\pi}{\omega} = 2\pi\sqrt{\frac{m}{k}} \qquad (9)$$

$$f = \frac{1}{T} = \frac{1}{2\pi}\sqrt{\frac{k}{m}} \qquad (10)$$

1.I

What are the units of ω, T and f?

ω has the unit rad/sec, T has the unit of sec and f has the unit cycles/sec or Hz (Hertz).

1.J

What is the physical meaning of the constant δ that appears in (3)?

δ is called the phase constant (or phase angle) and represents a constant which is adjusted to conform with the boundary conditions of the problem, that is, the values of x and v at t = 0.

1.K

Suppose that the block is released from *rest* at t = 0 from the *compressed* position, that is, x = -A at t = 0. What must δ equal to meet this condition?

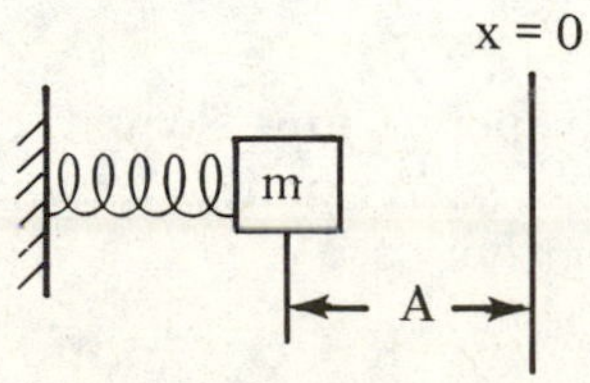

Using (3), we set t = 0 and x = -A. This gives

$$-A = A\cos\delta$$

$$\cos\delta = -1$$

$$\therefore \quad \delta = \pi$$

(or in general $\delta = n\pi$, where n is an odd integer).

1.L

Using the initial conditions given in 1.J and the value $\delta = \pi$ write a solution for x(t). Recall that

$$\cos(a + b) = \cos a \cos b - \sin a \sin b.$$

$$x = A \cos(\omega t + \delta)$$

$$= A[\cos \omega t \cos \pi - \sin \omega t \sin \pi]$$

$$x = -A \cos \omega t \qquad (11)$$

1.M

Show that (11) satisfies the initial conditions, namely, at t = 0, x = –A and $v_0 = 0$.

At t = 0, cos(0) = 1, so (11) gives x(0) = –A. From (4) we see that $v_0 = 0$, since $\sin(\pi) = 0$.

1.N

What is the acceleration of m at t = 0, assuming the conditions given in 1.J. Also, what is the acceleration as a function of time?

At t = 0, $\delta = \pi$. (5) gives

$$a(0) = -A\omega^2 \cos(\pi) = A\omega^2$$

That is, a(0) is a *maximum* at t = 0 and to the right, since the spring is *compressed.*

$$a(t) = A\omega^2 \cos \omega t$$

1.O

When |x| is a *maximum,* what can you say about v and |a|?

When |x| = A, the block comes to rest (at its turning points), so v = 0. Also, since $|a| = \omega^2 x$, then |a| is a maximum when |x| = A.

1.P

Write expressions for the potential and kinetic energies of the simple harmonic oscillator, using (3) and (4). Recall that $U = \frac{1}{2}kx^2$ and $K = \frac{1}{2}mv^2$.

$$U = \frac{1}{2}kx^2 = \frac{1}{2}kA^2 \cos^2(\omega t + \delta) \qquad (12)$$

$$K = \frac{1}{2}mv^2 = \frac{1}{2}mA^2\omega^2 \sin^2(\omega t + \delta) \qquad (13)$$

1.Q

Show that the *maximum* value of U equals the *maximum* value of K.

From (12), $U_{max} = \frac{1}{2}kA^2$. But $\omega^2 = k/m$, so from (13) we see that

$$K_{max} = \frac{1}{2}mA^2\omega^2 = \frac{1}{2}kA^2$$

$$\therefore U_{max} = K_{max}$$

1.R

What is the total energy of the simple harmonic oscillator?

The total energy must equal U_{max} (when K = 0 at x = ±A) or K_{max} (when U = 0 at x = 0). In either case,

$$E = K + U = \frac{1}{2}kA^2 \qquad (14)$$

1.S

What is E a constant of the motion?

We have neglected frictional (nonconservative) forces for the problem, so the total mechanical energy is conserved.

2 A mass of 50 g, attached to the end of a light spring whose force constant is 100 dynes/cm, executes simple harmonic motion along a frictionless horizontal surface. The mass is pulled a distance 10 cm to the right and released from rest at t = 0.

2.A

What is the period of the oscillating motion?

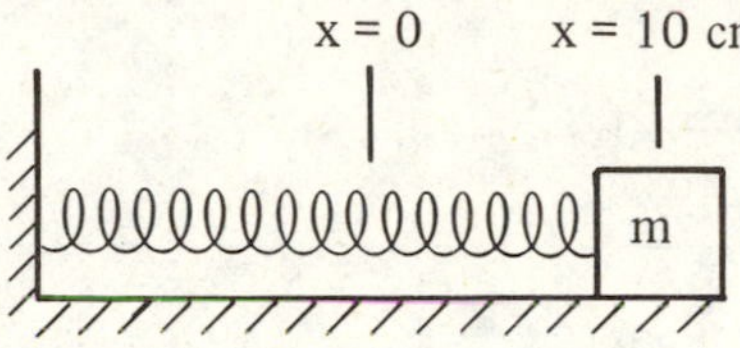

$$T = 2\pi\sqrt{\frac{m}{k}} = 2\pi\sqrt{\frac{50\ \text{g}}{100\ \text{dynes/cm}}}$$

$$T = \frac{2\pi}{\sqrt{2}} = \sqrt{2}\,\pi\ \text{sec} \qquad (1)$$

2.B

What is the boundary condition on x at t = 0?

At t = 0, we require that x = A = 10 cm.

2.C

What is the boundary condition on v at t = 0?

Since the mass is released from rest, then $v_0 = 0$ at t = 0.

2.D

What is the phase angle δ in the general solution $x = 10 \cos(\omega t + \delta)$, such that the condition x = 10 cm at t = 0 be met?

In our general solution, we set t = 0 and x = 10 cm. This gives $\cos \delta = 1$ or $\delta = 0$, (or in general even multiples of π).

2.E

Write the expression for x(t) for this example, using the condition that $\delta = 0$.

Since $\delta = 0$, our solution is $x = 10 \cos \omega t$, where $\omega = 2\pi/T = \sqrt{2}$ rad/sec.

$$\therefore x(t) = 10 \cos(\sqrt{2}\, t)\ \text{cm} \qquad (2)$$

---

†2.F

Find the velocity as a function of time for this example. Use the fact that v = dx/dt.

$$v = \frac{dx}{dt} = 10\frac{d}{dt}\cos(\sqrt{2}\, t)$$

$$v = -10\sqrt{2}\sin(\sqrt{2}\, t)\ \text{cm/sec} \qquad (3)$$

Note that $v_0 = 0$ at t = 0.

2.G

What is the *maximum* speed of the mass m? At what value of x is v a maximum?

From (3) we see by inspection that

$$|v_{max}| = 10\sqrt{2} \text{ cm/sec} \tag{4}$$

[This is simply $A\omega$]. The speed is a maximum at x = 0.

---

†2.H

Find the acceleration as a function of time, using the relation a = dv/dt.

$$a = \frac{dv}{dt} = -10\sqrt{2}\frac{d}{dt}\sin(\sqrt{2}\,t)$$

$$a = -20\cos(\sqrt{2}\,t) \text{ cm/sec}^2 \tag{5}$$

2.I

What is the maximum value of the acceleration? At what values of x is |a| a maximum?

From (5) we see that

$$|a_{max}| = 20 \text{ cm/sec}^2 \tag{6}$$

[This is simply $A\omega^2$.] |a| is a maximum at x = ±10 cm.

2.J

How long does it take the mass to first reach the position x = 5 cm?

When x = 5 cm, we can use (2) to find t. This gives

$$5 = 10\cos(\sqrt{2}\,t)$$

$$\therefore \quad \cos(\sqrt{2}\,t) = 1/2$$

$$t = \frac{\pi}{3\sqrt{2}} \text{ sec} \tag{7}$$

2.K

What is the velocity of the mass at the time $t = \frac{\pi}{3\sqrt{2}}$ sec?

We substitute this value of t into (3) and get

$$v = -10\sqrt{2}\sin\left(\frac{\pi}{3}\right) = -5\sqrt{6} \text{ cm/sec}$$

2.L

What is the acceleration of the mass at the time $t = \frac{\pi}{3\sqrt{2}}$ sec?

We substitute this value of t into (5) and get

$$a = -20\cos\left(\frac{\pi}{3}\right) = -10 \text{ cm/sec}^2$$

2.M

Find the *force* acting on the mass at $x = 10$, $x = 5$ and $x = 0$ cm.

$F = -kx$

$F(x = 10) = -100 \times 10 = -10^3$ dynes

$F(x = 5) = -100 \times 5 = -5 \times 10^2$ dynes

$F(x = 0) = -100 \times 0 = 0$

The minus sign indicates that F is to the *left* for these displacements.

2.N

Calculate the potential energy of the system as a function of time.

Using (2), we have

$$U = \frac{1}{2} kx^2 = \frac{1}{2} (100)(100) \cos^2 (\sqrt{2}\, t)$$

$$U = 5 \times 10^3 \cos^2 (\sqrt{2}\, t) \quad \text{ergs}$$

2.O

Calculate the kinetic energy of the system as a function of time.

Using (3), we have

$$K = \frac{1}{2} mv^2 = \frac{1}{2} (50)(100)\, 2 \sin^2 (\sqrt{2}\, t)$$

$$K = 5 \times 10^3 \sin^2 (\sqrt{2}\, t) \quad \text{ergs}$$

2.P

What is the *total* energy in the system?

$E = K + U$

$E = 5 \times 10^3 [\sin^2 (\sqrt{2}\, t) + \cos^2 (\sqrt{2}\, t)]$

$E = 5 \times 10^3$ ergs

3 A body oscillates in simple harmonic motion and its displacement varies in time according to the expression $x = 3 \cos (\pi t + \pi/6)$ cm.

3.A

What is the amplitude of motion for the body? Compare it with $x = A \cos (\omega t + \delta)$.

The amplitude is the maximum value of x, by inspection of the equation,

$$A = 3 \text{ cm}$$

3.B

What is the period of the motion?

By comparison with the general expression, we see that

$$\omega = \pi = \frac{2\pi}{T}$$

$$\therefore \qquad T = 2 \text{ sec}$$

3.C

What is the phase constant?

Again, by inspection,

$$\delta = \pi/6 \text{ rad}$$

3.D

Determine the displacement of the body at t = 0.

$$x(t) = 3 \cos\left(\pi t + \frac{\pi}{6}\right) \text{ cm} \tag{1}$$

$$x(t = 0) = 3 \cos \frac{\pi}{6} = \frac{3\sqrt{3}}{2} \text{ cm} \tag{2}$$

---

†3.E

What is the velocity of the body at any time t and at t = 0.

$$v(t) = \frac{dx}{dt} = -3\pi \sin\left(\pi t + \frac{\pi}{6}\right) \tag{3}$$

$$\therefore v(t = 0) = -3\pi \sin\left(\frac{\pi}{6}\right)$$

$$v(t = 0) = -\frac{3\pi}{2} \text{ cm/sec} \tag{4}$$

---

†3.F

Find the acceleration of the body at any time t and at t = 0.

$$a = \frac{dv}{dt} = -3\pi^2 \cos\left(\pi t + \frac{\pi}{6}\right) \tag{5}$$

$$a(t = 0) = -3\pi^2 \cos\left(\frac{\pi}{6}\right)$$

$$a(t = 0) = \frac{-3\sqrt{3}\,\pi^2}{2} \text{ cm/sec}^2 \tag{6}$$

3.G

Sketch the displacement, velocity and acceleration as functions of time. Note that the initial conditions are given by (2) and (4). Also, $v_{max} = 3\pi$ cm/sec and $a_{max} = -3\pi^2$ cm/sec^2.

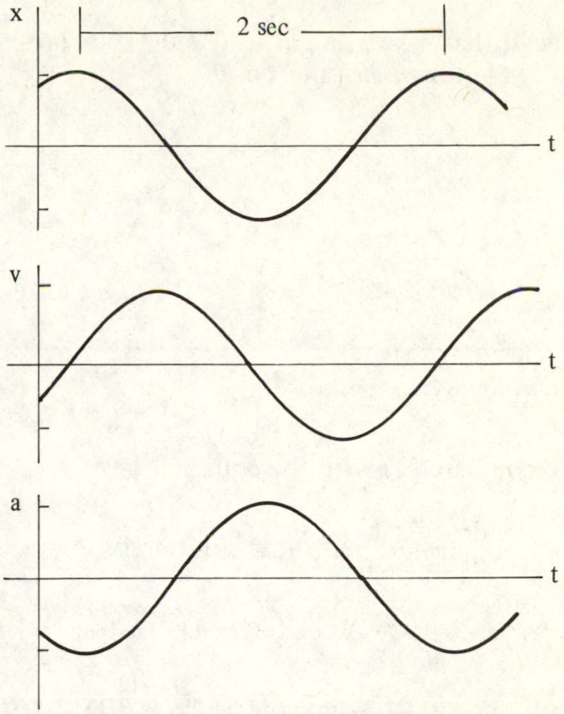

4 A mass m is suspended by a light string of length l. The upper end of the string is fixed, and the angle that the string makes with the vertical is θ.

4.A

Show the free body diagram for m.

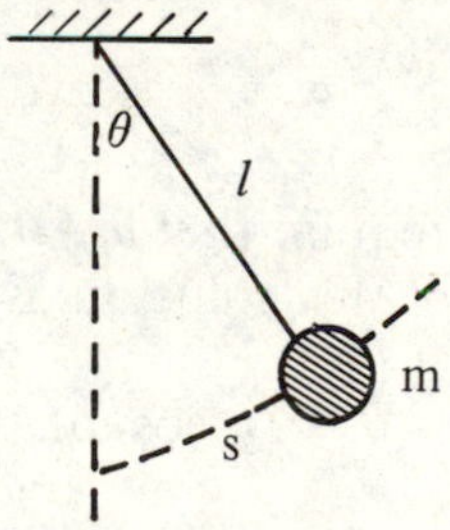

There are two forces, T and mg.

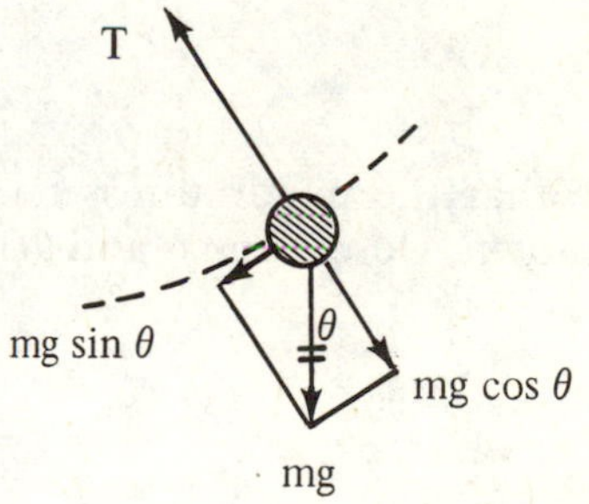

4.B

Write Newton's second law for the motion along the direction tangent to the circle.

$$\Sigma F_t = -mg \sin\theta = ma_t \qquad (1)$$

where the minus sign indicates that this force is a restoring force acting towards $\theta = 0$.

4.C

Recall that $s = l\theta$ and $a_t = d^2s/dt^2$. Now rewrite (1) for *small* values of θ.

For *small* θ, $\sin\theta \cong \theta$, so (1) becomes

$$-g\theta = \frac{d^2s}{dt^2} = l\frac{d^2\theta}{dt^2}$$

or

$$\frac{d^2\theta}{dt^2} = -\frac{g}{l}\theta \qquad (2)$$

4.D

We compare (2) with the equation

$\frac{d^2x}{dt^2} = -\frac{k}{m}x$ whose solution is

$$x = A\cos(\omega t + \delta)$$

with $\omega = \sqrt{\frac{k}{m}}$. What, therefore, is $\theta(t)$?

By comparison,

$$\theta(t) = \theta_0 \cos(\omega t + \delta) \qquad (3)$$

where

$$\omega = \sqrt{\frac{g}{l}} \qquad (4)$$

θ_0 and δ are constants, and θ is in radians.

4.E

What is the *period* of the simple pendulum?

$$T = \frac{2\pi}{\omega} = 2\pi\sqrt{\frac{l}{g}} \qquad (5)$$

4.F

Suppose that the pendulum is released from rest as shown. Determine δ and $\theta(t)$.

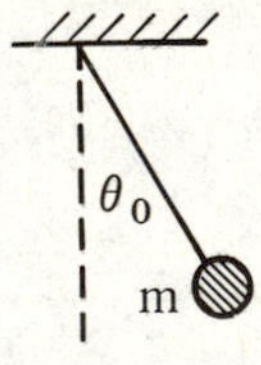

At $t = 0$, we require that $\theta = \theta_0$. $\therefore \cos\delta = 1$, or $\delta = 0$. So the solution is of the form

$$\theta = \theta_0 \cos\omega t \qquad (6)$$

4.G

How long does it take the mass to reach the vertical positon, where $\theta = 0$.

We set $\theta = 0$ in (6) and get $\cos\omega t = 0$, $\omega t = \pi/2$; $\therefore t = \pi/2\omega$. The value for ω is determined from (4). Note that $t = T/4$.

†4.H

Calculate the tangential velocity of the mass as a function of time, using the boundary conditions given in frame 4.F. Recall that

$$v = l\omega = l\frac{d\theta}{dt}$$

Using (6), we have

$$v = l\frac{d\theta}{dt} = l\theta_0 \frac{d}{dt} \cos \omega t$$

$$v = -l\omega\theta_0 \sin \omega t \qquad (7)$$

4.I

What is the velocity of the mass at t = 0 and at t = T/4 (when $\theta = 0$).

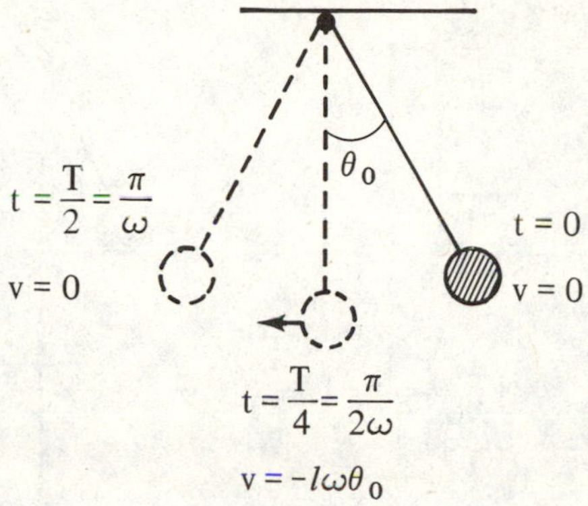

Using (7), we see that v = 0 at t = 0, since it is released from rest. At $t = T/4 = \pi/2\omega$, $v = -l\omega\theta_0 \sin(\omega\pi/2\omega) = -l\omega\theta_0$. That is, $|v|$ is a maximum at this time, when the mass is at $\theta = 0$.

4.J

Find the tangential acceleration of m as a function of time, using the boundary conditions given in 4.F. Assume θ is small, so $\sin\theta \cong \theta$.

From Newton's second law, the equation of motion for *small* θ is

$$ma_t = -mg\theta$$

$$\therefore \qquad a_t = -g\theta = -g\theta_0 \cos \omega t \qquad (8)$$

4.K

Does (8) make physical sense? Explain.

Yes. At t = 0 and $t = \pi/\omega$, $|a_t|$ is a maximum given by $g\theta_0$. This occurs when $|\theta|$ is a maximum. Likewise, $a_t = 0$ at $t = \pi/2\omega$, when $\theta = 0$.

4.L

If we analyze the free-body diagram in frame 4.A, the second law applied to the radial direction gives $T - mg = ma_r$. Use this result and the results of frame 4.I to find the maximum tension in the string for small values of θ_0.

Since $a_r = v^2/l$, we get

$$T - mg = \frac{mv^2}{l}$$

But for small θ_0, (7) gives $|v| = l\omega\theta_0$. This is the *maximum* value of v which occurs at $\theta = 0$. $\therefore$ T is a maximum when v is a maximum, or

$$T_{max} = m(g + l\omega^2\theta_0^2) \qquad (9)$$

4.M

We can make a comparison of the pendulum motion with the mass on the end of a spring with pictures. As an example, the systems at t = 0 look as follows:

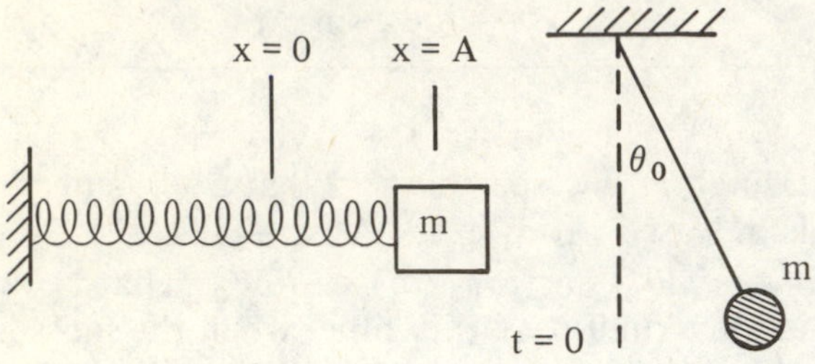

Complete the pictures for a complete cycle of the motions, giving a comparison for every quarter of a cycle.

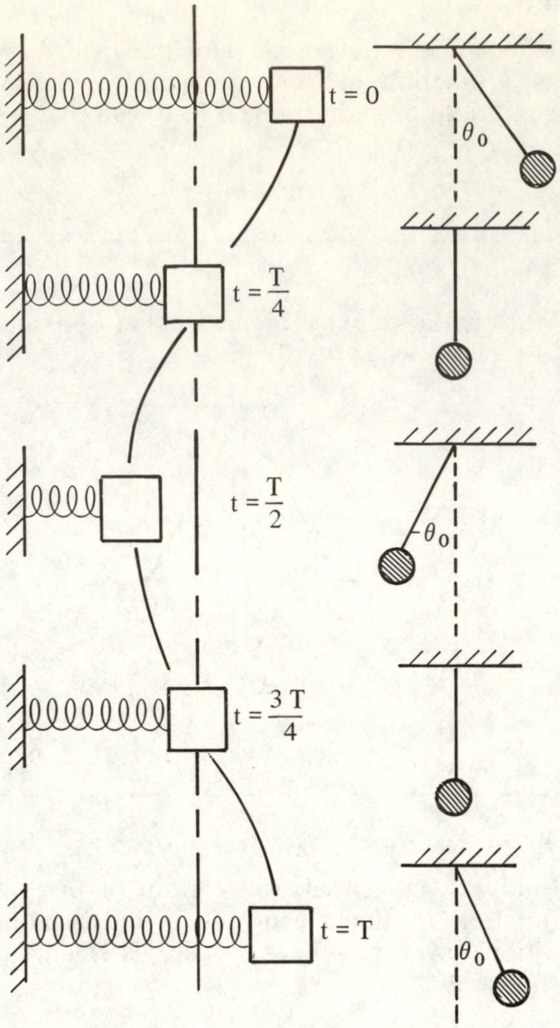

5 In this exercise, we demonstrate a more general approach to writing the solution to the simple harmonic oscillator problem. The complete solution is known if the *position* and *velocity* of the mass are specified at t = 0.

5.A

The general solution of the simple harmonic oscillator equation is given by $x = A\cos(\omega t + \delta)$. Write the expression in another form, using the identity

$$\cos(a + b) = \cos a \cos b - \sin a \sin b.$$

$$\cos(\omega t + \delta) = \cos \omega t \cos \delta - \sin \omega t \sin \delta$$

$$\therefore x = A \cos \delta \cos \omega t - A \sin \delta \sin \omega t$$

or,

$$x = B \sin \omega t + C \cos \omega t \qquad (1)$$

where

$$B = -A \sin \delta \qquad (2)$$

and

$$C = A \cos \delta \qquad (3)$$

5.B

Write a relationship between A, B and C in view of (2) and (3).

Squaring (2) and (3) and adding gives

$$A^2 = B^2 + C^2 \tag{4}$$

5.C

Write a relationship for δ in terms of B and C.

Dividng (2) by (3) gives

$$\tan \delta = -\frac{B}{C} \tag{5}$$

---

†5.D

Equation (1) is a form of the solution to the harmonic oscillator equation and is used in many texts. Suppose $x = x_0$ at $t = 0$ and $v = v_0$ at $t = 0$. Determine the constants B and C for these initial conditions. Note that x_0 is any arbitrary initial value of the position lying between $-A$ and A.

From (1) we have

$$x_0 = B \sin(0) + C \cos(0)$$

or,

$$x_0 = C \tag{6}$$

Since $v = dx/dt$, differentiation of (1) gives

$$v = \omega B \cos \omega t - \omega C \sin \omega t \tag{7}$$

$$v_0 = \omega B \cos(0) - \omega C \sin(0)$$

or,

$$B = \frac{v_0}{\omega} \tag{8}$$

5.E

Now write the solution of the harmonic oscillator equation for the initial conditions stated in frame 5.D.

Substituting (6) and (8) into (1) gives

$$x = \frac{v_0}{\omega} \sin \omega t + x_0 \cos \omega t \tag{9}$$

5.F

What is the phase constant δ for the initial conditions stated in frame 5.D?

Using (6) and (8) in (5) gives

$$\tan \delta = -\frac{v_0}{\omega x_0}$$

or,

$$\delta = \arctan\left(-\frac{v_0}{\omega x_0}\right) \tag{10}$$

5.G

What is the amplitude of the motion, A, in terms of x_0, v_0 and ω?

Substituting (6) and (8) into (4) gives

$$A = \left[\frac{v_0{}^2}{\omega^2} + x_0{}^2\right]^{1/2} \qquad (11)$$

5.H

From (10) and (11) we see that a complete solution is obtained if v_0, x_0 and ω are known, since

$$x = A \cos(\omega t + \delta)$$

Suppose that at $t = 0$, $v_0 = 4$ cm/sec $x_0 = 2$ cm and $\omega = 2$ rad/sec. Find A and δ, and write the general solution for x(t).

$$A = \left[\frac{4^2}{2^2} + 2^2\right]^{1/2} = 2\sqrt{2}\ \text{cm}$$

$$\delta = \arctan\left(-\frac{4}{2 \times 2}\right) = -\pi/4\ \text{rad}$$

$$\therefore x(t) = 2\sqrt{2}\cos(2t - \pi/4)\ \text{cm} \qquad (12)$$

10.7 SUMMARY

The general solution of the simple harmonic oscillator equation is

$$x = A \cos(\omega t + \delta) \tag{10.4}$$

where

$$\omega = \sqrt{\frac{k}{m}} \tag{10.3}$$

A is the amplitude of the oscillations and δ is a phase constant adjusted to fit the initial conditions of the problem. The *period* of the motion is

$$T = 2\pi \sqrt{\frac{k}{m}} \tag{10.5}$$

while the frequency is 1/T. The velocity and acceleration of the vibrating mass are given by

$$v = \frac{dx}{dt} = -A\omega \sin(\omega t + \delta) \tag{10.7}$$

$$a = \frac{d^2x}{dt^2} = -A\omega^2 \cos(\omega t + \delta) \tag{10.8}$$

The maximum velocity and maximum acceleration are given by

$$|v_{max}| = A\omega$$
$$|a_{max}| = A\omega^2 \tag{10.13}$$

The potential energy, total energy and kinetic energy of the simple harmonic oscillator are given by

$$U = \frac{1}{2}kx^2 \tag{10.18}$$

$$E = \frac{1}{2}kA^2 \tag{10.19}$$

and

$$K = E - U = \frac{1}{2}k(A^2 - x^2) \tag{10.21}$$

The *period* of motion of a simple pendulum of length l oscillating in simple harmonic motion for small displacements from equilibrium is given by

$$T = 2\pi \sqrt{\frac{l}{g}} \tag{10.25}$$

10.8 PROBLEMS

1. A certain spring stretches 3.9 cm when a 10 g mass is hung from it. If a 25 g mass attached to this spring oscillates in simple harmonic motion, as in Figure 10-1, calculate the period of the motion.

2. The frequency of vibration of the simple harmonic oscillator shown in Figure 10-1 is measured to be 5 Hz when a 4 g mass is connected to the spring. What is the force constant of the spring?

†3. A block oscillates with simple harmonic motion and its displacement varies in time according to the expression $x = 5.0 \cos (2t + \pi/6)$ cm. At $t = 0$ sec, find (a) the displacement, (b) the velocity and (c) the acceleration of the block. (d) Find the period of the motion.

†4. At $t = 0$, the displacement of a simple harmonic oscillator is given by $x_0 = 2$ cm, its velocity is $v_0 = -24$ cm/sec and the amplitude of the motion is 3 cm. If the frequency of oscillation is 2 Hz, find (a) the phase constant, (b) the displacement as a function of time, (c) the velocity and acceleration as functions of time, (d) the maximum speed and acceleration.

5. Verify by direct substitution that Equations (10.3) and (10.4) represent a general solution to Equation (10.2).

6. *The Physical Pendulum.* An arbitrarily shaped body pivoted at the axis through the point 0, as in Figure 10-7, is called a physical pendulum. For *small* angular displacements, θ, show that the period of oscillation is given by

$$T = 2\pi \sqrt{\frac{I}{Mgd}}$$

where I is the moment of inertia about 0 and d is the distance between 0 and the center of mass. *Hint:* The restoring torque about 0 is $-Mgd \sin \theta$, and $\tau = I\alpha = I d^2\theta/dt^2$.

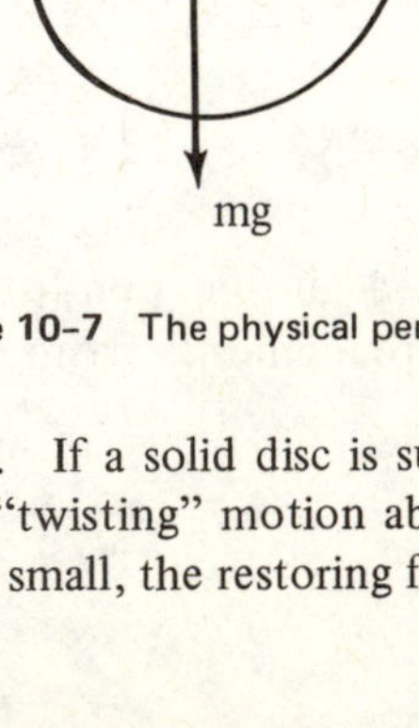

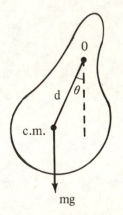

Figure 10-7 The physical pendulum.

†7. *The Torsional Pendulum.* If a solid disc is suspended from a wire, as in Figure 10-8, the disc will oscillate in a "twisting" motion about the vertical axis. If the angular displacement from equilibrium is small, the restoring force is given by $\tau = -\kappa\theta$, where κ is called the *torsional constant* of the wire. For small values of θ, show that the system oscillates in simple harmonic motion with a period given by

$$T = 2\pi\sqrt{\frac{I}{\kappa}}$$

where I is the moment of inertia about the axis of rotation. *Hint:* $\tau = I\alpha = Id^2\theta/dt^2$.

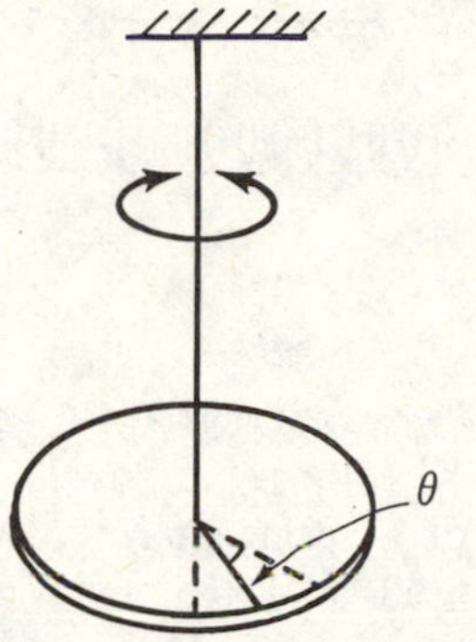

Figure 10-8 The torsional pendulum.

8. A heavy ball swings on the end of a light rope whose length is 4 m. (a) If the ball oscillates about the vertical position with a very small amplitude, determine its period of oscillation. (b) Determine the *change* in the period of motion if the maximum angular displacement from the vertical is 20 degrees.

9. A meter stick is pivoted at one end and oscillates in a vertical plane with simple harmonic motion. What is its period of oscillation for small amplitudes? *Hint:* The meter stick is a physical pendulum, as described in Problem 6, and has a moment of inertia $I = \frac{1}{3}Ml^2$ about an axis through one end.

10. A simple pendulum of length 2 m is released from rest at t = 0 when its angular displacement from the vertical is 5 degrees. The general solution of the equation of motion, Equation (10.23), is given by $\theta = \theta_0 \cos(\omega t + \delta)$. (a) Determine the appropriate constants θ_0, ω and δ for this situation. (b) What is the maximum speed of the suspended mass? (c) If a 3 kg mass is suspended, what is the tension in the rope at the instant the mass swings through the vertical position?

11. A bullet of mass m is shot into and becomes imbedded in a block of mass M. The block lies on a frictionless surface, connected to a light spring of force constant k, as in Figure 10-9. If the spring is compressed a total distance A, show that the initial speed of the bullet is given by

$$v_0 = \frac{\sqrt{k(m+M)}}{m}A$$

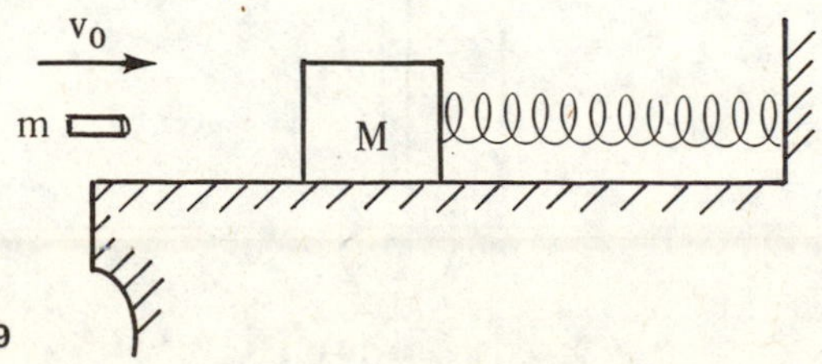

Figure 10-9

†12. A mass M connected to a spring of mass m oscillates in simple harmonic motion, as in Figure 10-10. The spring has a force constant k and an equilibrium length l. (a) Find the kinetic energy of the spring when the mass has a velocity v. (b) Find the period of oscillation. *Hint:* Assume all portions of the *spring* oscillate in phase, and that the velocity of a segment dx is proportional to the distance from the fixed end, that is, $v_x = \frac{x}{l} v$. Also, note that $dm = \frac{m}{l} dx$.

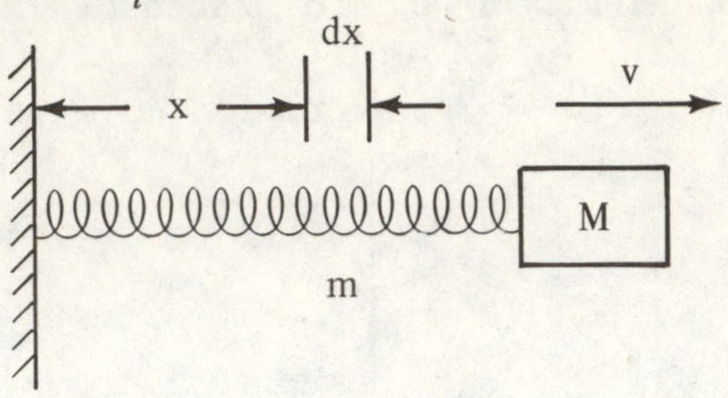

Figure 10-10

13. A mass on the end of a spring of force constant 50 N/m undergoes simple harmonic motion with an amplitude of 12 cm. (a) What is the total mechanical energy of the system? (b) When the mass is at a distance of 9 cm from equilibrium, what is its kinetic energy? (c) What is the potential energy of the system when x = 9 cm? (d) At what value of x is the energy equally divided between potential and kinetic energy?

14. A 50 g mass connected to a spring moves on a horizontal frictionless surface in simple harmonic motion with an amplitude of 16 cm and period of 4 sec. At t = 0, the mass is released from rest at a distance of +16 cm from its equilibrium position, as in Figure 10-2 [that is, the spring is initially *stretched* as in Figure 10-2]. Find (a) the coordinate of the mass at any time t, and the value of x at t = 0.5 sec; (b) the magnitude and direction of the force acting on the mass at t = 0.5 sec; (c) the minimum time required for the mass to move from its initial position to x = 8 cm; (d) the velocity at any time t and the speed when x = 8 cm; and (e) the total mechanical energy of the system and the force constant of the spring.

15. A mass M is supported by a uniform rod of mass M and length l as shown in Figure 10-11. The rod is supported at the top by a frictionless pivot. (a) Determine the tension in the rod at a point p a distance y from the bottom when the system is stationary. (*Note:* You must not neglect the mass of the rod.) (b) What is the tension at the top where the rod is supported? (c) Calculate the period of oscillation of the system (for small displacements) if l = 2 m. [Note that $I_{rod} = 1/3\ Ml^2$ about an axis through one end. Assume the lower mass is a point mass, and make use of the results of Problem 6.]

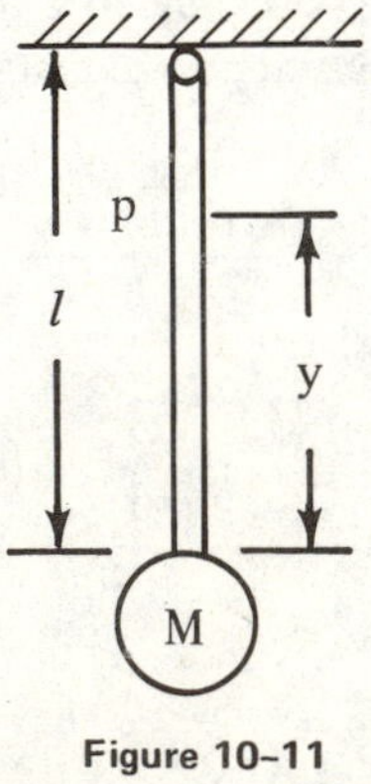

Figure 10-11

11

GRAVITATION

11.1 THE UNIVERSAL LAW OF GRAVITY

In 1687 Newton published his work on the universal law of gravitation. This theory was based on various astronomical observations. His analysis showed that *any* pair of particles attract each other with a force proportional to the product of their masses and inversely proportional to the square of their separation. This inverse square law of force has a magnitude given by

$$F = G\frac{m_1 m_2}{r^2} \tag{11.1}$$

where m_1 and m_2 are the particle masses, r is their separation and G is a universal constant, not to be confused with the acceleration of gravity g. The magnitude of this constant, as determined by experiment, is

$$G = (6.673 \pm 0.003) \times 10^{-11}\ \frac{Nm^2}{kg^2} \tag{11.2}$$

The attractive force given by Equation (11.1) always exists between two particles, *regardless* of the nature of the medium which separates them. Gravitational forces are the weakest forces in nature. For example, if we compare the gravitational force between the electron and proton of a hydrogen atom with the electric force between them (since they are both charged), we would find that the electric force is $\sim 10^{39}$ times greater than the gravitational force. Nuclear forces, which are very short-ranged ($\sim 10^{-15}$ m), are even stronger than electric forces. Since most large scale objects are electrically neutral, and nuclear forces are short-ranged, gravitational forces are readily experienced in everyday life. An electrically neutral object will normally accelerate towards the earth when dropped from some elevation. Massive bodies, such as planets and the sun, exhibit sizeable forces on each other, regardless of their large separations. Consequently, their motions are governed by these forces.

Another important aspect of the universal law of gravity is the fact that, by Newton's third law, the force of m_1 on m_2 is equal in magnitude and *opposite* in direction to the force of m_2 on m_1. This is demonstrated in Figure 11-1, where $\mathbf{F}_{12}$

means the force on 1 due to 2, while $\mathbf{F}_{21}$ means the force on 2 due to 1. In other words,

$$\mathbf{F}_{12} = -\mathbf{F}_{21} \tag{11.3}$$

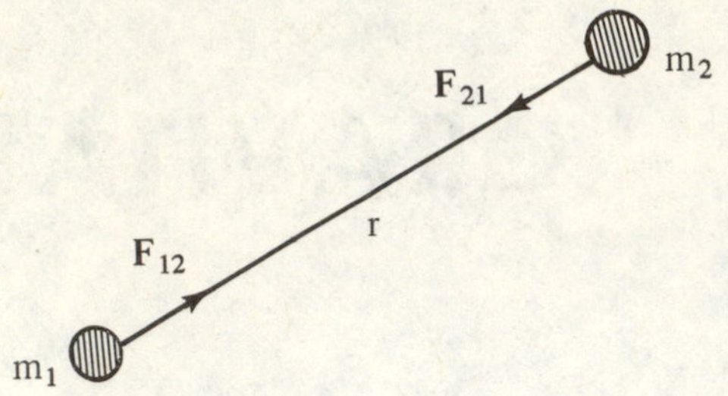

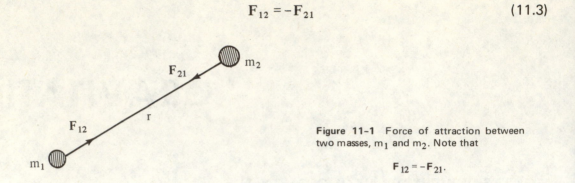

Figure 11-1 Force of attraction between two masses, m_1 and m_2. Note that

$$\mathbf{F}_{12} = -\mathbf{F}_{21}.$$

If more than two masses are present, we can use the *superposition* principle to find the resultant force on any one of the particles. This says that the resultant gravitational force on a body equals the *vector* sum of the individual forces of the various external bodies.

Example 11.1

Three point masses of 3 kg, 4 kg and 5 kg are placed as shown in Figure 11-2. Calculate the resultant gravitational force on the 4 kg mass. (Assume they are isolated from the rest of the universe.)

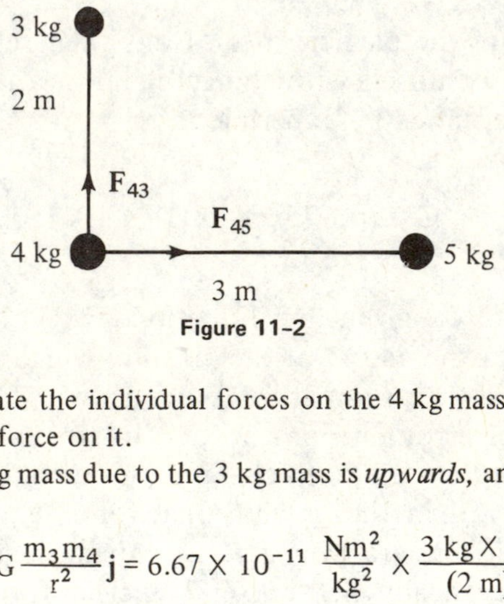

Figure 11-2

First, we will calculate the individual forces on the 4 kg mass, and then take a vector sum to get the resultant force on it.

The force on the 4 kg mass due to the 3 kg mass is *upwards,* and is given by

$$\mathbf{F}_{43} = G\frac{m_3 m_4}{r^2}\mathbf{j} = 6.67 \times 10^{-11}\ \frac{\text{Nm}^2}{\text{kg}^2} \times \frac{3\ \text{kg} \times 4\ \text{kg}}{(2\ \text{m})^2}\mathbf{j}$$

or,

$$\mathbf{F}_{43} = 2.0 \times 10^{-10}\ \mathbf{j}\ \text{N}$$

The force on the 4 kg mass due to the 5 kg mass is to the *right,* and is given by

$$\mathbf{F}_{45} = 6.67 \times 10^{-11}\ \frac{\text{Nm}^2}{\text{kg}^2} \times \frac{4\ \text{kg} \times 5\ \text{kg}}{(3\ \text{m})^2}\mathbf{i} = 1.5 \times 10^{-10}\ \mathbf{i}\ \text{N}$$

Therefore, the resultant force on the 4 kg mass is the *vector* sum of $\mathbf{F}_{43}$ and $\mathbf{F}_{45}$.

$$\mathbf{F} = \mathbf{F}_{43} + \mathbf{F}_{45} = (1.5 \times 10^{-10}\ \mathbf{i} + 2.0 \times 10^{-10}\ \mathbf{j})\ \mathrm{N}$$

The magnitude of this force is 2.5×10^{-10} N, which is equivalent to only 5.6×10^{-11} lb! The direction of F is given by arc tan $(2/1.5) = 59°$. That is, **F** makes an angle of 59° with the x axis.

Example 11.2

Assume the moon orbits the earth in a circular path with a period of 27 days, at a distance of 3.8×10^8 m. Using this information, calculate the mass of the earth.

The centripetal force acting on the moon is provided by the gravitational attraction between the moon and the earth. Therefore, Newton's second law applied to the moon, together with Equation (11.1), gives

$$F_m = G\frac{M_E M_m}{r^2} = M_m \frac{v^2}{r} \qquad (1)$$

where v is the orbital velocity of the moon. Since the moon travels a distance of $2\pi r$ in a time of 27 days, where $r = 3.8 \times 10^8$ m, we get

$$v = \frac{2\pi r}{T} = \frac{2\pi \times 3.8 \times 10^8\ \mathrm{m}}{27\ \mathrm{days} \times 8.64 \times 10^4\ \dfrac{\mathrm{sec}}{\mathrm{day}}} = 1.02 \times 10^3\ \mathrm{m/sec} \qquad (2)$$

Therefore, simplifying (1) for the mass of the earth gives

$$M_E = \frac{v^2 r}{G} = \frac{\left(1.02 \times 10^3\ \dfrac{\mathrm{m}}{\mathrm{sec}}\right)^2 \times 3.8 \times 10^8\ \mathrm{m}}{6.67 \times 10^{-11}\ \dfrac{\mathrm{Nm}^2}{\mathrm{kg}^2}}$$

or,

$$M_E \cong 5.93 \times 10^{24}\ \mathrm{kg}$$

This is to be compared with the more accurate value of 5.98×10^{24} kg. Using this value of M_E, and the fact that the moon's mass is about 7.34×10^{22} kg, the student should show that the force of attraction between the earth and the moon is equal to about 2.03×10^{20} N. This provides a centripetal acceleration for the moon equal to 2.8×10^{-8} m/sec^2.

It is useful to demonstrate the relationship between the acceleration of gravity and other relevant constants. Since the force between any object of mass m and the earth is simply the object's weight, mg, we can equate this force to Equation (11.1), giving

$$mg = G\frac{mM_E}{R^2}$$

where M_E and R are the mass and radius of the earth, respectively. Solving this for g gives

$$g = \frac{GM_E}{R^2} \qquad (11.4)$$

If we take $g = 9.80\ m/sec^2$, and the value $R = 6.37 \times 10^6$ m, M_E is calculated to be about 5.96×10^{24} kg. Of course, the value of g varies over the surface of the earth, owing to the bulging of the earth at the equator as well as to local variations in altitude.

11.2 THE GRAVITATIONAL FORCE BETWEEN BODIES OF FINITE EXTENT

The inverse square law of attraction between two masses as given by Equation (11.1) is generally valid only when the bodies are considered to be *particles,* with their respective masses concentrated at points in space. However, we can apply the same concept to evaluate the force between bodies having finite dimensions. To do this, we imagine dividing the body up into infinitesimally small segments of mass dM, as in Figure 11-3, where r is the distance from dM to a point P. The force on a particle of mass m placed at P due to the segment dM is given by

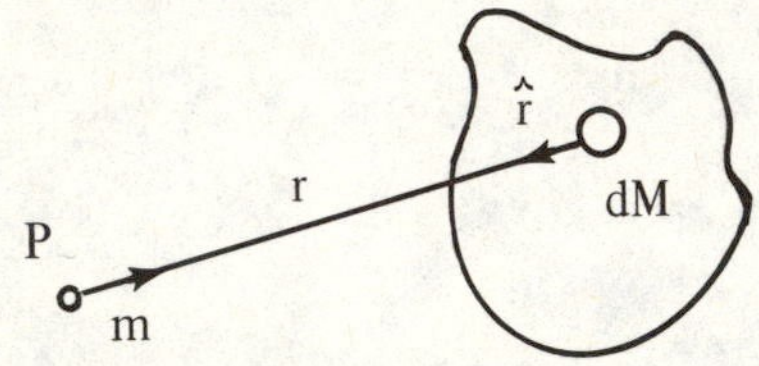

Figure 11-3 A particle of mass m near a body of finite extent having a mass M.

$$dF = -G\frac{mdM}{r^2}\hat{r} \qquad (11.5)$$

where $\hat{r}$ is a unit vector directed from dM to P. The *total* force on m due to the body is then given by the *integral* of Equation (11.5), which represents a *vector* sum over all segments making up the body. The integral may be cumbersome for some mass distributions; however, a few examples may help clarify this type of situation.

Example 11.3

A homogeneous bar of length *l,* mass M, is at a distance d from a point mass m, as in Figure 11-4. Calculate the force on m.

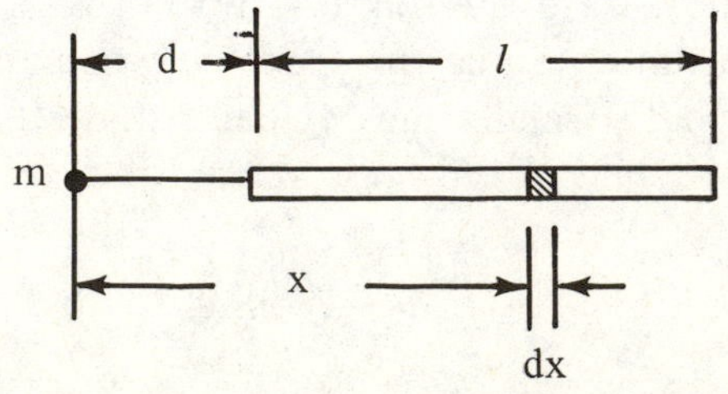

Figure 11-4

Let us assume m is at the origin. The shaded portion of the bar has a length dx and a mass dM. Since the mass per unit length is constant, then $dM = \frac{M}{l}\,dx$. Therefore, using Equation (11.5) and integrating gives

$$\mathbf{F} = Gm \int_{d}^{l+d} \frac{M}{l}\frac{dx}{x^2}\,\mathbf{i}$$

where the variable r is now x, and the force on m is in the +x direction for every segment. Since $\int \frac{dx}{x^2} = -\frac{1}{x}$, the expression above reduces to

$$\mathbf{F} = \frac{GmM}{d\,(l+d)}\,\mathbf{i}$$

Note that in the limit $l \to 0$, $F \sim 1/d^2$, which corresponds to the force between two point masses.

Various problems in gravitation involve the force between a particle and a sphere. Several facts will be stated here without proof, since the calculus is rather tedious.

1. If a particle of mass m is located *outside* a spherical shell of mass M, as in Figure 11-5, then the sphere attracts the particle as though the mass of the sphere were concentrated at its *center.* That is, the force of attraction is given by

$$|\mathbf{F}| = \frac{GMm}{r^2} \quad \text{for } r > R \tag{11.6}$$

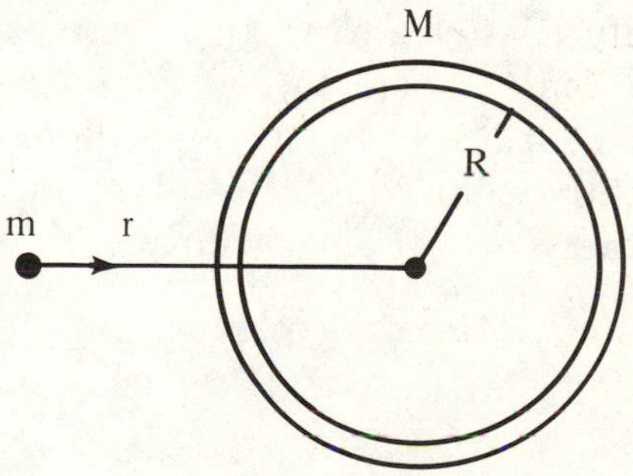

Figure 11-5 A particle of mass m near a spherical shell of mass M.

2. If a particle of mass m is located *inside* a thin spherical shell, the force on m is *zero.* That is, F = 0 for $r < R$.

3. If a particle of mass m is located *outside* a uniformly dense *solid* sphere of mass M, the sphere attracts m as though the mass of the sphere were concentrated at its center. Therefore, Equation (11.6) is valid for $r > R$.

4. If a particle of mass M is located *inside* a uniformly dense *solid* sphere of mass M, radius R, as in Figure 11-6, the force on m is due *only* to the mass contained *within* a spherical surface of radius r, represented by the dotted line in Figure 11-6. In this case, the force on m (directed towards the center of the sphere) is given by

$$|\mathbf{F}| = G\,\frac{mM'}{r^2} \tag{11.7}$$

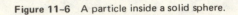

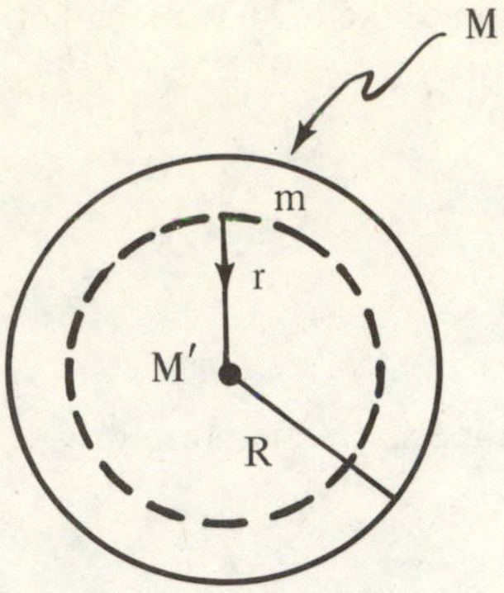

Figure 11-6 A particle inside a solid sphere.

where M′ is the mass of that portion of the sphere whose radius is r. Since the sphere has a uniform mass density ρ, we can write $\rho = M'/V'$, where V′ is the volume given by $\frac{4}{3}\pi r^3$. But we also have $\rho = M/V$, where $V = \frac{4}{3}\pi R^3$, the *total* volume. Taking the ratio gives

$$M' = \frac{r^3}{R^3} M$$

so Equation (11.7) can be written as

$$|\mathbf{F}| = \frac{GmM}{R^3} r \qquad r < R \text{ (solid sphere)} \tag{11.8}$$

5. Finally, if a particle of mass m is located *inside* a solid sphere whose mass density is *not* uniform, then we replace M′ in Equation (11.7) by an integral of the form $\int \rho dV$, where the integration is taken over the volume contained *within* the dotted surface in Figure 11-6. For spherical symmetry, that is, when $\rho = \rho(r)$, take the unit volume as $dV = 4\pi r^2\, dr$, the volume of a spherical shell of radius r, thickness dr. For example, *if* $\rho(r) = \rho_0 r$, where ρ_0 is a constant, then $M' = \int_0^r \rho_0 r\, 4\pi r^2\, dr = \pi \rho_0 r^4$.

11.3 GRAVITATIONAL POTENTIAL ENERGY

In Chapter 6 we found that the inverse square law of force is *conservative*. This being the case, a potential energy function was derived for a two particle system, given by

$$U_{12}(r) = -\frac{Gm_1 m_2}{r_{12}} \tag{11.9}$$

where r_{12} is the separation between the particles and $U(\infty) = 0$.

If there are more than two masses comprising a system of particles, the total potential energy is the *scalar sum* over all *pairs* of the system, each pair containing a term as in Equation (11.9).

Example 11.4

Calculate the total potential energy of the system of 3 particles shown in Figure 11-2.

For a system of three particles, summing over all *pairs* according to Equation (11.9) gives

$$U = U_{12} + U_{13} + U_{23}$$

$$U = -G\left[\frac{m_1 m_2}{r_{12}} + \frac{m_1 m_3}{r_{13}} + \frac{m_2 m_3}{r_{23}}\right]$$

Note that in Figure 11-2, $m_1 = 3$ kg, $m_2 = 4$ kg, $m_3 = 5$ kg, $r_{12} = 2$ m, $r_{13} = \sqrt{13}$ m and $r_{23} = 3$ m. Therefore,

$$U = -6.67 \times 10^{-11}\left[\frac{3\times 4}{2} + \frac{3\times 5}{\sqrt{13}} + \frac{4\times 5}{3}\right]\ \text{J} = -1.12 \times 10^{-9}\ \text{J}$$

11.4 PLANETARY MOTION

The laws that govern the motions of planets in the solar system are a consequence of Newton's universal law of gravity. Before Newton formulated this law, however, Kepler recognized three empirical laws, based on astronomical data. Briefly, these laws state the following: (1) Each planet of the solar system moves in elliptical orbits, with the sun at one focus. (2) The area swept out per unit time by the radius vector from the sun to any planet is constant. (3) The square of the period of motion for each planet is proportional to the *cube* of the semimajor axis, a.

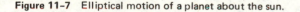

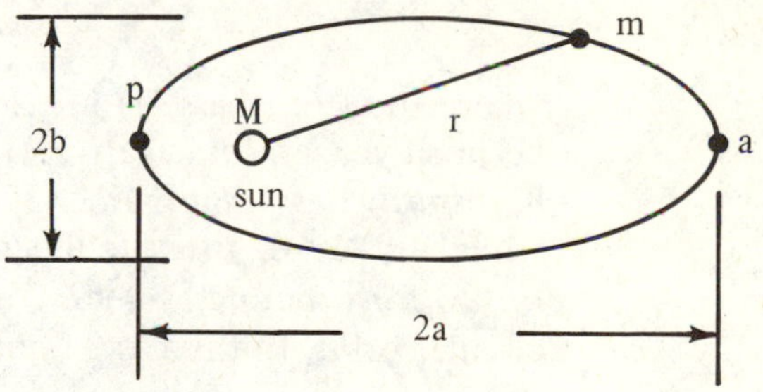

Figure 11-7 Elliptical motion of a planet about the sun.

Kepler's second law follows from the fact that the gravitational force is a *central* force (that is, F is a function only of r). For this special type of force, the torque on any planet due to the gravitational force of the sun is identically *zero*, since **r** || **F** $\boldsymbol{\tau} = \mathbf{r} \times \mathbf{F} = 0$ (Figure 11-7). Since $\boldsymbol{\tau} = d\mathbf{L}/dt$, it follows that the *orbital angular momentum*, **L**, of the planet relative to the sun is a *constant* of the motion. For a useful demonstration of this law, consider the closest and farthest separations of a planet from the sun—the *perihelion* and *aphelion* positions (labeled p and a in Figure 11-7). If the planet has speeds of v_p and v_a at these points, and the corresponding distances from the sun are r_p and r_a, conservation of angular momentum gives the expression

$$mv_a r_a = mv_p r_p \tag{11.10}$$

that is, the planet moves fastest when it is *closest* to the sun, and slowest when it is farthest from the sun. This also applies to the motion of satellites about the earth.

Kepler's third law can be derived for an elliptical orbit, but for simplicity we will assume the motion of the satellite is circular. In this case, Newton's second law applied to a planet of mass m, together with Equation (11.1), gives

$$F = \frac{GMm}{r^2} = m\frac{v^2}{r} \tag{11.11}$$

Since the speed of the planet is given by $2\pi r/T$, where T is the period, or the time for one revolution, Equation (11.11) becomes

$$\frac{GM}{r^2} = \frac{1}{r}\left(\frac{2\pi r}{T}\right)^2$$

or,

$$T^2 = \frac{4\pi^2}{GM}r^2 \tag{11.12}$$

Obviously, this law follows from the fact that $F \sim r^{-2}$. The student should note that Equation (11.12) *also* applies to an elliptical orbit, as in Figure 11–7, with r replaced by a, the semimajor axis.

Finally, let us consider the *energy* of a two-body system, such as the earth-sun combination. Since the gravitational force is conservative, the *total energy* of this two-body system is *constant.* We can express this constant of the motion as

$$E = K + U = \frac{1}{2}mv^2 - \frac{GMm}{r} = \text{constant} \tag{11.13}$$

where the sun (mass M) is assumed to be stationary, since $M \gg m$. Obviously, E can be positive, zero or negative. However, for a *bound* system such as the earth-sun, E is *necessarily less than zero.* [If $E = 0$, the satellite would move in a *parabolic* path, while if $E > 0$, the satellite would describe a *hyperbolic* path.] To establish that $E < 0$ for a bound system, consider a planet of mass m moving about the sun in a circular orbit. In this case, Equation (11.11) gives

$$\frac{1}{2}mv^2 = \frac{GMm}{2r} \tag{11.14}$$

If we substitute Equation (11.14) into Equation (11.13), we get

$$E = \frac{GMm}{2r} - \frac{GMm}{r} = -\frac{GMm}{2r} \tag{11.15}$$

The total energy is *negative* and equal to *half* the potential energy. If the orbit is an *ellipse,* the total energy is also given by Equation (11.15), with r replaced by a. In order to separate the two bodies by an *infinite* distance, we would have to supply at *least* the energy given by Equation (11.15).

Example 11.5

How much energy is required to separate the moon from the earth by an infinite distance?

Here, we simply evaluate $|E|$ as given by Equation (11.15), using the numerical values $M_E = 5.98 \times 10^{24}$ kg. $M_m = 7.34 \times 10^{22}$ kg and $r = 3.8 \times 10^8$ m. The result is $|E| = 3.85 \times 10^{30}$ J. The U.S. uses about 4×10^{18} J of electrical energy per year, so it would take $\sim 10^{12}$ years to generate this much energy by electrical means!

11.5 PROGRAMMED EXERCISES

1.A

Two *particles* of masses m_1 and m_2 are separated by a distance r as shown. What is the magnitude of the gravitational force between them?

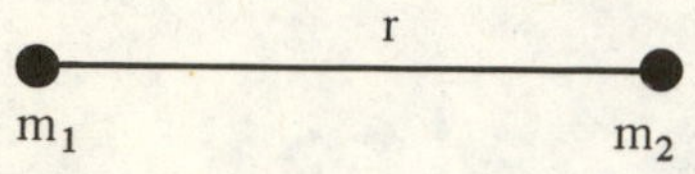

$$F = G\frac{m_1 m_2}{r^2} \qquad (1)$$

1.B

What law does (1) represent, and what is G?

(1) represents Newton's universal law of gravity.

$$G = 6.67 \times 10^{-11}\,\text{Nm}^2/\text{kg}^2$$

1.C

Is (1) valid for the force between *any* two bodies? Explain.

No. In general, the inverse square law applies only to the force between *point* masses.

1.D

Are gravitational forces attractive or repulsive? Show the direction of the force on m_1 due to m_2. Likewise, show the force on m_2 due to m_1.

Gravitational forces are *always* attractive.

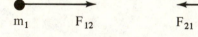

1.E

Show that the universal constant G, by definition, has units of Nm^2/kg^2 in the SI system.

From (1), we get

$$G = \frac{Fr^2}{m_1 m_2}$$

$$\therefore \quad [G] = \frac{\text{Nm}^2}{\text{kg}^2}$$

1.F

A man and a woman have masses of 70 kg and 50 kg, respectively, and are separated by 3 m. Assume they are point masses to estimate the force between them.

$$F = G\frac{m_1 m_2}{r^2}$$

$$F = 6.67 \times 10^{-11}\left[\frac{70 \times 50}{3^2}\right]$$

$$F \cong 2.6 \times 10^{-8}\ \text{N}$$

1.G

Compare the force calculated in frame 1.F with the *weight* of the man.

$$W = m_1 g = 70 \text{ kg} \times 9.8 \frac{\text{m}}{\text{sec}^2}$$

$$W \cong 6.9 \times 10^2 \text{ N}$$

$$\frac{W}{F} \cong 2.7 \times 10^{10}$$

1.H

The result of frame 1.G shows that the "weight" force overwhelms the attractive force between the man and the woman. What is the origin of the force we call weight?

This force arises from the attractive force between an object and the earth.

1.I

If an object of mass m is at the earth's surface, what is the force on it in terms of the earth's mass M and radius R?

If the earth is treated as a uniform sphere, its mass can be assumed to be concentrated at its *center* when considering bodies *external* to it.

$$\therefore \quad W = \frac{GmM}{R^2} \qquad (2)$$

1.J

Compare (2) with the more familiar formula for weight, that is, W = mg, and obtain a basic relation for g. Do *not* confuse g with G.

$$mg = G\frac{mM}{R^2}$$

$$\therefore \quad g = \frac{GM}{R^2} \qquad (3)$$

1.K

Does g remain constant as one goes away from the earth's surface? Explain.

No. Equation (3) only applies to a point on the earth's surface. For $r > R$, we get $g = GM/r^2$, so the "effective" g *decreases* as we progress away from the earth's surface.

1.L

An astronaut lands on planet x, whose mass is M_x and whose radius is R_x. What is the effective g on this planet?

Since (3) applies to the surface of *any* large sphere, it follows that

$$g_x = G\frac{M_x}{{R_x}^2}$$

1.M

In reality, the earth rotates about its axis with a period of 24 hours. Does this affect the value of g? Explain.

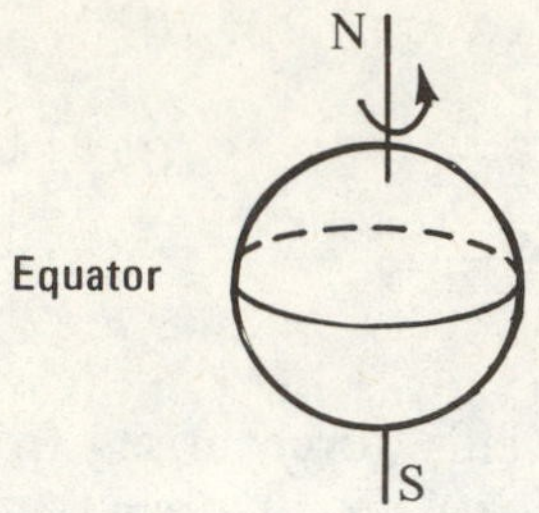

Yes. In a rotating system, the gravitational force given by (2) [the true weight] must *exceed* the apparent weight.

1.N

Assume the earth is a uniform sphere, and obtain g_e at the equator in terms of centripetal acceleration, a_r, experienced by the body at the equator.

Newton's second law gives

$$\frac{GMm}{R^2} - mg_e = ma_r$$

$$\therefore \quad g_e = \frac{GM}{R^2} - a_r$$

1.O

What is g at the North Pole? What can you conclude about the weight of a body at the North Pole compared to its weight at the equator?

Since a point on the North Pole is stationary relative to the earth's center ($a_r = 0$ here), g is simply GM/R^2. Consequently, a body weighs less at the equator than at the North Pole. The difference is small, but measurable, since $a_r \cong 0.034\ \text{m/sec}^2$ at the equator.

2 Three masses are placed at the corners of an equilateral triangle as shown below. Assume they are isolated from the rest of the universe (that is, free of other gravitational forces).

2.A

In a diagram, show the direction of the forces on particle 3 due to particles 1 and 2.

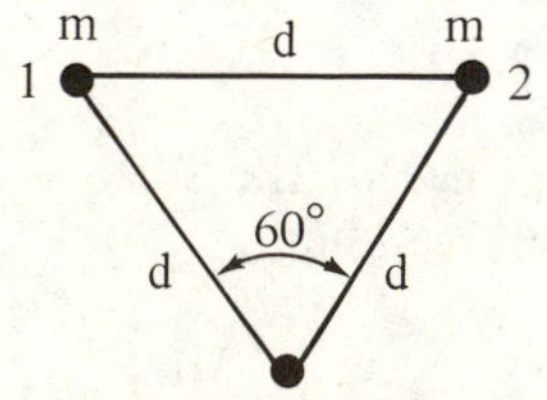

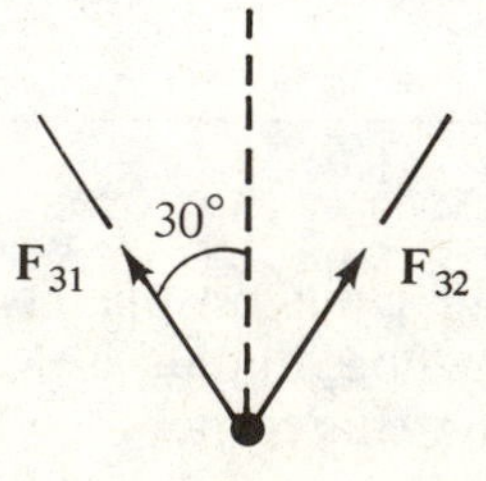

2.B

What are the magnitudes of $\mathbf{F}_{31}$ and $\mathbf{F}_{32}$?

Since particle 3 is equidistant from particles 1 and 2,

$$|\mathbf{F}_{31}| = |\mathbf{F}_{32}| = G\frac{2m^2}{d^2} \quad (1)$$

2.C

Write a *vector* expression for the *resultant* force on particle 3. *Note* the force diagram in frame 2.A.

From the force diagram we see that the resultant force in the x direction is *zero*, so the resultant force is in the y direction.

$$\therefore \quad \mathbf{F}_3 = (F_{31} \cos 30 + F_{32} \cos 30)\,\mathbf{j}$$

$$\mathbf{F}_3 = \left(G\frac{2m^2}{d^2}\frac{\sqrt{3}}{2} \times 2\right)\mathbf{j}$$

$$\mathbf{F}_3 = 2\sqrt{3}\,G\frac{m^2}{d^2}\,\mathbf{j} \quad (2)$$

2.D

What is the *acceleration* of particle 3 when the system is as shown in frame 2.A?

Using (2) and Newton's second law for particle 3 gives

$$\mathbf{F}_3 = 2m\mathbf{a} = 2\sqrt{3}\frac{m^2}{d^2}\,\mathbf{j}$$

$$\therefore \quad \mathbf{a} = \sqrt{3}\frac{m}{d^2}\,\mathbf{j} \quad (3)$$

2.E

In a diagram, show the approximate direction of the resultant force on particle 1 due to particles 2 and 3.

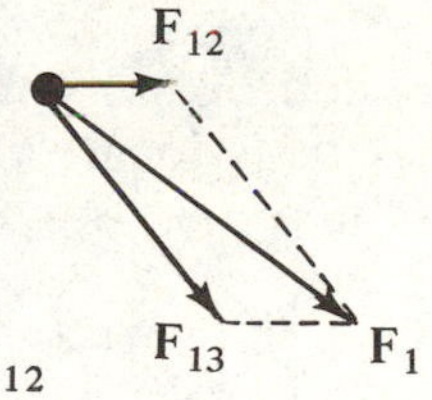

Note: $F_{13} = 2F_{12}$

$$\mathbf{F}_1 = \mathbf{F}_{12} + \mathbf{F}_{13}$$

2.F

Since the particles are "isolated" from the rest of the universe, what can you conclude about the motion of the center of mass? What about the momentum of the center of mass?

For an isolated system of particles, the center of mass is either stationary or moves with constant velocity. Therefore, the momentum of the center of mass is either zero or some constant.

2.G

In view of what was concluded in frame 2.F, what can you say about the resultant force on the system of particles?

Since $\mathbf{P}$ = constant for the system of particles,

$$\mathbf{F} = \frac{dp}{dt} = 0 \qquad (4)$$

2.H

What can you conclude about the *"internal"* forces acting between each particle, in view of (4)?

Since $\mathbf{F} = \mathbf{F}_1 + \mathbf{F}_2 + \mathbf{F}_3$, then the vector sum of the internal forces is *zero*, since $\mathbf{F} = 0$. *Note:* $\mathbf{F}_1$ means the *resultant* force on particle 1; likewise for $\mathbf{F}_2$ and $\mathbf{F}_3$.

2.I

Recall that the potential energy for any *pair* of particles separated by a distance r is given by $U = -Gm_1 m_2/r$. Use this to find the total potential energy of the system of particles shown in frame 2.A.

$$U = U_{12} + U_{13} + U_{23}$$

That is, the total P.E. is a *scalar* sum over all pairs.

$$U = -G\frac{m^2}{d} - G\frac{2m^2}{d} - G\frac{2m^2}{d}$$

$$U = -\frac{5Gm^2}{d} \qquad (5)$$

2.J

If the particles are stationary in the configuration shown in Frame 2.A, what would U represent physically?

U represents the *binding energy* of the system, or the energy required to separate the particles by an infinite distance from each other.

†3 A uniform rod of mass M is bent into the shape of a circular arc of radius R, as shown. A particle of mass m is located at the center of curvature. We wish to calculate the force on the particle due to the rod.

3.A

Is the force on the particle equal to GmM/R^2? Explain.

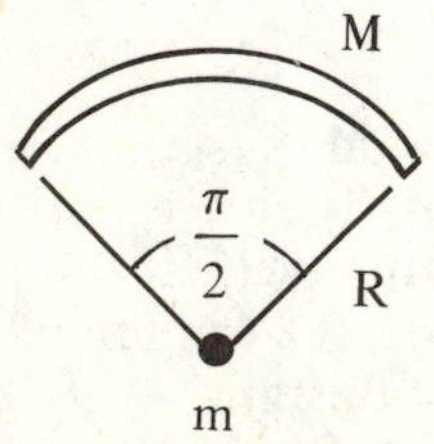

No. This equation would be valid if the bar were a point mass. This is obviously not the case.

3.B

In order to calculate the force on the particle, let us divide the bar up into segments of length ds, each segment having a mass dM. What is the force on m due to this segment?

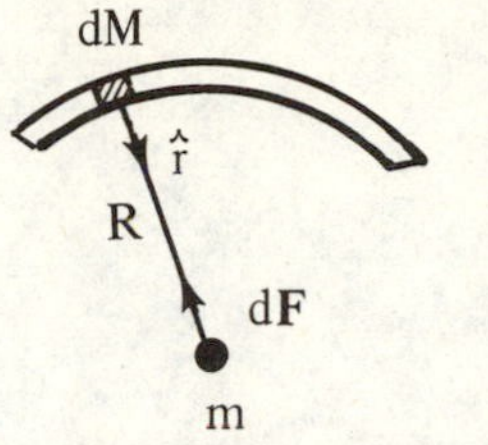

$$dF = -G\frac{mdM}{R^2}\hat{r} \qquad (1)$$

3.C

To get the total force on m, we have to integrate (1). Can this be done directly? Explain.

No. The expression is a *vector* equation and we must treat the integration with care.

3.D

Note the symmetry in the configuration about the y axis. From this observation, what can you conclude about the resultant force acting on m in the x direction?

$F_x = 0$, since the right half of the bar produces a force in the +x direction, which cancels the force in the −x direction due to the left half.

3.E

Note that we can write $dM = \frac{M}{l}\,ds$, where $l = \pi R/2$, the length of the arc. Write an expression for dF_y, that is, the force on m in the y direction due to dM.

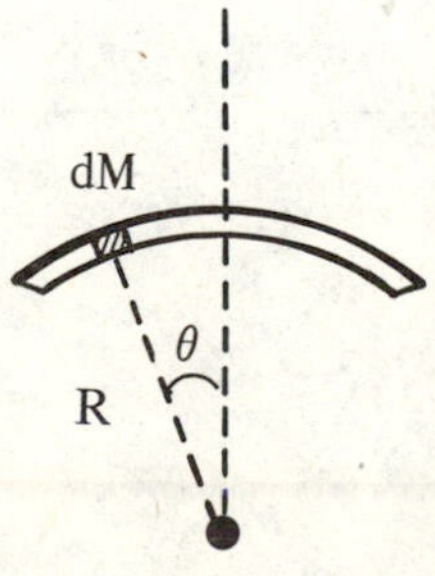

$$dF_y = dF\cos\theta$$

$$dF_y = G\frac{mdM}{R^2}\cos\theta$$

$$dF_y = \frac{GmM}{R^2 l}\cos\theta\,ds \qquad (2)$$

3.F

We are now ready to integrate (2) to get the total force on M. Since $s = R\theta$, $ds = Rd\theta$, the integral reduces to an integration over θ. Perform this integration, recalling that

$$\int \cos\theta\, d\theta = \sin\theta$$

$$F_y = \frac{GmM}{lR^2} R \int_{-\pi/4}^{\pi/4} \cos\theta\, d\theta$$

Since $l = \pi R/2$, we get

$$F_y = \frac{2\,GmM}{\pi R^2} [\sin\theta]_{-\pi/4}^{\pi/4}$$

$$F_y = \frac{2\sqrt{2}\,GMm}{\pi R^2} \qquad (3)$$

4 In this exercise, we will deal with the motion of an object along a tunnel, where the tunnel is assumed to be a straight hole dug between two points on the earth's surface. Neglecting frictional forces, the analysis will show that the object moves with simple harmonic motion between the two points, A and B.

4.A

When m is located as shown above, show a free body diagram for it, neglecting friction. Define your forces.

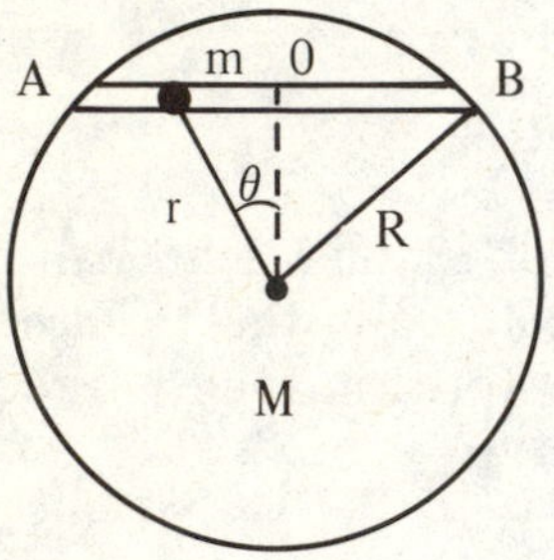

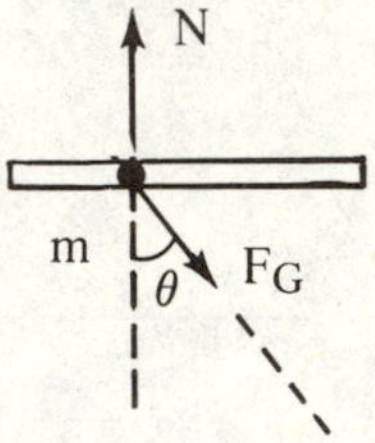

N = Normal Force

$$|F_G| = \frac{GM'm}{r^2} \qquad (1)$$

where $r < R$.

4.B

Equation (1) represents the gravitational force acting on m. What is M'?

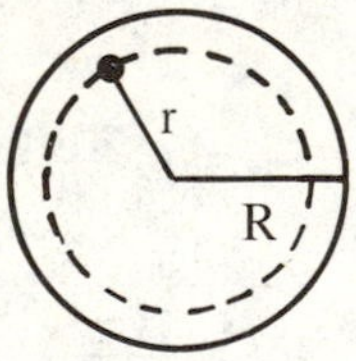

M' is that mass of the earth contained *within* a sphere of radius r. $\therefore M' < M$. In other words, as m descends below the earth's surface, the gravitational force *decreases.*

4.C

If we assume the density of the earth is *constant,* then $\rho = M'/V' = M/V$, where $V' = \frac{4}{3}\pi r^3$ and $V = \frac{4}{3}\pi R^3$. This shows that $M' = \frac{r^3}{R^3} M$. Use this fact, and rewriter (1).

$$|F_G| = \frac{GMm}{r^2}\frac{r^3}{R^3}$$

or

$$|F_G| = \frac{GMm}{R^3} r \qquad (2)$$

4.D

In the free body diagram, note that $N = F_G \cos\theta$, while $|F_x| = |F_G| \sin\theta$. The x-motion is of interest. Write an expression for F_x, using (2) and this last relation.

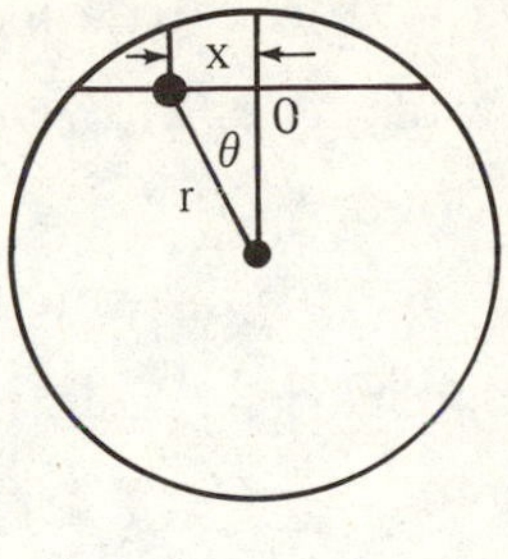

$$F_x = -F_G \sin\theta$$

Since $\sin\theta = x/r$,

$$F_x = -\frac{GMm}{R^3}x \qquad (3)$$

The *negative* sign in (3) is used here, since F_x is a *restoring* force, that is, always directed towards 0. Note that we are measuring the x displacement of m from the point 0.

4.E

Apply Newton's second law to the x-motion, using (3). What is the form of this expression?

$$F_x = -\frac{GMm}{R^3}x = m\frac{d^2x}{dt^2}$$

or

$$\frac{d^2x}{dt^2} = -\frac{GM}{R^3}x \qquad (4)$$

This expression is identical in form to the simple harmonic oscillator expression

$$\frac{d^2x}{dt^2} = -\frac{k}{m}x \qquad (5)$$

4.F

Since (4) is the oscillator equation, the mass will move along the tunnel with simple harmonic motion. What is the period of this motion? That is, how long does it take the mass m to travel from A to B and back to A?

If we compare (4) with (5), and recall that $\omega = \sqrt{k/m}$ for the simple harmonic oscillator, then

$$\omega = \sqrt{\frac{GM}{R^3}} = \frac{2\pi}{T}$$

$$\therefore \qquad T = 2\pi\sqrt{\frac{R^3}{GM}} \qquad (6)$$

4.G

Obtain a numerical value for T, using the values $R = 6.4 \times 10^6$ m and

$$M = 5.98 \times 10^{24} \text{ kg}$$

Inserting these values into (6), with $G = 6.67 \times 10^{-11}$ Nm²/kg² gives $T \cong 84$ min. Note that the result *does not* depend on the length of the tunnel!

4.H

Suppose the tunnel has a length $2x_0$. Determine the speed of the mass released from rest at A when it is at $x = 0$. Note that this is the point at which the speed is a maximum.

When the mass is at A, $x = x_0$, and it has an "effective" potential energy $\frac{1}{2}kx_0{}^2$, where $k = GMm/R^3$. [Compare (4) and (5).] At $x = 0$, the mass has kinetic energy $\frac{1}{2}mv^2$, while its P.E. is zero relative to the point at 0.

$$\therefore \quad \frac{1}{2}mv^2 = \frac{1}{2}\frac{GMm}{R^3}x_0{}^2$$

or

$$v = \left(\frac{GM}{R^3}\right)^{1/2} x_0 \qquad (7)$$

4.I

Suppose the tunnel has a length of 6.4×10^5 m (~400 mi). Determine the speed of m at $x = 0$. [Note that we can also write (6) as $v = \omega x_0$, where $\omega = 2\pi/T$, and $T \cong 84$ min.]

Using (7) with $x_0 = 3.2 \times 10^5$ m (half the length of the tunnel), together with the numerical values in frame 4.G, gives

$$v \cong 400 \text{ m/sec}$$

This is equal to about 880 mi/hr! Do you think this idea would be practical for a public transportation system?

5 A satellite of mass m orbits the earth in a *circular* path or radius r with a speed v. Let us analyze the motion of this system.

5.A

Since the satellite moves in a circular orbit, some force must provide the centripetal acceleration. What is it?

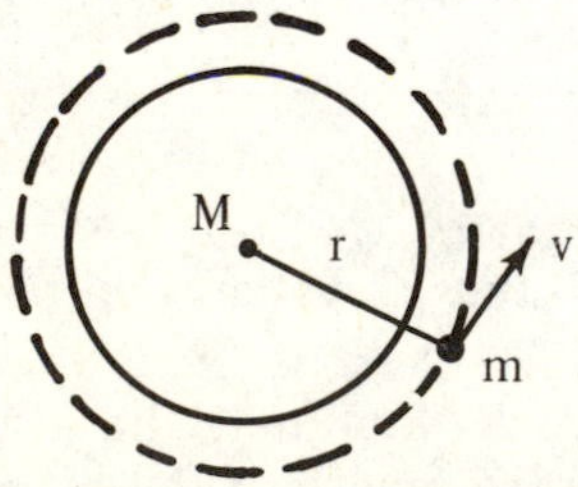

The gravitational force provides the centripetal acceleration

$$F = \frac{GMm}{r^2} \qquad (1)$$

5.B

The direction of F acting on m is *towards the center of motion.* Write Newton's second law, equating (1) to ma_r, and determine an expression for v.

$$F = \frac{GMm}{r^2} = ma_r$$

But $a_r = v^2/r$, so

$$\frac{v^2}{r} = \frac{GM}{r^2} \qquad (2)$$

or

$$v = \left(\frac{GM}{r}\right)^{1/2} \qquad (3)$$

5.C

If the period of motion is T (the time for one complete orbit), write an expression for v in terms of r and T.

Since the satellite moves a distance of $2\pi r$ in a time T, then

$$v = \frac{2\pi r}{T} \qquad (4)$$

5.D

Equate (3) to (4), and determine the relation between T and r. What law does this correspond to?

$$\frac{2\pi r}{T} = \left(\frac{GM}{r}\right)^{1/2}$$

or,

$$T^2 = \left(\frac{4\pi^2}{GM}\right) r^3 \qquad (5)$$

This is *Kepler's third law* applied to circular orbits.

5.E

By Newton's third law, the force given by (1) also acts on the *earth.* Determine the acceleration of the *earth;* show that it is *negligible* for typical values such as $m = 10^4$ kg, and $r = 8 \times 10^6$ m.

$$\frac{GMm}{r^2} = Ma_r$$

$$a_r = \frac{Gm}{r^2} \text{ (earth)}$$

$$a_r = \frac{6.67 \times 10^{-11} \times 10^4}{(8 \times 10^6)^2}$$

$$a_r \cong 10^{-20} \frac{m}{sec^2}$$

Of course, a_r is not negligible if one considers the moon as a satellite of the earth, since m is then not negligible.

5.F

Since a_r is negligibly small for the earth, its speed, and hence its kinetic energy, are small. Write an expression for the *total* energy of the satellite-earth system. What is the physical significance of E?

$$E = \frac{1}{2}mv^2 - \frac{GMm}{r} \qquad (6)$$

This represents the energy required to separate the satellite from the earth by an infinite distance.

5.G

From (2) we see that $v^2 = GM/r$. Use this in (6) and obtain another expression for E, eliminating v.

$$E = \frac{1}{2}m\left(\frac{GM}{r}\right) - \frac{GMm}{r}$$

or

$$E = -\frac{GMm}{2r} \qquad (7)$$

5.H

Note that (7) gives a *negative* value for E. What is the physical significance of this sign?

This means that the earth-satellite system represents a *bound* system. In other words, the *negative* potential energy is necessarily twice as large in magnitude as the positive kinetic energy for circular motion.

5.I

What is the trend for v as r gets smaller or larger?

As r gets smaller, v gets *larger* according to (3). As $r \to \infty$, $v \to 0$. This is consistent with the fact that $U \to 0$ and $K \to 0$ as $r \to \infty$. $\therefore E \to 0$ as $r \to \infty$, as predicted by (7).

5.J

How much energy is required to take the satellite from an orbit of radius r to one of radius 3r?

From (7) we have

$$E(r) = -\frac{GMm}{2r}$$

$$E(3r) = -\frac{GMm}{2(3r)}$$

$\therefore$ The required energy is

$$E(3r) - E(r) = -\frac{GMm}{6r} + \frac{GMm}{2r}$$

$$E = \frac{GMm}{3r}$$

5.K

Let us go back to Kepler's third law, that is, frame 5.D. Rewriting (5) as $T^2/r^3 = 4\pi^2/GM$, we note that the ratio T^2/r^3 is a *constant* for all satellites of the *earth*. Evaluate this constant, and note that M is the mass of the earth.

$$G = 6.67 \times 10^{-11}\ \frac{Nm^2}{kg^2}$$

$$M = 5.98 \times 10^{24}\ kg$$

$$\frac{4\pi^2}{GM} = \frac{4\pi^2}{6.67 \times 10^{-11} \times 5.98 \times 10^{24}\ \frac{Nm^2}{kg}}$$

$$\frac{4\pi^2}{GM} \cong 9.9 \times 10^{-14}\ \frac{sec^2}{m^3} \qquad (8)$$

5.L

A typical satellite of the earth moves in an orbit 100 miles above the earth. What is the period of this orbit? Note that $r = R + h$, where R is the radius of the earth given by 6.37×10^6 m. Also, 1 mile = 1.61×10^3 m.

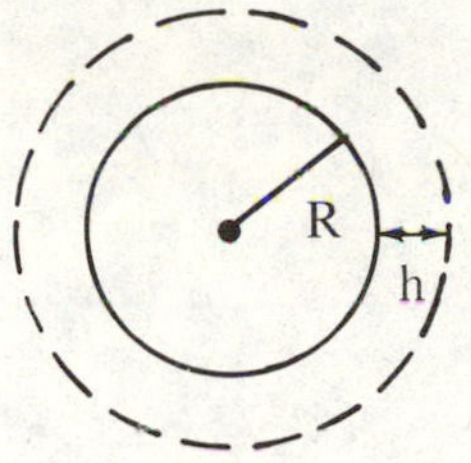

$h = 100$ mi

$h = 1.16 \times 10^5$ m

$r = R + h = 6.53 \times 10^6$ m

Using (5) and (8) gives

$$\frac{T^2}{r^3} = 9.9 \times 10^{-14}\ \frac{sec^2}{m^3}$$

$$T^2 = 9.9 \times 10^{-14}\ \frac{sec^2}{m^3}\ (6.53 \times 10^6)^3\ m^3$$

$$T^2 = 27.6 \times 10^6\ sec^2$$

$$T = 5.25 \times 10^3\ sec \approx 88\ min$$

5.M

Evaluate the orbital speed of the satellite described in frame 5.L. Obtain an answer in mi/hr, and note that 1 m/sec = 2.24 mi/hr.

$$v = \frac{2\pi r}{T} = \frac{2\pi \times 6.53 \times 10^6\ m}{5.25 \times 10^3\ sec}$$

$$v = 7.8 \times 10^3\ \frac{m}{sec} \cong 17{,}500\ \frac{mi}{hr}$$

5.N

With some modifications, (5) can be applied to any satellite of the sun, that is, one of the planets. What are these modifications?

First, the mass M is now the mass of the sun. Second, since the orbits are *elliptical*, r is replaced by the semimajor axis, a. The student should show that for Mars, where $a = 2.28 \times 10^{11}$ m, the period is 5.94×10^7 sec.

5.O

Calculate the *minimum* speed a rocket orbiting the earth must have to *escape* the gravitational field of the earth. Note that in order for the rocket to just reach ∞, its *total* energy must be *zero,* since U (∞) = 0 and K (∞) = 0. Assume h = 100 miles, so r = R + h = 6.53 × 10⁶ m.

$$E = \frac{1}{2} mv^2 - \frac{GMm}{r} = 0$$

$$\therefore \; v_{escape} = \sqrt{\frac{2\,GM}{r}}$$

$$v_{escape} = \sqrt{\frac{2 \times 6.67 \times 10^{-11} \times 5.98 \times 10^{24}}{6.53 \times 10^6}}$$

$$v_{escape} = 1.1 \times 10^4 \text{ m/sec}$$

$$v_{escape} \cong 24{,}600 \text{ mi/hr}$$

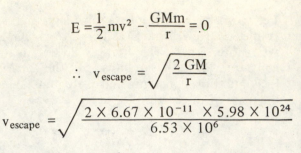

11.6 SUMMARY

The force of attraction between any two *particles* of masses m_1 and m_2 separated by a distance r is given in magnitude by

$$F = G\frac{m_1 m_2}{r^2} \tag{11.1}$$

where $G = 6.673 \times 10^{-11}\ Nm^2/kg^2$. Note that Equation (11.1) does *not* apply to bodies of finite extent. For finite sized bodies, one must use integral expressions to get the resultant force.

The gravitational potential energy of a system of two particles of masses m_1 and m_2 separated by r is

$$U = -G\frac{m_1 m_2}{r} \tag{11.9}$$

For those problems involving elliptical (or circular) motions of planets about the sun (or satellites about the earth), the gravitational forces are *central* and *conservative.* Consequently, the *angular momentum* of any planet relative to the sun (or any satellite relative to the earth) is *constant.*

Second, the period of motion of any planet (or any satellite of the earth) is given by

$$T^2 = \frac{4\pi^2}{GM} r^3 \tag{11.12}$$

Finally, the total energy of these two-body systems is constant, and is given by

$$E = \frac{1}{2}mv^2 - \frac{GMm}{r} = \text{constant} \tag{11.13}$$

where $M \gg m$, and the mass M is assumed to be stationary. For bound systems, the total energy is *negative* and is given by

$$E = -\frac{GMm}{2r} \tag{11.15}$$

When a problem involves *elliptical* orbits, replace r by a in Equations (11.12) and (11.15), where a is the semimajor axis.

11.7 PROBLEMS

1. The masses of the earth and sun are 5.98×10^{24} kg and 1.97×10^{30} kg, respectively. Assuming the orbit of the earth is circular, with a period of 365 days, determine (a) the radius of the earth's orbit about the sun and (b) the force between the earth and sun.

2. Two stars of masses M and 4M are separated by a distance d. Determine the position of a point measured from M where the net force on a third mass is zero.

3. Four particles are positioned at the corners of a rectangle as shown in Figure 11-8. Determine the *components* of the resultant force acting on the particle whose mass is m.

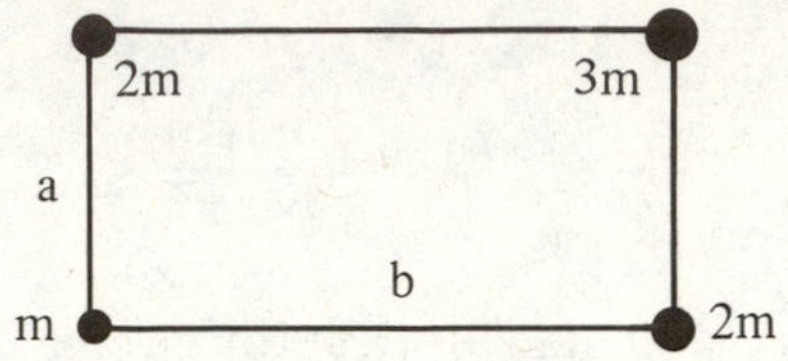

Figure 11-8

4. Calculate the total potential energy for the system of four particles shown in Figure 11-8.

5. An experiment which demonstrates the variation of g with altitude is the following: A simple pendulum is measured to have a period of 1 sec at sea level, where $g = 9.80 \text{ m/sec}^2$. The pendulum is taken to the top of a mountain, where its period is measured to be 1.01 sec. (a) What is effective gravitational acceleration at the top of the mountain? (b) What is the height of the mountain? Take the radius of the earth to be 6.37×10^6 m.

†6. A uniform rod of M is bent into the shape of a semicircle of radius R as shown in Figure 11-9. What is the force between this object and a point mass m placed at 0?

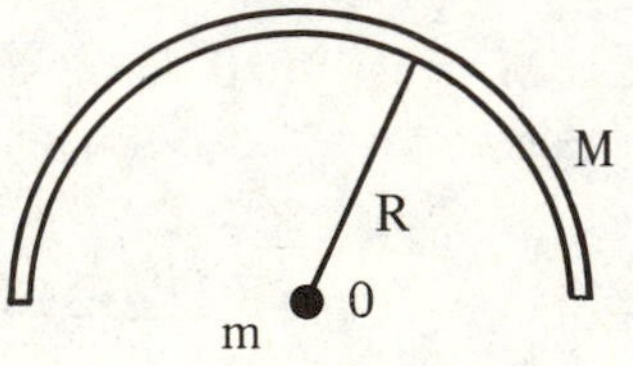

Figure 11-9

†7. The density of a certain sphere of mass M and radius R varies with the distance from its center according to the expression $\rho = \rho_0 r^2$, for $0 \leqslant r \leqslant R$. (a) What is the constant ρ_0 in terms of M and R? [Recall that $M = \int_0^R \rho dV$.] (b) Determine the force on a particle of mass m when placed *outside* the sphere. (c) Determine the force on the particle if it is *inside* the sphere.

8. A satellite orbits the earth in an elliptical path in a time of 6×10^3 sec. If the perigee of the orbit is at a distance of 2×10^5 m from the earth's *surface,* how far is the apogee from the earth? *Hint:* First calculate the length of the semimajor axis. The radius of the earth is 6.4×10^6 m.

9. Two astronauts, each of the same mass M, are situated *opposite* each other in a gravity free orbiting space station. The room they are in is a cylinder whose radius is R, as in Figure 11-10. What is the *minimum* angular speed of the cylinder which will keep the astronauts from moving towards each other? (Assume they are unfastened to their seats, but their seats are tied down.)

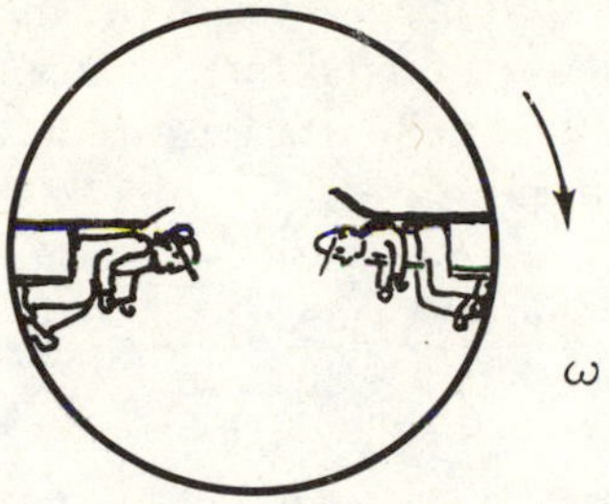

Figure 11-10

10. In a certain attempted landing on a peculiar *hollow* planet a vertically descending unmanned space vehicle loses its power while under the influence of the planet's gravity. This unfortunate incident happens just above a large hole in the planet, as in Figure 11-11. Describe the fate of the space vehicle.

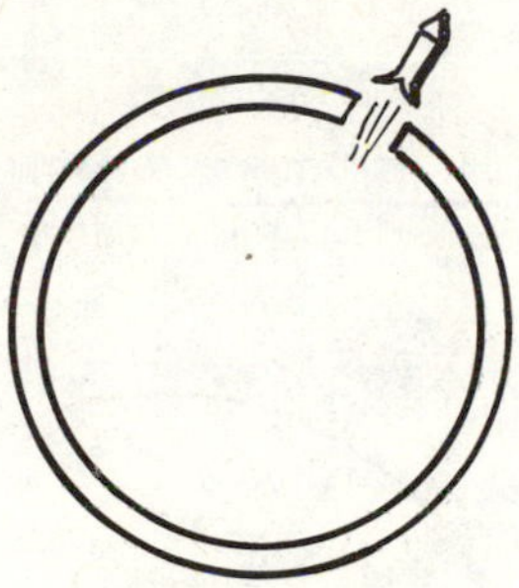

Figure 11-11

11. What is the acceleration of gravity at a point R_E above the earth's surface, where R_E is the radius of the earth?

12. A satellite of mass 2000 kg orbits the earth in a circular path at a distance of 0.2×10^6 m above the surface of the earth. The mean radius of the earth is 6.4×10^6 m and the mass of the earth is 5.98×10^{24} kg. Determine the period of the satellite from this data.

13. What *total* energy must be supplied to the satellite described in Problem 12 to move it an infinite distance from the earth?

14. All planets revolve around the sun in elliptical orbits with the sun at one focus. Kepler's third law states that $T^2/a^3 = 4\pi^2/GM_s$ = constant, where a is the length of the semimajor axis and M_s is the mass of the sun given by 1.97×10^{30} kg. (a) Determine the value of the constant $4\pi^2/GM_s$. (b) What is the period of revolution of Mercury, for which $a = 5.79 \times 10^{10}$ m?

15. Calculate the escape velocity from Mars. The mass of Mars is $0.106\ M_E$, and its radius is 3.43×10^6 m.

16. A particle of mass 2.0 kg is separated by 4×10^5 m from a uniform spherical body of mass 1.0×10^{24} kg and radius 1×10^4 m. Both bodies are initially at rest. Determine the speed of the 2 kg mass when it is 2×10^4 m from the large mass. [*Hint:* Neglect the motion of the large mass and use energy methods.]

17. Prove Kepler's second law, which states that the radius vector from the sun to each planet sweeps out equal areas in equal times. Use the fact that the gravitational force between any two point masses is a central force, hence the angular momentum L of the planet is conserved. In particular, show that if ΔA is the area swept out by r in a time Δt, and m is mass of the planet, then

$$\lim_{\Delta t \to 0} \frac{\Delta A}{\Delta t} = \frac{dA}{dt} = \frac{L}{2m} = \text{constant}$$

18. Two stars of comparable masses M and m, separated by a distance d, rotate in circular orbits about their center of mass as shown in Figure 11-12. Show that each star has a period given by

$$T^2 = \frac{4\pi^2}{G(M + m)} d^3$$

Hint: The center of mass condition requires that $Mr_1 = mr_2$. Also, from Newton's second law, and the universal law of gravity,

$$\frac{Mv_1^2}{r_1} = \frac{mv_2^2}{r_2} = \frac{GMm}{d^2}$$

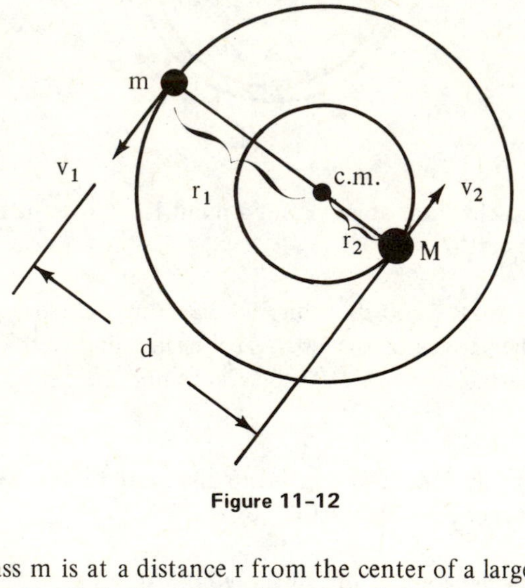

Figure 11-12

19. A point mass m is at a distance r from the center of a large uniform solid sphere of radius R and mass M. Find the potential energy of the system for (a) $r > R$, (b) $r = R$ and (c) $r < R$. Make use of the definition

$$U(r) - U(\infty) = -\int_{\infty}^{r} \mathbf{F} \cdot d\mathbf{r}, \text{ where } U(\infty) = 0$$

Also, note that for $r < R$, F is given by Equation (11.8).

20. A circular ring of mass M, radius a, lies in the yz plane, as in Figure 11-13. A point mass m is at a distance d from the ring along the x axis as shown. Prove that the force on m is given by

$$\mathbf{F} = \frac{GMmd}{(a^2 + d^2)^{3/2}}\,\mathbf{i}$$

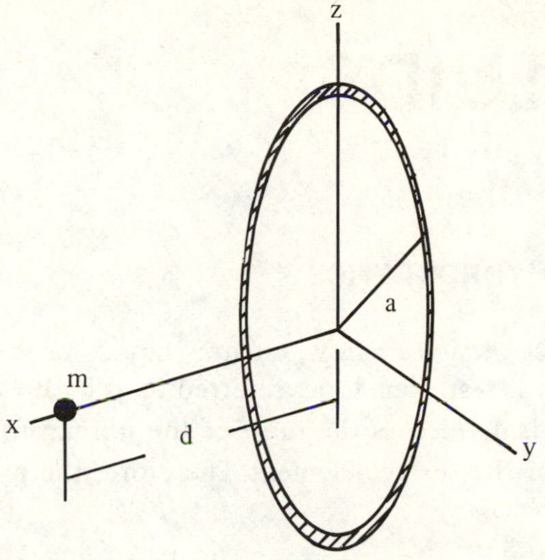

Figure 11-13

21. A planet of mass M has three moons of equal mass m located in an equilateral triangle and rotating in circular orbits of radius R, as in Figure 11-14. (a) Find the total potential energy of the system. (b) Determine the orbital speed of each moon such that they each remain in the orbit of radius R.

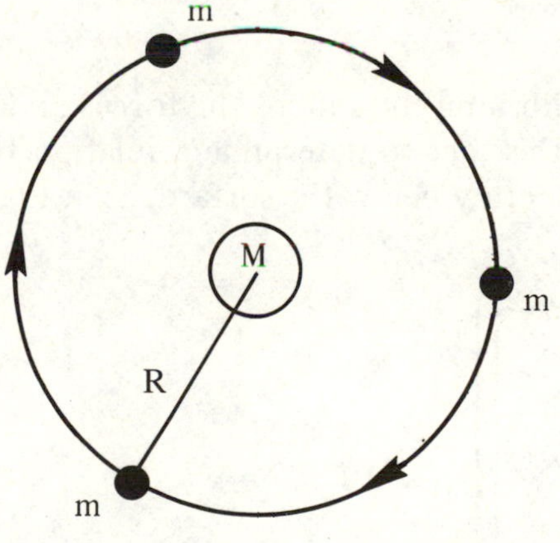

Figure 11-14

12

PHYSICS OF FLUIDS

12.1 FLUIDS AT REST—HYDROSTATICS

Fluids are any substances that can flow, such as liquids or gases. We will first describe the nature of fluids at rest, sometimes referred to as hydrostatics.

The *pressure* of a fluid is defined as the ratio of the normal force per unit area ΔA, where ΔA is the area of the surface element. Therefore, the pressure at a given point on the surface is

$$p = \lim_{\Delta A \to 0} \frac{\Delta F}{\Delta A} = \frac{dF}{dA} \qquad (12.1)$$

Pressure has the unit force/area, or N/m^2 in the SI system.

The *density* ρ of any homogeneous substance is defined as its mass per unit volume.

$$\rho = \frac{m}{V} \qquad (12.2)$$

When a fluid is in equilibrium, the sum of the forces on any element of the fluid must be zero. We can use this fact to determine a relation between the pressure at any point in terms of the depth y below the surface, as in Figure 12-1. If the weight

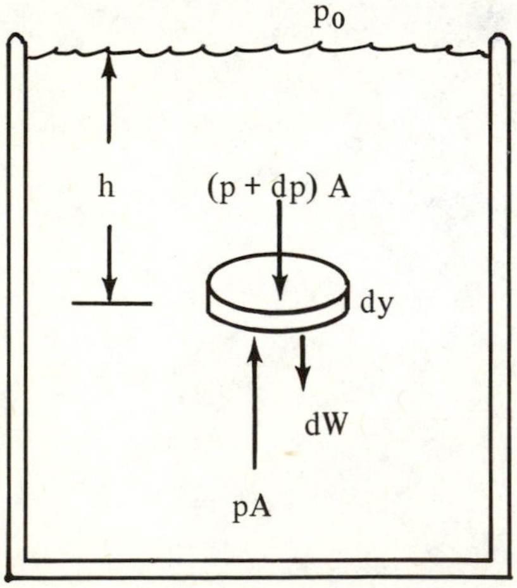

Figure 12-1 A fluid element in equilibrium.

of the cylindrical element is dW, and the difference in pressure between the lower face and the upper face is dp, then the sum of the forces on the element gives

$$pA - (p + dp)A - dW = 0 \tag{12.3}$$

But the volume of the element is Ady; therefore, $dW = \rho A dy$, so Equation (12.3) reduces to

$$\frac{dp}{dy} = -\rho g \tag{12.4}$$

The negative sign in Equation (12.4) indicates that the pressure is larger at the bottom of the element than at the top. If h is taken to be the depth of the element below the surface, and p_0 is taken to be atmosphere pressure ($14.7\ \text{lb/in}^2 \cong 1.01 \times 10^5\ \text{N/m}^2$), integration of Equation (12.4) gives

$$p = p_0 + \rho g h \tag{12.5}$$

In other words, the *absolute* pressure p at a depth h below the surface of a liquid is greater than atmospheric pressure by an amount $\rho g h$. The pressure difference $p - p_0$ is referred to as *gauge pressure.*

Example 12.1

(a) Calculate the absolute pressure at an ocean depth of 1000 m (about 0.62 miles). The density of water is about $1.0 \times 10^3\ \text{kg/m}^3$. (b) Calculate the total force which would be exerted on a 30 cm diameter window of a submarine at this depth.

Solution

(a) $p = p_0 + \rho g h = 1.01 \times 10^5 \frac{N}{m^2} + 1 \times 10^3 \times 9.8 \times 10^3 \frac{N}{m^2}$

$p \cong 9.9 \times 10^6 \frac{N}{m^2}$

(b) $F = pA = p\pi r^2 = 9.9 \times 10^6 \frac{N}{m^2} (\pi \times 0.15^2)\ m^2$

$F \cong 7.0 \times 10^5\ N \cong 157{,}000$ lb!

The calculation of F assumes that the pressure on the inner face of the window is zero. However, even if we take the inner pressure to be 1 atmosphere, F is still *very large.* Obviously, the design and construction of windows and vessels to withstand such high pressures is not a trivial matter.

There is a useful fact called *Pascal's law* which states that *pressure applied to an enclosed fluid is transmitted undiminished to every portion of the fluid and the walls of the containing vessel.* As an example, if atmospheric pressure in Equation (12.5) changes by an amount Δp_0, the change in pressure at a depth h is $\Delta p = \Delta p_0$. The principle is useful in lifting heavy weights, such as automobiles, using a hydraulic press. A small force is applied to a piston of small cross-section. This applied pressure is transmitted uniformly through an enclosed fluid to a large cross-section piston, which in turn lifts the heavy load.

Another useful law of hydrostatics is *Archimedes' principle,* which states that *when an object is partially or fully submersed in a fluid, the fluid exerts a force on the object equal to the weight of the fluid displaced by the object.* This *buoyant* force acts vertically upward through the center of gravity of the object if it is at *rest.* If the object is not at rest, the buoyant force acts through the center of gravity of the fluid before it is displaced.

Example 12.2

A boy weighing 100 lb makes a raft out of solid Styrofoam whose dimensions are 1 m X 1 m X 0.06 m. If the Styrofoam "just" supports the boy in water, as shown in Figure 12-2, determine the density of the Styrofoam.

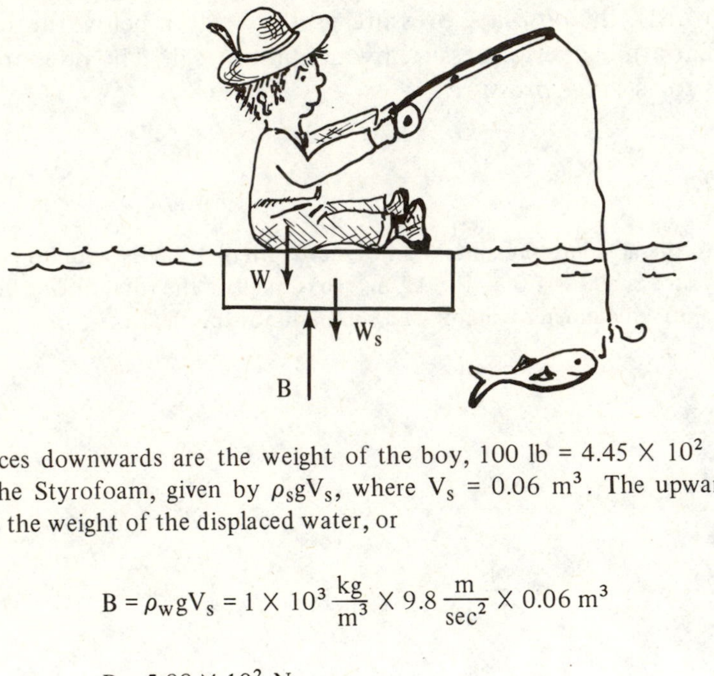

Figure 12-2

The forces downwards are the weight of the boy, 100 lb = 4.45×10^2 N, and the weight of the Styrofoam, given by $\rho_s g V_s$, where $V_s = 0.06\ m^3$. The upward buoyant force equals the weight of the displaced water, or

$$B = \rho_w g V_s = 1 \times 10^3 \frac{kg}{m^3} \times 9.8 \frac{m}{sec^2} \times 0.06\ m^3$$

$$B = 5.88 \times 10^2\ N$$

Setting the sum of the forces equal to zero gives

$$B - W - W_s = 0$$

$$\therefore\ 5.88 \times 10^2 - 4.45 \times 10^2 - \rho_s g V_s = 0$$

$$\rho_s g V_s = 1.43 \times 10^2\ N \qquad \text{or} \qquad \rho_s = 2.44 \times 10^2 \frac{kg}{m^3}$$

That is, the density of the Styrofoam is about one fourth the density of water.

12.2 FLUIDS IN MOTION–HYDRODYNAMICS

The treatment of fluids in motion, or hydrodynamics, can be a rather complex subject, unless some simplifying assumptions are made. For the present treatment of hydrodynamics we will assume the following for our "ideal liquid": (1) the flow is *steady,* that is, the fluid velocity at a given point is constant in time; (2) the flow is *irrotational,* that is, there is no vortex motion; (3) the fluid flow is *nonviscous,* so that internal frictional forces are neglected; and (4) the fluid is *incompressible,* hence there are no variations in its density. We are now in a position to discuss the flow of this so-called streamline flow as illustrated in Figure 12-3.

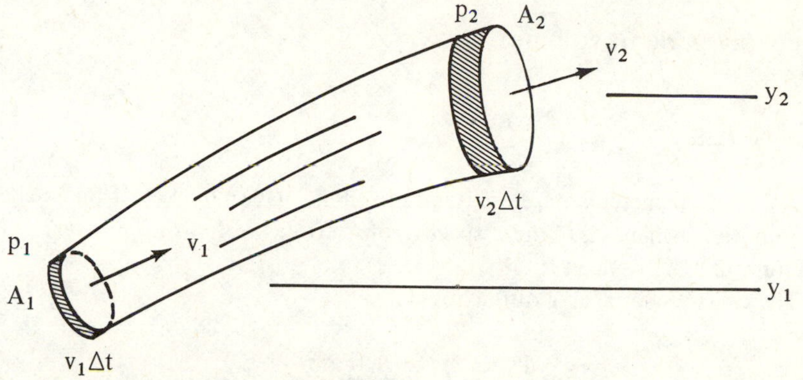

Figure 12-3 Streamline flow of an "ideal fluid" through a tube.

There are two basic laws one can obtain for this type of flow. The first is the *equation of continuity* which is a consequence of the fact that mass is conserved; that is, whatever mass passes through the cross-section A_1 in a time Δt also passes through A_2 in the same time. Since the mass crossing A_1 is $\rho_1 A_1 v_1 \Delta t$, while the mass crossing A_2 is $\rho_2 A_2 v_2 \Delta t$, we get

$$\rho_1 A_1 v_1 \Delta t \cong \rho_2 A_2 v_2 \Delta t$$

The equality is exact as $\Delta t \to 0$. Therefore,

$$\rho_1 A_1 v_1 = \rho_2 A_2 v_2 = \text{constant} \tag{12.6}$$

If the fluid is *incompressible,* then $\rho_1 = \rho_2$ and Equation (12.6) becomes

$$Av = \text{constant} \tag{12.7}$$

That is, the product Av is a *constant everywhere* along the tube. In other words, v is *large* at points where the tube is constricted, and vice versa. Note that Av has units of volume/time, or it represents a *flow rate,* sometimes given the symbol Q.

The second law of hydrodynamics for our "ideal fluid" is *Bernoulli's equation,* perhaps the most fundamental relation in fluid mechanics. This equation is a statement of *conservation of mechanical energy* as applied to a fluid, and is given as

$$p + \frac{1}{2}\rho v^2 + \rho g y = \text{constant} \tag{12.8}$$

where p is the absolute pressure at an arbitrary point in the tube, v is the fluid velocity at that point and y is the elevation of that point above (or below) some reference level. If we apply this expression to our fluid flowing through the tube in Figure 12–3, we get

$$p_1 + \frac{1}{2}\rho v_1{}^2 + \rho g y_1 = p_2 + \frac{1}{2}\rho v_2{}^2 + \rho g y_2 \qquad (12.9)$$

Note that if the fluid is at rest, $v_1 = v_2 = 0$, and Equation (12.9) reduces to Equation (12.5), which is our basic law of hydrostatics. Therefore, the pressure $p + \rho gy$ is referred to as the *static pressure*, while the term $\frac{1}{2}\rho v^2$ is called the *dynamic pressure* (or kinetic energy per unit volume).

Example 12.3

A 1 inch diameter water hose is used to fill a 5 gallon bucket. If the bucket is filled in 1 minute, what is the speed of efflux of the water? Note that 1 gallon = 2.31 in³ = 1.34 ft³.

The cross-sectional area of the hose is

$$A = \pi \frac{d^2}{4} = \frac{\pi}{4}\left(1 \text{ in } \frac{1 \text{ ft}}{12 \text{ in}}\right)^2 = 5.5 \times 10^{-3} \text{ ft}^2$$

The flow rate is given by Equation (12.7).

$$Av = 5.5 \times 10^{-3} \text{ ft}^2 \text{ (v)} = \frac{5 \text{ gal} \times 1.34 \text{ ft}^3/\text{gal}}{60 \text{ sec}}$$

$$\therefore \qquad v \cong 20 \text{ ft/sec}$$

Example 12.4

A tank containing a liquid of density ρ has a small hole in it at a distance y above the bottom, as in Figure 12-4. The atmosphere at the top portion of the tank is maintained at a pressure p. Determine the velocity at which the fluid leaves the hole when the liquid level is a distance h above the bottom.

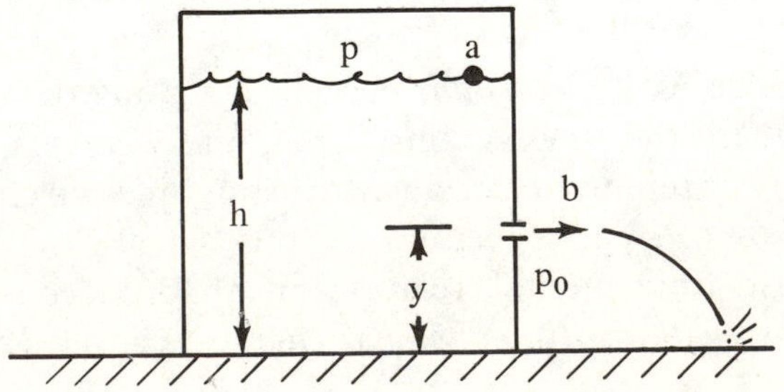

Figure 12–4

If we assume the tank is large, compared to the hole, the fluid will be approximately at rest at the top. Applying Bernoulli's equation to the points a and b (where b is at atmospheric pressure, p_0), gives

$$p + \rho gh = p_0 + \rho gy + \frac{1}{2}\rho v^2$$

$$\therefore \quad v = \sqrt{\frac{2(p - p_0)}{\rho} + 2g(h - y)}$$

If the tank is open to atmosphere at the top, then $p = p_0$ and $v = \sqrt{2g(h - y)}$. This is known as *Torricelli's law*. If the area of the hole is A, the discharge rate (or leak rate) is simply Av. The student should show that the horizontal distance which the water travels before hitting the ground is $2\sqrt{yh}$ when $p = p_0$.

12.3 PROGRAMMED EXERCISES

1 A U-shaped tube containing a liquid (usually mercury or oil) with one end opened to the atmosphere is commonly used in the laboratory to measure absolute pressures. One side of such a device is connected to a vacuum pump as shown below.

1.A

Before the valve is opened, each side of the manometer is at atmospheric pressure p_0, so the fluid level is the same on either side. Describe what happens when the valve is opened.

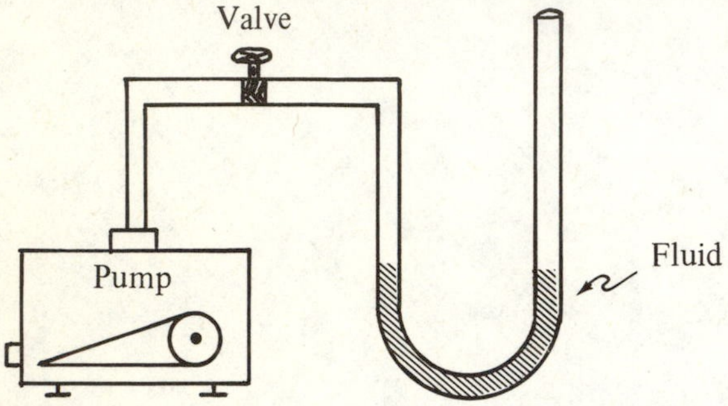

The pump reduces the pressure on the left side to $p < p_0$, so the liquid rises on the left and falls on the right. The valve should be opened *slowly* so as not to draw the liquid into the pump.

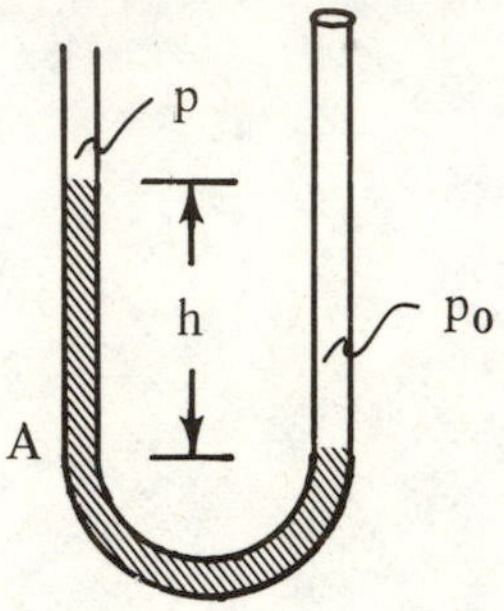

1.B

What is the absolute pressure at the point A, a distance h below the liquid at the left?

The pressure at A is p_0, since it is at the same elevation as the liquid at the right.

1.C

Determine the absolute pressure p on the left in terms of p_0, ρ and h. Assume the fluid is in equilibrium.

When the fluid is *at rest,* we have

$$p_A = p + \rho g h \qquad (1)$$

But $p_A = p_0$, so

$$p = p_0 - \rho g h \qquad (2)$$

1.D

When mercury is used as the liquid, it is convenient to express the absolute pressure in mm of Hg (where $p/\rho g$ has this unit). Atmospheric pressure is equal to 760 mm of Hg. What would p be if h = 760 mm?

p = 0. This is the principle of the mercury barometer. Of course, in any real system, p is never identically zero, although vacuums have been obtained as good as $\sim 10^{10}$ mm Hg.

1.E

What would the absolute pressure be on the left if h = 750 mm Hg? Express your answer in lb/in^2, and recall that 1 atm = 14.7 lb/in^2 = 760 mm Hg.

From (2), $\frac{p}{\rho g} = \frac{p_0}{\rho g} - h$

$$\frac{p}{\rho g} = (760 - 750) \text{ mm Hg}$$

$$\frac{p}{\rho g} = 10 \text{ mm Hg}$$

This corresponds to a pressure

$$p = \frac{10}{760} \times 14.7 \frac{lb}{in^2} = 0.19 \frac{lb}{in^2}$$

2 A large balloon, whose mass is 20 kg, is filled with helium gas until its volume is 400 m^3. Helium has a density of 0.178 kg/m^3, while the air is assumed to have a density of 1.30 kg/m^3.

2.A

Show all the external forces acting on the balloon. *Neglect* air resistance (a gross over-simplification).

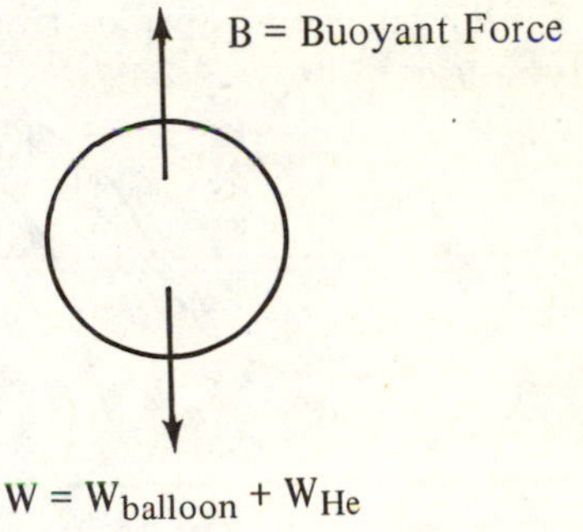

2.B

Calculate the buoyant force, that is, the force of the air on the balloon, or the so-called "lift."

B = weight of displaced air

$$B = \rho_{air} V_{balloon} g$$

$$B = 1.30 \frac{kg}{m^3} \times 4 \times 10^2 \, m^3 \times 9.8 \frac{m}{sec^2}$$

$$B = 5.1 \times 10^3 \text{ N}$$

2.C

Determine the resultant force on the balloon. Note that the weight of the helium is $\rho_{He} V g = 0.7 \times 10^3$ N, while the weight of the balloon is $\sim 0.2 \times 10^3$ N.

$$B - W = (5.1 \times 10^3 - 0.9 \times 10^3) \text{ N}$$

$$F \cong 4.2 \times 10^3 \text{ N upwards}$$

that is, the buoyant force is larger than W, so the balloon will *rise*.

2.D

Suppose the "payload" (added weight) is to be such that a = g/4 upwards. What must the mass of the payload be? Call the mass of the payload M. Note that:

$$m = m_{balloon} + m_{He} = \frac{W}{g}$$

$$m = \frac{0.9 \times 10^3 \text{ N}}{9.8 \text{ m/sec}^2}$$

$$m \cong 92 \text{ kg}$$

$$B - (m + M)\,g = (m + M)\,a$$

$$B = (m + M)\,(a + g)$$

$$m + M = \frac{B}{a + g} = \frac{B}{5g/4}$$

$$\therefore 92 + M = \frac{5.1 \times 10^3}{5\left(\frac{9.8}{4}\right)} = 4.2 \times 10^2$$

$$M \cong 328 \text{ kg}$$

An average man has a mass of around 70 kg, so he has plenty of room for supplies and ballast.

3.A

An "ideal fluid" flows through a constricted tube as shown below. What do we mean by "ideal fluid" in our simple treatment of hydrodynamics?

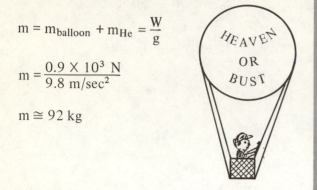

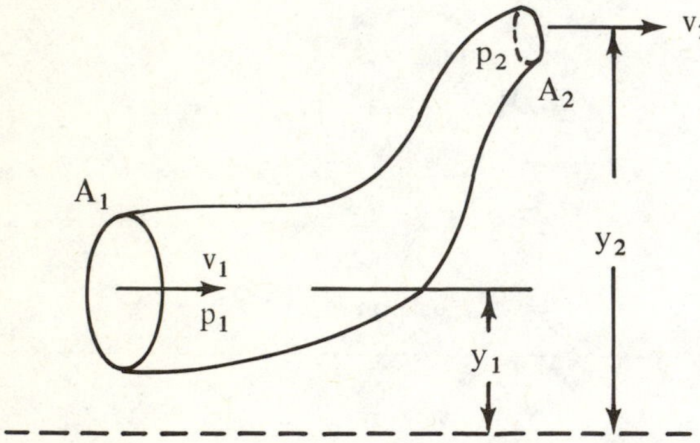

We mean that the fluid is *nonviscous, incompressible, irrotational* and hence moves in steady flow along so-called streamlines. That is, the velocity of each point over a given cross-section is the same.

3.B

If the velocity of the fluid is v_1 where the cross-sectional area is A_1, what is the velocity v_2 at a point where the area is A_2? What law does this expression represent?

$$A_1 v_1 = A_2 v_2 \qquad (1)$$

$$\therefore \quad v_2 = \frac{A_1}{A_2} v_1 \qquad (2)$$

Equation (1) represents *conservation of mass,* that is, the mass of fluid that passes A_1 in a time Δt must equal the mass that passes A_2 in the same time.

3.C

Show that the unit of Av is volume/time, that is, Av represents the *rate of flow* of fluid.

$$[Av] = L^2 \times \frac{L}{T} = \frac{L^3}{T}$$

3.D

Write Bernoulli's equation as applied to the lower and upper portions of the tube shown in frame 3.A.

In general,

$$p + \frac{1}{2}\rho v^2 + \rho g y = \text{constant} \qquad (3)$$

$$p_1 + \frac{1}{2}\rho v_1{}^2 + \rho g y_1 = p_2 + \frac{1}{2}\rho v_2{}^2 + \rho g y_2 \qquad (4)$$

3.E

Discuss the physical significance of Equation (3) and the meaning of each term in the expression.

Equation (3) is a statement of *conservation* of *mechanical energy* for our "ideal fluid." p is the absolute pressure, $\frac{1}{2}\rho v^2$ represents the kinetic energy per unit volume (or dynamic pressure) and $\rho g y$ is the potential energy per unit volume.

3.F

Pressure has units of force/area, (or N/M^2 in SI). Show that the terms $\frac{1}{2}\rho v^2$ and $\rho g y$ also have units of force/area. Note that

$$\left[\frac{\text{Force}}{\text{Area}}\right] = \frac{ML/T^2}{L^2} = \frac{M}{LT^2}$$

$$\left[\frac{1}{2}\rho v^2\right] = \frac{M}{L^3} \times \frac{L^2}{T^2} = \frac{M}{LT^2}$$

$$[\rho g y] = \frac{M}{L^3} \times \frac{L}{T^2} \times L = \frac{M}{LT^2}$$

3.G

A constricted pipe has two vertical tubes open to the *atmosphere* as shown. If the fluid velocity at the left is 5 m/sec, what is the fluid velocity at the right and what is the discharge rate?

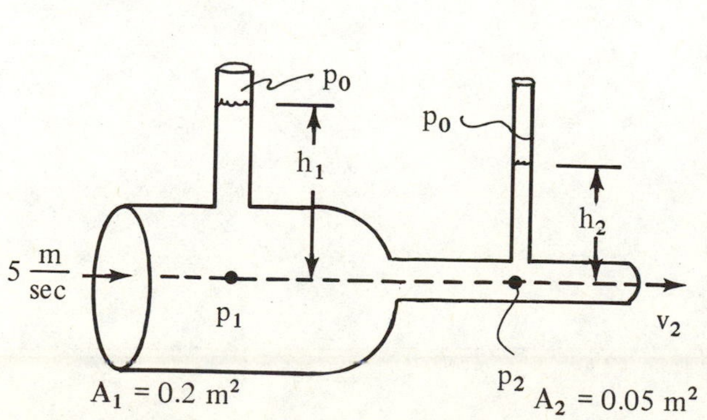

Applying (2) gives

$$v_2 = \frac{A_1}{A_2} \times v_1 = \frac{0.2}{0.05} \times 5\ \frac{m}{sec}$$

$$v_2 = 20\ m/sec$$

Discharge rate = $A_1 v_1 = A_2 v_2$

$$Q = 5 \times 0.2 \frac{m^3}{sec} = 1\ \frac{m^3}{sec}$$

3.H

Assuming the fluid is water ($\rho = 1 \times 10^3$ kg/m^3), determine the absolute pressure p_1 if the *gauge pressure* at the constricted part is 3.01×10^5 N/m^2. Recall that atmospheric pressure is 1.01×10^5 N/m^2.

Gauge pressure equals $p_1 - p_2$; therefore, $p_1 = 2.0 \times 10^5$ N/m^2. Since the point in question is at the same level as the constricted part, (4) gives

$$p_1 + \frac{1}{2}\rho v_1^2 = p_2 + \frac{1}{2}\rho v_2^2$$

$$p_1 = p_2 + \frac{1}{2}\rho (v_2^2 - v_1^2)$$

$$p_1 = 2.0 \times 10^5 + \frac{1 \times 10^3}{2}(20^2 - 5^2)$$

$$p_1 = 3.87 \times 10^5 \text{ N/m}^2$$

3.I

To what height h_1 will the water rise in the left vertical tube? Assume the water in the vertical tube to be at rest. This devise, called a *Venturi meter,* is used to measure flow velocities of liquids.

Since the tube is open to atmosphere, we have

$$p_1 = p_0 + \rho g h_1$$

$$\therefore \quad h_1 = \frac{p_1 - p_0}{\rho g} = \frac{2.86 \times 10^5}{1 \times 10^3 \times 9.8}$$

$$h_1 \cong 29 \text{ m}$$

Likewise, the student should show that $h_2 = 20.4$ m.

12.4 SUMMARY

The absolute pressure p at a point a distance h below the surface of a stationary fluid is given by

$$p = p_0 + \rho gh \tag{12.5}$$

where p_0 is the pressure at the surface of the liquid and ρ is its density.

Archimedes' principle states that an object partially or fully submerged in a fluid is buoyed up by a force equal to the weight of the displaced fluid.

For steady streamline flow of an "ideal fluid," that is, irrotational, steady and nonviscous flow of an incompressible fluid, we can write the laws of conservation of mass and conservation of energy as

$$Av = \text{constant} \tag{12.7}$$

and

$$p + \frac{1}{2}\rho v^2 + \rho gy = \text{constant} \tag{12.8}$$

Equation (12.8), the law of conservation of energy, is commonly referred to as *Bernoulli's equation.*

12.5 PROBLEMS

1. The U-shaped tube shown in Figure 12-5 contains colored water. The left side is connected to a pressurized tank, while the right side is open to atmosphere. What is the absolute pressure on the left if h = 20 cm? What is the gauge pressure?

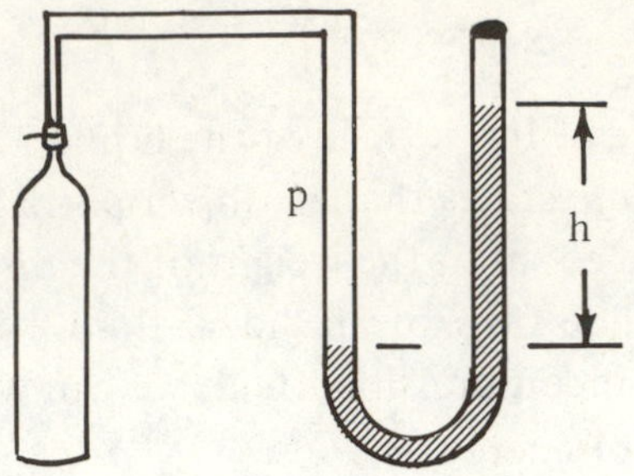

Figure 12-5

2. A hollow plastic ball has a radius of 5 cm and a mass of 10 g. The ball has a tiny hole at the top through which lead shot can be inserted. How many grams of lead (density ~11.3 g/cm^3) can be put into the ball before it sinks in the water? Assume the ball does not leak.

3. The Eisenhower lock at the St. Lawrence Seaway has a depth of about 50 ft, a width of 100 ft and a length of 500 ft (these are estimates). (a) When the lock is full of water, what is the total force on the end walls where the gates are located? (*Hint:* The pressure varies with depth, so the total force must be obtained by integration of dF = pdA.) (b) the lock empties (and fills) in about 10 min. What flow rate does this correspond to in ft^3/sec? How many gal/sec does this correspond to? (*Note:* 1 gal = 0.134 ft^3.)

4. An aluminum boat of mass M is designed with a compartment containing Styrofoam of density ρ_s. Show that the minimum *volume* of Styrofoam necessary to just keep the boat from sinking when filled with water is given by $V_s = M/(\rho_w - \rho_s)$, where ρ_w is the density of water. *Note:* The total volume of the boat equals the volume occupied by the Styrofoam plus the volume occupied by the water.

5. A mercury barometer, a device used to measure atmospheric pressure, is shown in Figure 12-6. The bulb above the mercury column is evacuated and sealed. Show that at standard atmospheric conditions, the mercury will rise to a height of 76 cm. That is, a 76 cm column of Hg exerts a pressure of 1 atmosphere at A. *Note:* The density of mercury is 13.6 g/cm^3.

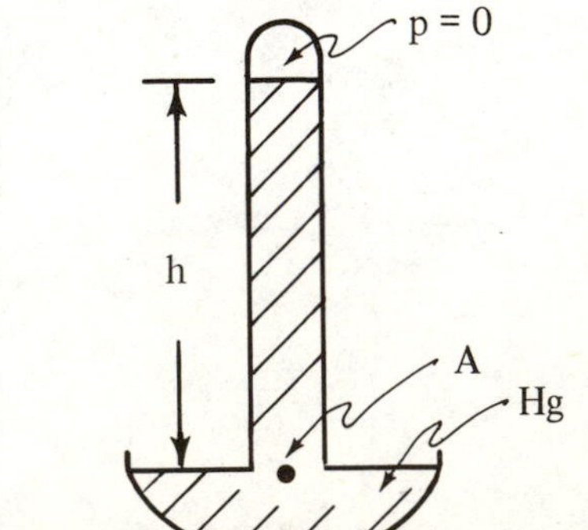

Figure 12-6 Barometer.

6. A block of metal of dimensions 8 cm × 10 cm × 10 cm is submersed in water and suspended from a string, as in Figure 12-7. The "true" mass of the block is 2 kg. (a) What is the reading of the spring scale? (b) What are the forces on the top and bottom of the block? (c) Show that the difference in the forces calculated in (b) equals the buoyant force. (Take $p_0 = 1.01 \times 10^5$ N/m^2).

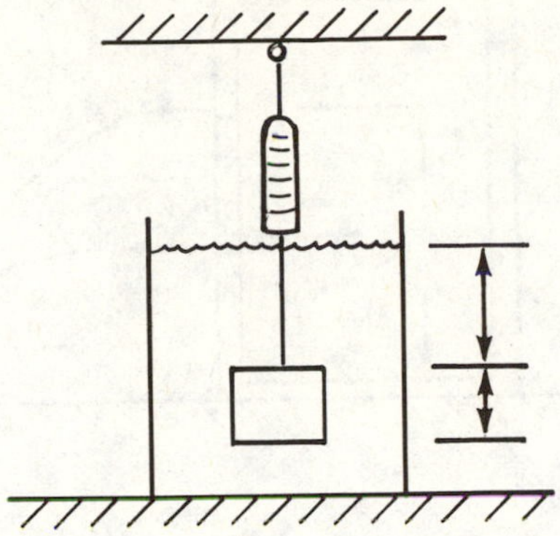

Figure 12-7

7. Assume that the decrease in atmospheric pressure with altitude is given by Equation (12.4) and that the density ρ is proportional to the absolute pressure; show that the variation of pressure with altitude if $p = p_0 e^{-\alpha y}$, where $\alpha = g\frac{\rho_0}{p_0}$, p_0 is atmospheric pressure at sea level and ρ_0 is the corresponding atmospheric density.

8. The water supply of a building is fed through a 3 in diameter pipe. A 1 in diameter faucet located 30 ft above the 3 in main line is observed to fill a 5 gallon container in 2 minutes. (a) What is the speed of the water as it leaves the faucet? (b) What is the *gauge pressure* in the 3 in main line? Assume that the faucet is the only "leak" in the building.

9. A *syphon*, a device which allows fluid to flow out of a tank under atmospheric pressure, is shown in Figure 12-8. Show that the velocity of the fluid leaving the pipe is $v = \sqrt{2gh}$. That is, v is independent of y. Of course, the flow must be initiated by a partial vacuum in the tube, say by suction as in a straw. Also, flow will not persist if y is so large that the absolute pressure at the top of the tube falls to zero. [Incidentally, it has been told that gasoline tastes terrible.]

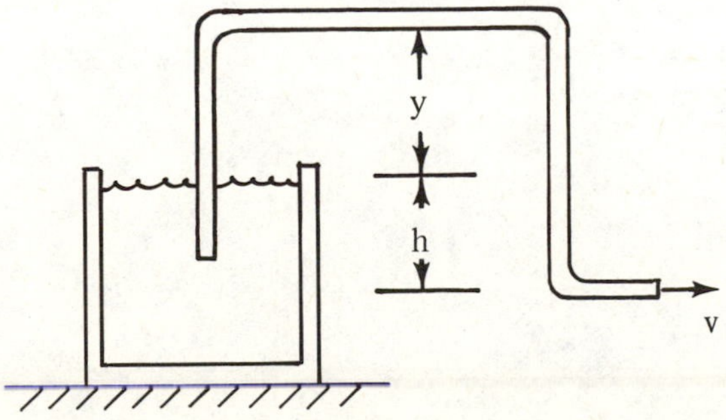

Figure 12-8 A syphon.

10. A large tank contains two holes, one directly above the other as in Figure 12-9. The upper hole has an area 2A, while the lower one has an area A. (a) Determine the velocities of efflux from the two holes. (b) Calculate the distances x_1 and x_2, measured from the base of the tank, at which each stream hits the floor. (c) What is the total discharge rate from the two holes?

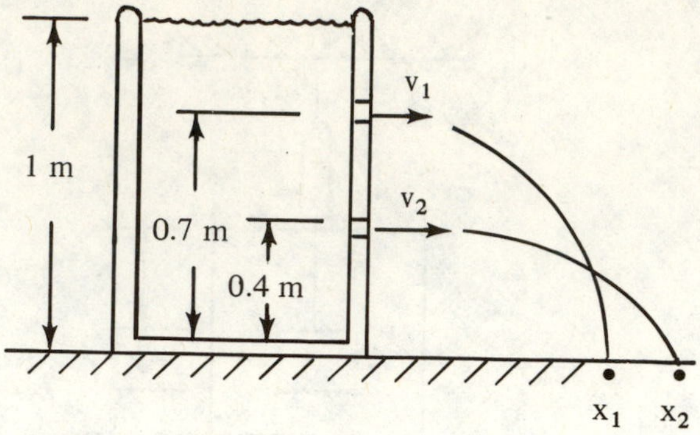

Figure 12-9

13

THERMAL PHYSICS

Thermal physics is the study of the behavior of solids, liquids and gases, using the concepts of heat and temperature. Two approaches are commonly used in the treatment of this science. The first is a *macroscopic* approach, or the field of *thermodynamics,* whereby the bulk thermal properties of matter are explained. These macroscopic properties include pressure, volume and temperature. The second is a *microscopic* approach, or *statistical mechanics,* whereby thermal properties are explained on an atomic scale. Here, the description includes the velocities, masses and energies of the particles of the system. Both approaches require some basic definitions, such as the concepts of temperature and heat. On the atomic scale, we will see that thermal energy (or heat energy) is actually the mechanical energy of a large number of particles making up the system. In other words, all thermal phenomena are manifestations of the laws of mechanics as we have learned them.

13.1 TEMPERATURE AND THE ZEROTH LAW OF THERMODYNAMICS

The concept of the *temperature* of a system can be understood in connection with a measurement, such as the reading of a thermometer. Temperature, a scalar quantity, is a property which is definable only when the system is in thermal equilibrium with another system. Thermal equilibrium implies that two (or more) systems are at the same temperature. The *zeroth law of thermodynamics* states that if two systems are in thermal equilibrium with a third *system,* they are in thermal equilibrium with each other. The third system is usually a calibrated *thermometer* (of which there are many types) whose reading determines whether or not other systems are in thermal equilibrium.

The gas *thermometer* is a standard device used for the definition of temperature. A low density gas is placed in a flask, and its volume is kept constant while it is heated (the constant volume gas thermometer). The pressure is measured as the gas is heated or cooled. Experimentally, the pressure rises as heat is applied, and vice versa. If we define temperature to be proportional to the absolute pressure, then

$$T = \alpha p \tag{13.1}$$

This expression defines the *absolute temperature* if the gas behaves in an "ideal" fashion. By ideal, we mean that the average distance between gas molecules is large compared to molecular dimensions, and the potential energy of interaction is negligible. At very low densities, all gases behave as ideal gases. The constant α that appears in SI units in Equation (13.1) has the value

$$\alpha = \frac{273.15°K}{p_t} \tag{13.2}$$

where p_t is the gas pressure at the *triple point* of water (the point where all three phases of water coexist in equilibrium). Different gases have different values of p_t. The temperature 273.15°K (where °K stands for degrees Kelvin) is the temperature of the triple point of water.

Other temperature scales include the Celsius temperature scale (°C) and the Fahrenheit scale (°F). The Celsius scale is related to the Kelvin scale by the expression

$$T(°C) = T(°K) - 273.15 \tag{13.3}$$

where 0°C corresponds to 273.15°K.

The Fahrenheit scale is related to the Celsius scale through the transformation

$$T(°F) = \frac{9}{5}T(°C) + 32 \tag{13.4}$$

where 0°C = 32°F and 100°C = 212°F. The Kelvin and Celsius scales are used most often in research.

As an exercise, the student should show that *absolute zero,* or 0°K, corresponds to −273.15°C or −459.67°F. Also, show that 40°C = 313.15°K = 104°F.

13.2 PROPERTIES OF SOLIDS AND LIQUIDS

When solids or liquids are heated, they expand as a result of the change in equilibrium separations between the atoms or molecules of the substance. If the temperature changes are small, then the *change in volume* of the substance is proportional to the change in temperature according to the relation

$$\Delta V = \bar{\beta} V_0 \Delta T \tag{13.5}$$

where V_0 is the original volume and $\bar{\beta}$ is the *average volume coefficient of thermal expansion.* This coefficient β is sensitive to changes in temperature, but at some temperature T, the *volume coefficient of thermal expansion* is given as

$$\beta = \frac{1}{V}\left(\frac{dV}{dT}\right)_p \tag{13.6}$$

where the subscript p means that the coefficient is measured at constant pressure. We see from Equation (13.6) that β has units of inverse temperature.

Example 13.1

Mercury has an average volume expansion coefficient of $\bar{\beta} = 18 \times 10^{-5}\ (°C)^{-1}$. Therefore, if we heat 20 cm^3 of Hg from 27°C (room temperature) to 37°C, $\Delta T = 10°C$, and Equation (13.5) gives $\Delta V = 3.6 \times 10^{-2}\ cm^3$. This corresponds to a *fractional* change in volume of $\Delta V/V = 1.8 \times 10^{-3}$.

If a body is shaped like a long rod with small cross-section, then its change in length over a small temperature interval is given as

$$\Delta L = \bar{\alpha} L_0 \Delta T \tag{13.7}$$

where $\bar{\alpha}$ is the *average linear coefficient of thermal expansion* and L_0 is the original length. At some temperature T, α is defined as

$$\alpha = \frac{1}{L}\frac{dL}{dT} \tag{13.8}$$

If the substance is isotropic, so that the *linear coefficient of thermal expansion,* α, is the same in all directions, then β and α are related by the expression

$$\beta = 3\alpha \tag{13.9}$$

Likewise, it follows that the *change* in *area* of an isotropic solid for a given change in temperature is given by

$$\Delta A = 2\alpha A \Delta T \tag{13.10}$$

Example 13.2

A thin copper plate has a length and width of 12 cm and 8 cm, respectively, at 27°C. Copper has a linear expansion coefficient of $17 \times 10^{-6}\ (°C)^{-1}$. If the plate is heated from 27°C to 100°C, its *change* in area, from Equation (13.10), is given by

$$\Delta A = 2\alpha A \Delta T = 2 \times 17 \times 10^{-6}\ (°C)^{-1} \times 96\ cm^2 \times 73°C$$

$$\Delta A = 0.24\ cm^2 \quad \textit{or} \quad \frac{\Delta A}{A} \cong 2.5 \times 10^{-3}$$

The *compressibility,* κ, of a material is a constant which is a measure of the *change in volume* of the material for a given *change in pressure.* It is defined as

$$\kappa = -\frac{1}{V}\frac{dV}{dp} \tag{13.11}$$

where the negative sign insures that κ is always a positive number. That is, if the pressure increases, dp is positive, while dV is negative (the material shrinks), so κ is positive. Likewise, if the pressure decreases, dp is negative, dV is positive, so κ is still positive. The *inverse* of κ is called the *bulk modulus*, B. The units of κ are $(\text{pressure})^{-1}$; B has units of pressure.

Example 13.3

The compressibility of mercury is $3.7 \times 10^{-11}\ m^2N^{-1}$. Therefore, if we subject 50 cm^3 of Hg to an *increase* in pressure of 1 atm $\cong 1.01 \times 10^5\ N/m^2$, the *decrease* in its volume from Equation (13.12) is calculated to be $\Delta V = -\kappa V \Delta p$, or

$$\Delta V = -3.7 \times 10^{-11}\ m^2N^{-1} \times 5 \times 10^{-5} m^3 \times 1.01 \times 10^5\ \frac{N}{m^2}$$

$$\Delta V = -1.87 \times 10^{-10}\ m^3$$

This corresponds to a fractional change of $\Delta V/V \cong 3.74 \times 10^{-5}$.

13.3 THE CONCEPT OF HEAT

When two systems at different temperatures are in contact with each other, energy will transfer between the two systems until they are at the same temperature (that is, when they are in equilibrium with each other). This energy is called heat, and the "flow of heat" implies an energy transfer as a consequence of the temperature difference. On an atomic scale, heat is transferred as a result of multiple atomic or molecular collisions. For example, when a hot liquid is mixed with a cold liquid, a "cold" molecule gains energy, while a "hot" molecule loses energy when they undergo a collision with each other.

The unit of heat is defined as the calorie(cal), where 1 cal = 4.184J = 4.184×10^7 erg. The kcal is more commonly used, where the conversion is 1 kcal = 4.184×10^3 J. The ratio 4.184×10^3 J/kcal is referred to as the *mechanical equivalent of heat.* The kcal is approximately equal to the heat required to raise the temperature of 1 kg of water from 14.5 to 15.5°C. The student should not confuse the term calorie as it is used in designating food ratings. One food calorie = 1 kcal.

Example 13.4

A man has a dinner which is rated at 2000 food calories. He wishes to "work this off" by lifting a 50 kg mass in the gym. How many times must he raise and lower the weight to expend this much energy? Assume he raises the weight a distance of 1 m each time.

First, the work required is

$$W = 4.184 \times 10^3\ \frac{J}{kcal} \times 2000\ kcal = 8.368 \times 10^6\ J$$

The work done by the man against gravity each time he lifts and lowers the weight (through a total distance of 2 m each time) is

$$w = mgd = 50\ kg \times 9.8\ \frac{m}{sec^2} \times 2\ m = 9.8 \times 10^2\ J$$

Therefore, the number of times he performs this feat to do an equivalent work W is

$$n = \frac{W}{w} = \frac{8.368 \times 10^6}{9.8 \times 10^2} \cong 8540 \text{ times}$$

Assuming he has a lot of stamina, and does this every 2 sec, it would take him 4270 sec ≅ 71 min to work 2000 food calories off in this manner. Obviously, it's much easier to lose weight by dieting.

The *heat capacity* C of a substance is the ratio of the heat ΔQ supplied to its temperature increase ΔT.

$$C = \frac{\Delta Q}{\Delta T} \tag{13.12}$$

The *specific heat* c of a substance whose mass is m is defined as the heat capacity per unit mass.

$$c = \frac{C}{m} = \frac{1}{m}\frac{\Delta Q}{\Delta T} \tag{13.13}$$

Since the specific heat is in general a function of temperature, the total heat that must be supplied to a substance to raise its temperature from T_0 to T can be obtained by integrating Equation (13.13) in the lim $\Delta T \rightarrow 0$. This gives

$$Q = m \int_{T_0}^{T} c dT \tag{13.14}$$

By definition, heat capacity has units of cal/°C, while specific heat has units of cal/g°C.

Example 13.5

An unknown substance has a mass of 100 g. It is raised to a temperature of 200°C, and dropped into a calorimeter (a device for measuring heat capacities). The calorimeter is made of aluminum, whose mass is 200 g, and is filled with 500 g of water. If the temperature of the calorimeter increases from 25°C to 30°C, determine the specific heat of the unknown. Assume there is no heat loss to the surroundings, and the specific heats for aluminum and water are 0.215 cal/g°C and 1.00 cal/g°C, respectively.

Since the heat lost by the unknown must equal the heat gained by the aluminum and water (conservation of energy), we have

$$Q_x = Q_{Al} + Q_{H_2O}$$

$$(mc\Delta T)_x = (mc\Delta T)_{Al} + (mc\Delta T)_{H_2O}$$

$$100\, c_x\, (200 - 30) = [200 \times 0.215 \times (30 - 25) + 500 \times 1.0 \times (30 - 25)] \text{ cal}$$

or,

$$c_x = 0.16 \text{ cal/g°C}$$

A *phase change* in a substance is accompanied by heat transfer and a change in volume. Examples of phase changes include vaporization (liquid to gas), fusion (liquid to solid), sublimation (solid to gas) and changes in solid state structure. The energy associated with the vaporization process is called *heat of vaporization,* while the energy associated with melting of a solid (or the reverse) is called *heat of fusion.* The terminology *latent heat* or *heat of transformation* is used to designate both heats of fusion and vaporization, using the symbol L. Therefore, the heat absorbed or liberated by a substance of mass m during a change of phase is

$$Q = mL \tag{13.15}$$

Example 13.6

Liquid helium has a very low boiling point, 4.2°K, and a very small heat of vaporization, 5.0 kcal/kg. A constant power of 10 watts (where 1 watt = 1 J/sec) is supplied to a resistor immersed in the helium fluid. How long does it take to boil away 1 kg of helium at this rate? [Liquid helium has a density of 0.1251 g/cm^3, so 1 kg corresponds to 8000 cm^3 or 8 liters of liquid.]

We have to supply 5.0 kcal of energy to 1 kg of liquid helium to boil it away. The mechanical equivalent of 5 kcal is

$$5.0 \text{ kcal} = 5.0 \times 4.184 \times 10^3 \text{ J} = 2.1 \times 10^4 \text{ J}$$

Therefore, since 1 watt = 1 J/sec, the time it would take is

$$t = \frac{2.1 \times 10^4 \text{ J}}{10 \text{ J/sec}} = 2.1 \times 10^3 \text{ sec} \cong 35 \text{ min}$$

Since 1 kg corresponds to about 8 liters of liquid, this means a "boil-off" rate of about 0.23 liters per minute! By comparison, 1 kg of liquid nitrogen, which has a boiling point of 77°K and a heat of vaporization of 48 kcal/kg, would take about 3.4 hours to boil away at the rate of 10 J/sec. One kg of boiling water (L = 539 kcal/kg) would require about 38 hours!

13.4 HEAT TRANSFER PROCESSES

There are three basic processes of heat transfer. These are (1) conduction, (2) convection and (3) radiation. We will give a rather qualitative description of these processes.

Conduction is a heat transfer process which occurs when there is a *temperature gradient* across the body. Hence, conduction of heat occurs only when the body is not at a uniform temperature. For example, if we heat a metal rod at one end with a Bunsen burner, heat will flow from the hot end to the colder end. If the heat flow is along x (that is, along the rod), and we define the *temperature gradient* as dT/dx, then a quantity of heat dQ will flow in a time dt along the rod. The rate of flow of heat along the rod, or the *heat current,* H, is given by

$$H = -kA\frac{dT}{dx} \tag{13.16}$$

where A is the cross-sectional area of the rod and k is a positive constant called the *thermal conductivity*. In steady-state, H is the same for all points along the rod. Also, if the rod has a uniform cross-section, dT/dx is constant along the rod; therefore, the temperature gradient is $(T_2 - T_1)/l$, where l is the length of the rod and T_1, T_2 are the temperatures at its ends.

When heat transfer occurs as the result of motion of material, such as in the mixing of hot and cold fluids, the process is referred to as *convection*. Other examples of convection heat are hot-air and hot-water heating systems. Convection currents are also common in the "mixing" of warm and cold air regions of the atmosphere, which in turn produces changes in weather conditions.

Heat transfer by *radiation* is a form of heat transfer common to all bodies. *Radiant energy* is continually emitted by all bodies in the form of electromagnetic waves of varying wavelength. For example, when we see that the heating element on the electric range is "red hot," we are observing electromagnetic radiation emitted by the hot element. The energy which reaches the earth from the sun is radiant energy, and techniques for harnessing this solar energy for practical purposes are of current interest. The rate of emission of radiant energy from the surface of a body is given by *Stefan's law*, which states

$$R = e\sigma T^4 \tag{13.17}$$

R is the *radiant emittance*, which represents the rate of energy per unit area ($J/sec\ m^2$), σ is a constant equal to 5.6699×10^{-5} in SI units, T is the temperature in °K and e is a constant called the *emissivity*. The value of e can vary between zero and unity, depending on the surface. In reality, a body will both radiate and abosrb electromagnetic radiation. If it is at equilibrium with its environment, it will radiate and absorb energy at the same rate, so its temperature doesn't change. If it is hotter than its surroundings, it will radiate more energy than it absorbs, so it cools. An *ideal radiator* (or ideal black body, $e \cong 1$), is one which absorbs all of the energy incident on it (and reflects none). Therefore, it is also a good emitter of radiant energy. On the other hand, a highly reflecting surface ($e \cong 0$) is a poor absorber and a poor emitter of radiation. For these reasons, a white house is cooler in the summer, and light colored clothing is preferable in the summer. On the other hand, dark colors are preferable in the winter if warmth is desired.

13.5 MACROSCOPIC DESCRIPTION OF AN IDEAL GAS

If a *dilute* gas occupies a volume V, the relation between the macroscopic variables p, V and T at equilibrium obeys the equation of state for an *ideal gas:*

$$pV = \mu RT \tag{13.18}$$

where μ is the number of moles of gas, and R is the *universal gas constant* given by

$$R = 8.3149\ J/mole\ °K \tag{13.19}$$

The value given for R in Equation (13.19) assumes that μ is expressed in g-moles. If kg-moles are used for μ, then R = 8.3149 × 10^3 J/kg-mole °K. To a good approximation, all real gases follow the behavior of an ideal gas at low densities and pressures.

Example 13.7

Pure oxygen gas is admitted into a container whose volume is 0.3 m^3. The absolute pressure and temperature are measured to be 500 N/m^2 and 300°K, respectively. (a) How many kg of oxygen are in the container? (b) If the volume of the container is decreased by means of a piston to 0.2 m^3, and the pressure increases to 900 N/m^2, what is the final temperature of the gas? (Assume it behaves like an ideal gas, and no gas escapes.)

Solution

(a) The number of moles of gas is given by $\mu = m/M$, where m = the mass of the gas and M is its molecular weight. Since the molecular weight of oxygen is 32 (each molecule contains two atoms), substitution into Equation (13.18) gives

$$500 \frac{N}{m^2} \times 0.3\ m^3 = \frac{m}{32\ g/mole} \times 8.3149 \frac{J}{mole\ ^\circ K} \times 300^\circ K$$

or,

$$m = 1.92\ g = 1.92 \times 10^{-3}\ kg$$

(b) Since we assume that no gas escapes, μ remains constant, and we have

$$\frac{pV}{T} = \frac{p_0 V_0}{T_0}$$

where p_0, V_0 and T_0 are initial values. Solving for T gives

$$T = \frac{pV}{p_0 V_0} T_0 = \frac{900 \times 0.2}{500 \times 0.3} \times 300^\circ K = 360^\circ K$$

13.6 MICROSCOPIC DESCRIPTION OF AN IDEAL GAS

An ideal gas can be quantitatively described on a microscopic scale using the assumptions that: (1) there are a large number of atoms or molecules making up the gas; (2) the molecules obey Newton's laws and move in a random fashion; (3) forces act on the molecules only during collisions; and (4) the collisions are elastic in nature and short in duration. If each molecule has a mass m, and there are N molecules occupying a volume V, the pressure they exert on the walls of the container due to collisions with the wall is given by

$$p = \frac{2}{3}\frac{N}{V} \times \frac{1}{2} m < v^2 > \qquad (13.20)$$

The derivation of this formula is given in most standard physics texts. One obtains the result by calculating the impulsive force per unit area, due to the collisions of the molecules with the walls. In other words, the pressure is proportional to the average kinetic energy per molecule, $\frac{1}{2}m < v^2 >$, where $< v^2 >$ means the average of the square of the velocity. Since $N \times \frac{1}{2} m < v^2 >$ is the *total* internal energy, U, we can write Equation (13.20) as

$$pV = \frac{2}{3}U \tag{13.21}$$

If we compare Equation (13.21) with $pV = \mu RT$ for an ideal gas, we get for the internal energy

$$U = \frac{3}{2}\mu RT \tag{13.22}$$

Consequently, the average kinetic energy per molecule is

$$\frac{1}{2}m < v^2 > = \frac{U}{N} = \frac{3}{2}\frac{\mu RT}{N} \tag{13.23}$$

But $\mu = N/N_0$, where N_0 is Avogadro's number, $N_0 = 6.02 \times 10^{23}$ molecules/mole. If we define Boltzmann's constant k as

$$k = \frac{R}{N_0} = 1.38 \times 10^{-23}\ \text{J/°K} \tag{13.24}$$

we can write Equation (13.23) as

$$\frac{1}{2}m < v^2 > = \frac{3}{2}kT \tag{13.25}$$

This important result says that the average kinetic energy per molecule depends *only* on T. It also provides a physical description of temperature as the average translational kinetic energy per molecule times the constant 2/3k.

The root mean square speed, v_{rms}, can be obtained from Equation (13.25) directly.

$$v_{rms} = \sqrt{< v^2 >} = \sqrt{\frac{3kT}{m}} \tag{13.26}$$

The student should show that v_{rms} can also be expressed as $v_{rms} = \sqrt{3p/\rho}$, where ρ is the density of the gas given by $\rho = Nm/V$.

Example 13.8

A tank of volume 2 m^3 contains 5 moles of helium gas at a temperature of 0°C. (a) Determine the total internal energy of the system. (b) What is the average kinetic energy per atom? (c) Determine the rms speed of the helium atoms. (d) What is the absolute pressure of the gas?

Solution

(a) From Equation (13.22), with $\mu = 5$, $T = 273\ ^\circ K$, we get

$$U = \frac{3}{2}\mu RT = \frac{3}{2} \times 5 \text{ moles} \times 8.3149 \frac{J}{\text{mole}^\circ K} \times 273^\circ K = 1.70 \times 10^4 \text{ J}$$

(b) The average kinetic energy per atom, from Equation (13.23), is

$$\frac{1}{2} m < v^2 > = \frac{U}{N} = \frac{U}{\mu N_0} = \frac{1.70 \times 10^4 \text{ J}}{5 \times 6.02 \times 10^{23}} = 5.65 \times 10^{-21} \text{ J}$$

(c) Since the helium atom has a mass number of 4, its mass is four times greater than the mass of hydrogen.

$$\therefore\ v_{rms} = \sqrt{\frac{3kT}{m}} = \sqrt{\frac{3 \times 1.38 \times 10^{-23} \times 273}{4 \times 1.67 \times 10^{-27}}} \cong 1.30 \times 10^3 \frac{m}{sec}$$

(d) From Equation (13.21) and (a), we get

$$p = \frac{2}{3}\frac{U}{V} = \frac{2}{3}\left(\frac{1.70 \times 10^4 \text{ J}}{2 \text{ m}^3}\right) = 5.67 \times 10^3 \text{ N/m}^2 \cong 0.056 \text{ atm}$$

13.7 MOLAR SPECIFIC HEAT OF A GAS AT CONSTANT VOLUME

We define the *molar specific heat* c_v of a gas at *constant volume* as the quantity of heat which must be transferred to one mole of a gas to increase its temperature by one degree. Following the definition of specific heat, we get

$$\Delta Q = \mu c_v \Delta T \qquad (13.27)$$

The heat that is added to the gas goes into increasing the internal energy of the gas. Thus, for μ moles of gas, the change in internal energy of the gas at constant volume is also

$$\Delta U = \Delta Q = \mu c_v \Delta T \qquad (13.28)$$

Note that for an *ideal gas*, $U = \frac{3}{2}\mu RT$, so c_v can be written as

$$c_v = \frac{1}{\mu}\frac{dU}{dT} = \frac{1}{\mu}\frac{d}{dT}\left(\frac{3}{2}\mu RT\right) = \frac{3}{2}R \qquad (13.29)$$

13.8 WORK DONE BY A GAS

Suppose gas is contained in a cylinder which can expand, as in Figure 13-1. If the absolute pressure is p, and the area of the piston is A, the force on the piston is F = pA. By definition, the work done by the gas on the piston for a displacement dx is

$$dW = Fdx = pAdx = pdV \tag{13.30}$$

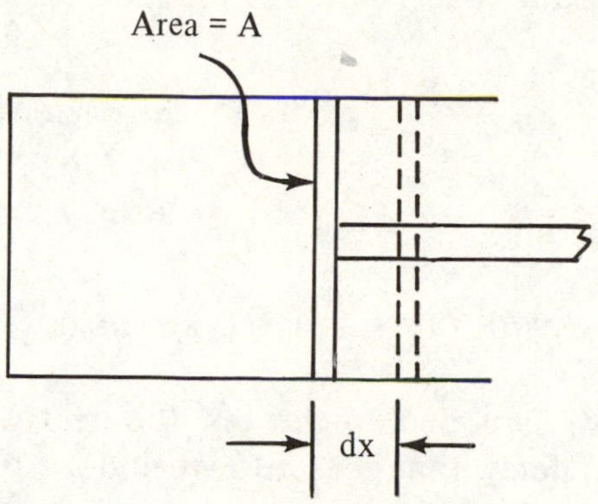

Figure 13-1

where dV is the change in volume of the contained gas. If the volume of the cylinder changes from V_1 to V_2, the *total work done* is

$$W = \int_{V_1}^{V_2} pdV \tag{13.31}$$

The pressure is not necessarily constant, so one must exercise care in evaluating W using Equation (13.31). In general, the work done is the *area* under the pV curve bounded by V_1 and V_2, and the function p as in Figure 13-2. This curve is for an arbitrary process.

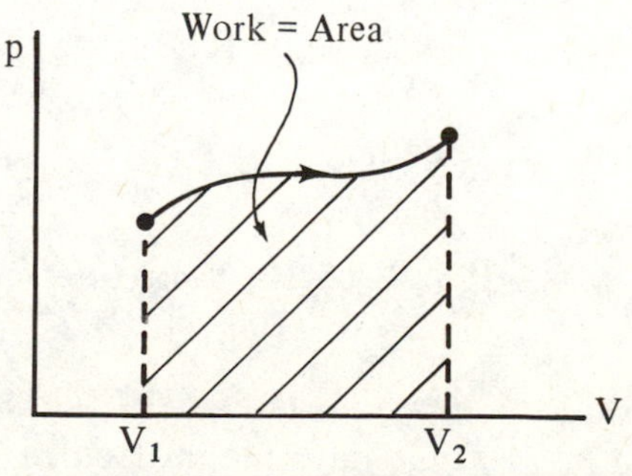

Figure 13-2 Work done by a gas is the area under the pV curve.

If the gas is compressed, $V_2 < V_1$, and work is done *on* the gas by the piston, so W is *negative.* If the gas expands, $V_2 > V_1$, and the gas does work on the piston, so W is *positive.*

If a gas expands at *constant pressure,* the process is referred to as an *isobaric process.* In this case, p can be taken out of the integrand of Equation (13.31).

$$W = p(V_2 - V_1) \quad \text{(p = constant) Isobaric Process} \tag{13.32}$$

For this special case, the equation of state for the ideal gas gives

$$V_2 - V_1 = \frac{\mu R}{p}(T_2 - T_1) \quad \text{(p = constant)}$$

so Equation (13.32) reduces to

$$W = \mu R(T_2 - T_1) \quad \text{(p = constant)} \tag{13.33}$$

If an amount of heat ΔQ is added to the gas, the heat will go into increasing the internal energy plus the energy that goes into work done by the expanding gas. This is a statement of the *first law of thermodynamics* which will be discussed in more detail in the next chapter. We can think of this law as a conservation of energy expression applied to a thermodynamic system:

$$\Delta Q = \Delta U + \Delta W \quad \text{First Law} \tag{13.34}$$

Substituting Equations (13.28) and (13.33) into (13.34) gives

$$\Delta Q = \mu c_v \Delta T + \mu R \Delta T \tag{13.35}$$

It is convenient to define the *molar specific heat at constant pressure* as

$$c_p = \frac{\Delta Q}{\mu \Delta T} \tag{13.36}$$

Using this definition, together with Equation (13.35), we get

$$c_p = c_v + R \tag{13.37}$$

Since $c_v = \frac{3}{2}R$ for *ideal gases,* c_p becomes

$$c_p = \frac{3}{2}R + R = \frac{5}{2}R \quad \text{Ideal Gases} \tag{13.38}$$

These values for c_p and c_v are in good agreement with monatomic gases such as argon, helium and neon at p = 1 atm. Diatomic molecules (or, in general, polyatomic molecules) have rotational and vibrational degrees of freedom, and correspondingly larger values of specific heats.

13.9 THERMODYNAMIC PROCESSES

We have already discussed one type of thermodynamic process, the *isobaric process,* which takes place at constant pressure. Other processes of interest are the *isothermal process* (constant temperature), the *adiabatic process* (no heat transfer) and the *isochoric process* (constant volume).

In the *isothermal process,* since T remains constant, we can conclude that for an *ideal gas,* $\Delta U = 0, \therefore \Delta Q = \Delta W$. That is, the heat put into the system goes into work done by the gas.

The *isochoric process,* where V = constant, corresponds to *no* work done, since $dV = 0$. Therefore, from the first law, $\Delta Q = \Delta U$, that is, all the added heat goes into *increasing the internal energy of the gas.*

Finally, the *adiabatic process* corresponds to $\Delta Q = 0$, that is, no heat is added to or taken from the system. In this case, the first law gives $\Delta U = -\Delta W$. Later, we will show in an exercise that this process, when applied to an ideal gas, gives the relation

$$pV^{\gamma} = \text{constant} \tag{13.39}$$

where $\gamma = c_p/c_v$. For an ideal gas, it follows that $\gamma = 1.67$. For diatomic gases such as O_2 and N_2, $c_v = \frac{5}{2}R$ and $c_p = \frac{7}{2}R$; therefore, $\gamma = 1.4$ at standard temperature and pressure.

13.10 PROGRAMMED EXERCISES

1 A long uniform metal rod of mass M has a length L_0 at a temperature T_0. It is suspended from one end to form a simple pendulum as shown.

1.A

What is the moment of inertia of the rod about the pivot?

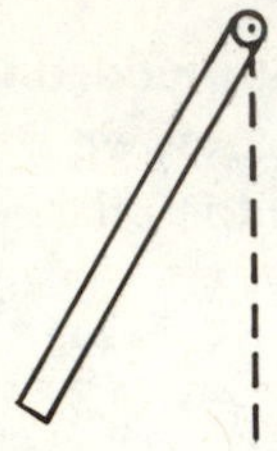

Using $I = \int r^2\, dM$ gives

$$I_0 = \frac{1}{3} M L_0{}^2 \qquad (1)$$

(See Chap. 8 for details.) If you have forgotten this result, do not be too concerned.

1.B

If the temperature is raised from T_0 to T, what is the new length of the rod? Assume it has an average linear expansion coefficient α.

By definition,

$$\Delta L = \bar{\alpha} L_0 \Delta T$$

$$L - L_0 = \bar{\alpha} L_0 (T - T_0)$$

$$\therefore \quad L = L_0 [1 + \bar{\alpha} \Delta T] \qquad (2)$$

1.C

What is the moment of inertia of the rod at the temperature T? Assume that ΔT is *small* ($T \cong T_0$), so ΔT^2 is negligible. Recall that for *small* x,

$$(1 + x)^n \cong 1 + nx + \ldots$$

$$I = \frac{1}{3} M L^2$$

or,

$$I = \frac{1}{3} M L_0{}^2 [1 + \bar{\alpha} \Delta T]^2$$

$$\therefore I \cong I_0 [1 + 2\bar{\alpha} \Delta T] \qquad (3)$$

1.D

Determine the *change* in the moment of inertia of the rod.

Subtracting (1) from (3) gives

$$\Delta I = I - I_0 = I_0 (2 \bar{\alpha} \Delta T)$$

$$\Delta I = \frac{1}{3} M L_0{}^2 (2 \bar{\alpha} \Delta T) \qquad (4)$$

1.E

Recall that the period of oscillation of the physical pendulum for small oscillations is $t = 2\pi\sqrt{I/Mgd}$, where d is the distance from the pivot to the center of mass. What is the change in period of our pendulum? Use the approximation $(1 + x)^{1/2} \cong 1 + \frac{1}{2}x$ for *small* x.

$$t_0 = 2\pi\sqrt{\frac{1/3\ ML_0{}^2}{MgL_0/2}} = \sqrt{\frac{2\ L_0}{3\ g}}$$

Likewise, $t = \sqrt{\frac{2\ L}{3\ g}} = \sqrt{\frac{2\ L_0}{3\ g}(1 + \bar{\alpha}\Delta T)}$

$$\therefore \Delta t = \sqrt{\frac{2\ L_0}{3\ g}}\ [(1 + \bar{\alpha}\Delta T)^{1/2} - 1]$$

or,

$$\Delta t \cong t_0\left(\frac{\alpha\Delta T}{2}\right) \qquad (5)$$

2 An upright cylinder of cross-section A contains one mole of an *ideal gas,* initially at atmospheric pressure p_0. At this time, a piston of mass M is held a distance h above the bottom as shown.

2.A

What is the initial temperature of the gas?

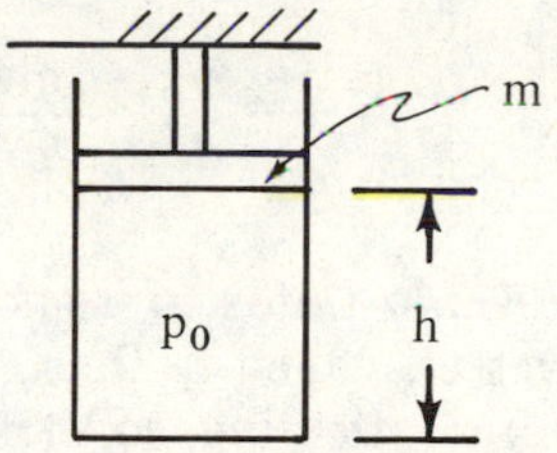

Since the gas is ideal, and $\mu = 1$,

$$p_0 V_0 = RT_0 \qquad (1)$$

but $V_0 = Ah$, $\therefore$

$$T_0 = \frac{p_0 Ah}{R} \qquad (2)$$

2.B

The piston is allowed to *slowly* drop until it reaches equilibrium under its own weight. If this position is as shown, and *no* gas escapes, what is the final pressure of the gas? Neglect friction between the piston and cylinder.

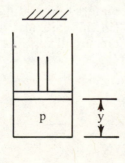

Since the piston is in equilibrium, and atmospheric pressure is also exerted from above, then,

$$pA = p_0 A + mg$$

or,

$$p = p_0 + \frac{mg}{A} \qquad (3)$$

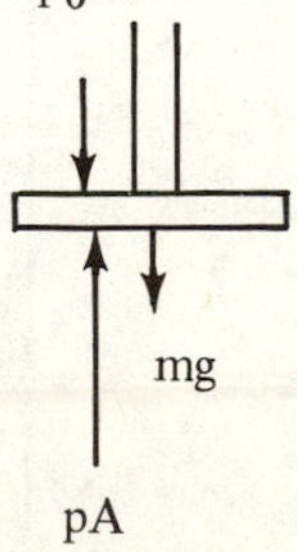

2.C

Since no gas escapes, what can we conclude about the final temperature of the gas?

From the equation of state, we get

$$p_0 V_0 = RT_0 \qquad \text{and} \qquad pV = RT$$

so

$$\frac{p_0 V_0}{T_0} = \frac{pV}{T}$$

or

$$T = \left(\frac{pV}{p_0 V_0}\right) T_0 \qquad (4)$$

2.D

Substitute (3) into (4) to obtain another expression for T. (Note that V = Ay.) The result shows that $y \propto T$! Why?

$$T = \frac{\left(p_0 + \frac{mg}{A}\right) Ay}{p_0 Ah} T_0$$

or,

$$T = \left(1 + \frac{mg}{p_0 A}\right) \frac{y}{h} T_0 \qquad (5)$$

3 A vessel for containing liquid nitrogen or other cryogenic fluids consists of two silvered glass walls separated by a vacuum, as shown below. This vessel, called a *Dewar,* is simply a sturdy thermos bottle. We will assume that the walls are perfectly reflecting, so there is no heat loss due to radiation. Convection currents will also be neglected. We wish to find the heat current if the vessel develops a leak, and one atmosphere of air is admitted in the vacuum jacket.

3.A

The cross-section of the vessel is circular. If the vessel is *long* compared to a or b, the heat current will be radial *inwards.* What is the differential expression for heat current in the radial direction? Define your terms.

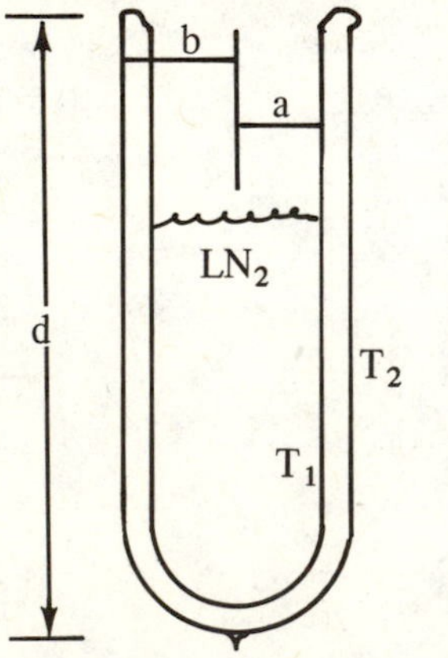

$$H = -kA\frac{dT}{dr} \qquad (1)$$

where k is the *thermal conductivity* of the material separating the inner and outer walls (in this case, air), and A is the cross-sectional area of all surfaces within $b \leqslant r \leqslant a$.

†3.B

Note that the area whose cross-section is represented by the dotted line is given by $2\pi rd$. Call the temperatures of the inner and outer walls T_1 and T_2, and integrate (1). We will *assume* here that the inner wall is kept at T_2, while the outer wall remains at T_1, that is, conduction takes place through the air. *Note:* the length of the dewar is d.

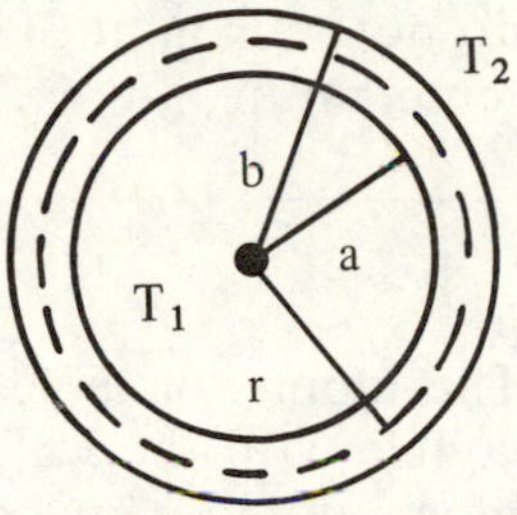

Cross-Section

$$H = -k2\pi rd \frac{dT}{dr}$$

$$\therefore \quad H \int_a^b \frac{dr}{r} = -2\pi kd \int_{T_1}^{T_2} dT$$

$$\therefore \quad H\, ln\left(\frac{b}{a}\right) = -2\pi kd\,(T_2 - T_1) \qquad (2)$$

3.C

Solve (2) for the heat current, $H = dQ/dt$. What does the negative sign in H imply?

$$H = \frac{dQ}{dt} = -\frac{2\,\pi kd\,(T_2 - T_1)}{n\,(b/a)} \qquad (3)$$

Since $T_2 > T_1$, H is *negative,* which implies that the heat flow is radially *inwards.* This corresponds to flow from the hotter surface to the colder one.

3.D

Let us assume that b = 20 cm, a = 18 cm, d = 100 cm, T_1 = 77°K (liquid nitrogen temperature), T_2 = 300°K (room temperature) and, for air, $k = 5.7 \times 10^{-5}$ cal/sec cm°C. Calculate the heat current.

$$H = -\frac{2\pi \times 5.7 \times 10^{-5} \times 10^2\ (223)}{ln\left(\frac{20}{18}\right)}$$

$$\frac{dQ}{dt} = -76 \text{ cal/sec} \qquad (4)$$

3.E

Liquid nitrogen has a *heat of vaporization* equal to 48 cal/g. What is the *total heat* necessary to vaporize 1000 g of nitrogen?

$Q = mL$

$Q = 1000 \text{ g} \times 48 \text{ cal/g}$

$$Q = 4.8 \times 10^4 \text{ cal} \qquad (5)$$

3.F

Assuming the heat current (or heat input) is constant, and given by (4), find the total time it takes to vaporize 1000 g of nitrogen in this vessel. [The density of liquid nitrogen is about 0.8 g/cm^3 at 77°K, so 1000 g corresponds to about 1.25 liters.]

Using (4) and (5) gives

$$t = \frac{Q}{76} = \frac{4.8 \times 10^4 \text{ cal}}{76 \text{ cal/sec}}$$

$$t \cong 630 \text{ sec}$$

This, of course, is an estimate, since in reality the glass conducts and the outer walls do not remain at 300°K, but at some lower temperature.

4.A

In the kinetic theory description of an ideal gas, what is the origin of the pressure exerted by the gas on the walls of its container?

The atoms (or molecules) making up the gas collide with the walls and the component of momentum perpendicular to the wall is changed. To conserve momentum, the atoms transfer momentum to the walls, hence a force is exerted.

4.B

We can write the absolute pressure of the ideal gas containing N molecules, each of mass m, speed v, in a volume V as

$$p = \frac{2}{3}\frac{N}{V} \times \frac{1}{2} m < v^2 > \qquad (1)$$

From this equation, write an expression for the pressure in terms of the density, ρ, of the gas.

By definition, ρ = Nm/V, where Nm is the *total* mass of the gas.

$$\therefore \quad p = \frac{1}{3} \rho < v^2 > \qquad (2)$$

where $< v^2 >$ is the average of the square of the speed of all particles in the container.

4.C

Use (2) to obtain an expression for the root mean square speed of the constituents of an ideal gas.

$$v_{rms} = \sqrt{< v^2 >} = \sqrt{\frac{3p}{\rho}} \qquad (3)$$

4.D

Consider nitrogen gas, N_2, at STP (standard temperature and pressure), that is, 0°C and p = 1 atm. What is its density?

The volume of one g-mole of an ideal gas at STP is 22.415×10^{-3} m^3. Since nitrogen is diatomic, one mole has a mass of 28 g, so

$$\rho = \frac{28 \times 10^{-3} \text{ kg}}{22.415 \times 10^{-3} \text{ m}^3} = 1.25 \text{ kg/m}^3$$

4.E

Use this value of ρ and (3) to find the rms speed of a nitrogen molecule at STP. $p = 1.01 \times 10^5\ N/m^2$.

$$v_{rms} = \sqrt{\frac{3 \times 1.01 \times 10^5\ N/m^2}{1.25\ kg/m^3}} = 492\ m/sec$$

The student should verify that the units of v_{rms} are m/sec.

4.F

In our original expression (1) for p in frame 4.B, what does $\frac{1}{2} m < v^2 >$ represent? What about $\frac{Nm}{2} < v^2 >$?

$\frac{1}{2} m < v^2 >$ represents the average kinetic energy per molecule of the gas. $\frac{Nm}{2} < v^2 >$ is the *internal* energy of the system of N molecules, U.

4.G

We can write (1) as $pV = \frac{2}{3} U$, where U is the internal energy of the system. Compare this with the equation of state of an ideal gas, and show that U is proportional to T.

$$pV = \frac{2}{3} U; \quad pV = \mu RT$$

$$\therefore \quad U = \frac{3}{2} \mu RT \tag{4}$$

where μ = the number of moles of gas and $R = 8.3149\ J/mole°K$.

4.H

Use (4) and the fact that $U = \frac{Nm}{2} < v^2 >$ to obtain another expression for the average kinetic energy per molecule. Note that $\mu = N/N_0$, where N_0 is Avogadro's number

$$N_0 = 6.02 \times 10^{23}\ \text{molecules/mole}$$

$$N \frac{m}{2} < v^2 > = \frac{3}{2} RT$$

$$\frac{m}{2} < v^2 > = \frac{3}{2} \frac{RT}{N}$$

Defining Boltzmann's constant as $k = R/N_0 = 1.38 \times 10^{-23}\ J/°K$ gives

$$\frac{m}{2} < v^2 > = \frac{3}{2} kT \tag{5}$$

4.I

In view of (5), give a description of temperature of an ideal gas.

The result says the temperature is proportional to the average kinetic energy per molecule.

4.J

We have concluded that $< E > = \frac{3}{2} kT$ for our ideal gas. Since a monatomic gas such as helium has three translational degrees of freedom, what can we conclude about the energy of each degree of freedom?

Since $< E > = \frac{1}{2} m < v^2 > = \frac{3}{2} kT$, where $< v^2 > = < v_x{}^2 > + < v_y{}^2 > + < v_z{}^2 >$, each degree of freedom has $\frac{1}{2} kT$ of energy associated with it.

†4.K

What is the *molar specific heat* of an ideal gas at constant volume?

By definition,

$$c_v = \frac{1}{\mu}\frac{dU}{dT} \tag{6}$$

where $U = \frac{3}{2}\mu RT$.

$$\therefore \quad c_v = \frac{3}{2}R \tag{7}$$

4.L

On the basis of the fact that each degree of freedom has $\frac{1}{2}kT$ of energy associated with it, do you think that the specific heat of a diatomic gas such as O_2 should have a molar specific heat of $\frac{3}{2}R$? Explain.

No. Diatomic molecules have more than simply translational degrees of freedom. They can also rotate (one degree) and vibrate (two degrees, corresponding to potential and kinetic energies). This predicts a total of 6 degrees of freedom, or $c_v = 3R$. In reality, the vibrational degrees of freedom are only important at very high temperatures.

5.A

Give a general definition of the *work done* by a gas whose volume changes from V_1 to V_2.

$$W = \int_{V_1}^{V_2} p\,dV \tag{1}$$

where p is the absolute pressure, which *may vary* during the process.

5.B

What is the work done by a gas in terms of quantities that can be obtained from a pV diagram?

Work done is the *area* under the curve as shown at the right. This is an arbitrary process.

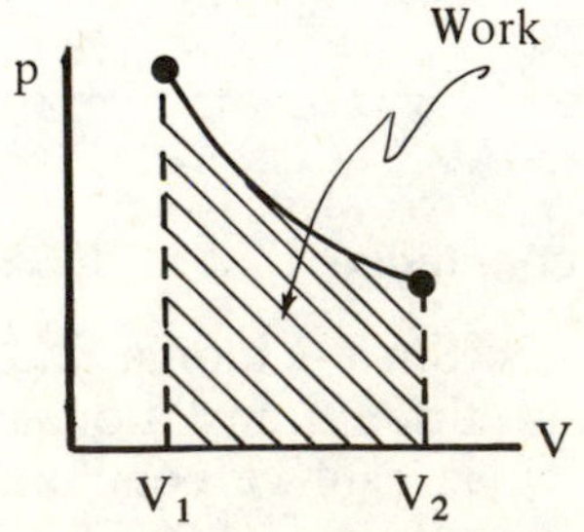

5.C

If p is constant = p_0, the process is called ______. If V_1 and V_2 are the initial and final volumes, W = ______? Draw a pV diagram.

Isobaric

$W = p_0 (V_2 - V_1)$

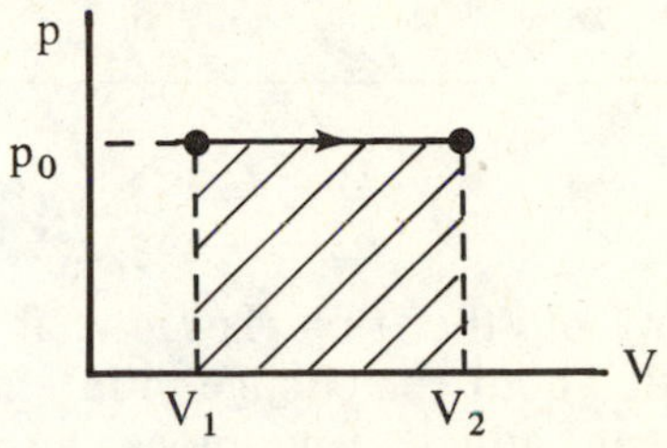

5.D

If T is constant, the process is called ______. Make use of the equation of state for an ideal gas, and sketch the pV diagram.

Isothermal

Since pV = constant, the equation is that of a *hyperbola.*

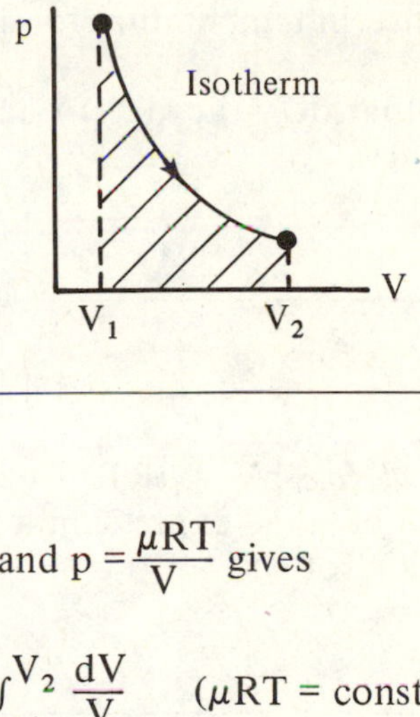

---

†5.E

Calculate W for an ideal gas whose volume changes from V_1 to V_2 in an *isothermal* process. Note that

$$pV = \mu RT = \text{constant}$$

therefore, $p = \frac{\mu RT}{V} = \frac{\text{constant}}{V}$ can be used in (1).

Using (1) and $p = \frac{\mu RT}{V}$ gives

$$W = \mu RT \int_{V_1}^{V_2} \frac{dV}{V} \qquad (\mu RT = \text{constant})$$

$$W = \mu RT\, ln \left(\frac{V_2}{V_1}\right) \qquad \textit{(isothermal process)} \; (2)$$

5.F

Describe the *first law* of thermodynamics. What conservation law does it represent?

The first law is a statement of *conservation of energy,* that is, $\Delta U = \Delta Q - \Delta W$. The change in internal energy of a system equals the heat put in (or taken out) of the system minus the work done by the system (which can be positive or negative).

5.G

Describe an *adiabatic process.*

An adiabatic process is one for which *no heat* is transferred between the system and its environment. That is, the system is thermally insulated from its environment.

5.H

In view of the first law, what can you conclude about the change in internal energy of a gas which undergoes an adiabatic process?

Since $\Delta Q = 0$ for an adiabatic process,

$$\Delta U = -\Delta W \qquad (3)$$

(adiabatic process)

5.I

Use (3) to obtain an expression for dT, that is, the change in temperature for differential changes.

Recall that $dU = \mu c_v dT$ for an ideal gas, and $dW = pdV$.

$$\mu c_v dT = -pdV$$

$$dT = -\frac{p}{\mu c_v} dV \qquad (4)$$

(adiabatic process)

---

†5.J

For an *ideal gas,* $pV = \mu RT$. Take the total differential of this expression to find another relation for dT.

$$d(pV) = pdV + Vdp = \mu RdT$$

$$\therefore \quad dT = \frac{pdV + Vdp}{\mu R} \qquad (5)$$

---

†5.K

The student should show that equating (4) and (5), and noting that $c_p - c_v = R$ for an ideal gas, gives

$$\frac{dp}{p} + \frac{c_p}{c_v}\frac{dV}{V} = 0 \qquad (6)$$

Let $\frac{c_p}{c_v} = \gamma$, and show that integration of (6) gives pV^γ = constant.

$$\int \frac{dp}{p} + \gamma \int \frac{dV}{V} = \text{constant}$$

$$\ln p + \gamma \ln V = \text{constant}$$

$$pV^\gamma = \text{constant} \qquad (6)$$

(adiabatic process)

By substituting $p = \mu RT/V$ into (7), it also follows that $TV^{\gamma-1}$ = constant.

5.L

What is the process called for which V remains constant? What is W in this case? Sketch a pV diagram.

When V is constant, the process is *isochoric.* Since V = constant, dV = 0, ∴ W = 0. Applying the first law gives

$$\Delta U = \Delta Q.$$

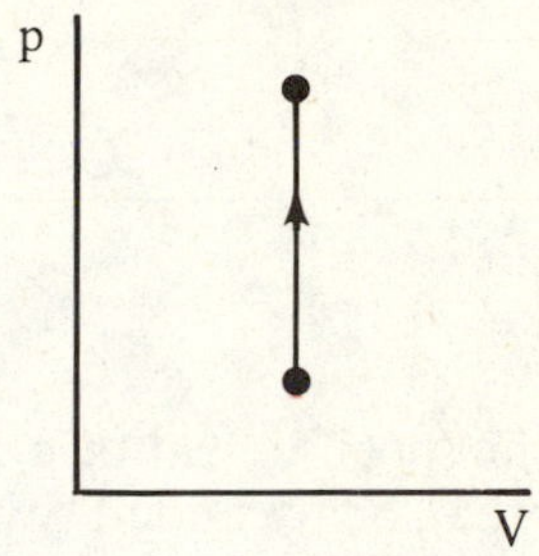

6 An ideal monatomic gas is contained in a vessel of *constant* volume 0.2 m^3. The initial temperature and pressure of the gas are 300°K and 5 atm, respectively. We wish to calculate the final pressure and temperature of the gas if 4000 calories of heat energy are supplied to the gas.

6.A

Calculate the number of moles of gas in the vessel, using $pV_0 = \mu RT_0$. Recall that 1 atm = 1.013 X 10^5 N/m^2, and R = 8.31 J/mole °K.

$$\mu = \frac{pV_0}{RT_0}$$

$$\mu = \frac{5 \times 1.01 \times 10^5 \frac{N}{m^2} \times 0.2\ m^3}{8.31\ J/mole\ ^\circ K \times 300\ ^\circ K}$$

$$\mu \cong 40 \text{ moles} \qquad (1)$$

6.B

The molar specific heat of an ideal gas at constant volume is $c_V = \frac{3}{2}R$. What is the *total* specific heat of our ideal gas?

$$C = \mu c_V = \mu \frac{3}{2} R$$

$$C = 40 \text{ moles} \times \frac{3}{2} \times 8.31\ J/moles\ ^\circ K$$

$$C \cong 500\ J/^\circ K \qquad (2)$$

6.C

In this problem, what is the work done by the gas during the heating process? Explain.

W = 0, since the volume of the gas remains constant. That is, dV = 0, so $W = \int p dV = 0$.

6.D

The heat added to the gas, Q = 4000 cal. What is the change in internal energy of the gas?

From the first law of thermodynamics,

$$\Delta Q = \Delta U + \Delta W$$

but $\Delta W = 0$ in this case, ∴

$$\Delta U = \Delta Q = 4000 \text{ cal} \qquad (3)$$

6.E

Recall that the specific heat of a substance is related to the heat added through the relation $\Delta Q = C\Delta T$. Use this expression together with (2) and (3), and find the *change* in temperature of the ideal gas in this problem.

$$\Delta T = \frac{\Delta Q}{C} = \frac{4000 \text{ cal}}{500 \text{ J/°K}}$$

but 1 cal = 4.184 J, ∴

$$\Delta T = \frac{4000 \times 4.184 \text{ J}}{500 \text{ J/°K}}$$

$$\Delta T = 32\text{°K} \qquad (4)$$

Note: We could also calculate ΔT using the fact that $U = \frac{3}{2}\,\mu RT$ for an ideal gas, so $\Delta U \sim \Delta T$

6.F

Now determine the final temperature and pressure of the gas. Note that using $pV = \mu RT$ implies that the initial and final states are *equilibrium* states.

$$\Delta T = T - T_0 = 32\text{°K}$$

$$T = 32\text{°K} + 300\text{°K} = 332\text{°K}$$

Using $pV = \mu RT$, where p, V and T are final values gives

$$p = \frac{\mu RT}{V} = \frac{40 \times 8.31 \times 332}{0.2}\ \frac{\text{N}}{\text{m}^2}$$

$$p \cong 5.7 \times 10^5\ \frac{\text{N}}{\text{m}^2} = 5.6 \text{ atm}$$

7 Two vessels of volumes V_0 and $3V_0$ are at an initial temperature T_0. The vessels contain a total of μ moles of an ideal gas, and are connected together by an insulating tube which permits the vessels to be at the same pressure, but at different temperatures. We wish to find the final pressure in the vessels if the vessel of volume V_0 is heated to a temperature T, while the vessel of volume $3V_0$ is maintained at T_0.

7.A

What is initial pressure in the system?

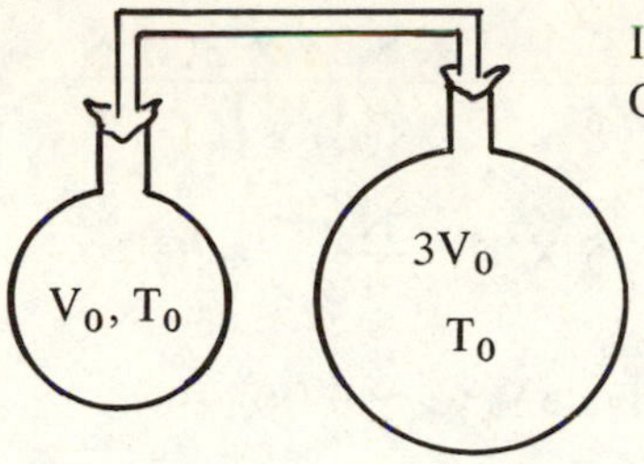

The *total* volume is $4V_0$, and since there are μ moles of gas,

$$p_0 V = \mu R T_0; \qquad V = 4V_0$$

$$p_0 = \frac{\mu R T_0}{4V_0} \tag{1}$$

7.B

As the vessel on the left is heated, what quantity in the system remains constant? Assume that no gas leaks from the system.

The *total* number of moles, μ, remains constant. If μ_1 is the number of moles in the vessel of volume V_0, and μ_2 is the number in the vessel of volume $3V_0$, then

$$\mu = \mu_1 + \mu_2 = \text{constant} \tag{2}$$

Note: μ_1 and μ_2 *do not* remain constant as the vessel is heated.

7.C

In the final equilibrium state shown below, write the equations of state for the two vessels.

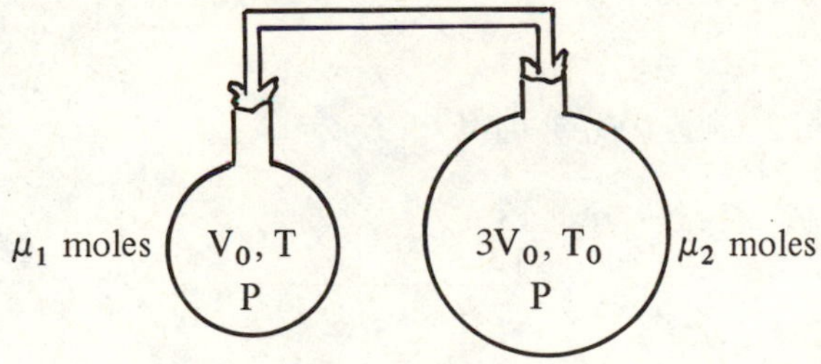

For the vessel of volume V_0,

$$pV_0 = \mu_1 RT \tag{3}$$

For the vessel of volume $3V_0$,

$$p3V_0 = \mu_2 RT_0 \tag{4}$$

7.D

Use (3) and (4) to find the *ratio* μ_1/μ_2. Note that the ratio only depends on the initial and final temperatures.

Dividing (3) by (4) gives

$$\frac{\mu_1}{\mu_2} = \frac{T}{3T_0} \tag{5}$$

7.E

Suppose the total number of moles is $\mu = 1$, $T_0 = 300°K$, $V_0 = 0.1\ m^3$ and $T = 500°K$. Find the *initial* pressure p_0.

Using (1), we have

$$p_0 = \frac{\mu R T_0}{4V_0} = \frac{1 \times 8.31 \times 300}{4\ (0.1)} \frac{N}{m^2}$$

$$p_0 \cong 6.25 \times 10^3 \frac{N}{m^2} \qquad (6)$$

(which is equivalent to ~ 0.06 atm).

7.F

What is the ratio μ_1/μ_2 for the data given in 7.E?

From (5),

$$\frac{\mu_1}{\mu_2} = \frac{T}{3T_0} = \frac{500}{3 \times 300} = \frac{5}{9} \qquad (7)$$

7.G

Calculate the number of moles μ_1 in the vessel of volume V_0 when its temperature is 500°K. *Hint:* Recall that

$$\mu = \mu_1 + \mu_2 = \text{constant},$$

and use (7).

Using (7) gives

$$\mu_2 = \frac{9}{5}\mu_1$$

Since $\mu_1 + \mu_2 = \mu$

$$\mu_1 + \frac{9}{5}\mu_1 = 1 \text{ mole}$$

$$\mu_1 = \frac{5}{14} \text{ mole} \qquad (8)$$

7.H

Now calculate the final pressure in the system using (3) and (8), and the data in frame 7.E.

$$pV_0 = \mu_1 RT$$

$$p = \frac{\frac{5}{14} \times 8.31 \times 500}{0.1} \frac{N}{m^2}$$

$$p \cong 14.9 \times 10^3\ N/m^2 \qquad (9)$$

or

$$p \cong 2.4\ p_0$$

The student should show that (9) also follows from (4), with $\mu_2 = 9/14$ moles.

13.11 SUMMARY

Absolute temperature, T, is defined to be proportional to the absolute pressure of an ideal gas according to the expression

$$T = \frac{273.15}{p_t} p \qquad (13.1)$$

where p_t is the gas pressure at the triple point of water.

The change in volume of an isotropic solid or liquid is proportional to the change in temperature.

$$\Delta V = \bar{\beta} V_0 \Delta T \qquad (13.5)$$

where $\bar{\beta}$ is the *average volume expansion coefficient* and V_0 is the initial volume. The *volume expansion coefficient* β is three times the *linear expansion coefficient* α.

The change in volume of a substance undergoing a change in pressure Δp is given by

$$\Delta V = -\kappa V_0 \Delta p$$

where κ is the *compressibility*. The inverse of κ is called the *bulk modulus*, B.

The *heat capacity* C of a substance is defined as the ratio of the heat supplied ΔQ to the corresponding temperature increase ΔT. The *specific heat*, c, of a substance whose mass is m is given by

$$c = \frac{1}{m} \frac{\Delta Q}{\Delta T} \qquad (13.13)$$

The heat liberated or absorbed by a substance of mass m undergoing a *phase change* is given by

$$Q = mL \qquad (13.15)$$

where L is the *latent heat* which depends on the type of phase change and the nature of the substance.

The transfer of heat can occur by three processes: (1) conduction, (2) convection and (3) radiation.

The equation of state of an ideal gas is given by

$$pV = \mu RT \qquad (13.18)$$

where p, V and T are pressure, volume and temperature, μ is the number of moles of gas and R = 8.3149 J/mole °K.

The *internal energy* of μ moles of an ideal gas at a temperature T is

$$U = \frac{3}{2} \mu RT \qquad (13.22)$$

The *average kinetic energy per molecule* of an ideal gas is

$$\frac{1}{2}m < v^2 > = \frac{3}{2}kT \tag{13.25}$$

where $k = 1.38 \times 10^{-23}$ J/°K is Boltzmann's constant. The *root mean square speed* is defined by $\sqrt{< v^2 >}$.

The work done by a gas whose volume changes from V_1 to V_2 is given by

$$W = \int_{V_1}^{V_2} pdV \tag{13.31}$$

An *isobaric* process is one which occurs at constant pressure. An *isothermal* process occurs at constant temperature. An *isochoric* process occurs at constant volume. An *adiabatic* process is one for which no heat enters or leaves the system.

The *first law of thermodynamics* states that the change in internal energy of the system equals the heat added (or taken) from the system minus the work done by the system. It is a statement of *conservation of energy.*

$$\Delta U = \Delta Q - \Delta W \tag{13.34}$$

13.12 PROBLEMS

1. Calculate the *fractional* change in volume of a copper bar which undergoes a change in temperature of 20°C. Copper has a linear expansion coefficient of $\alpha = 1.7 \times 10^{-7}\ (°C)^{-1}$. Recall that $\beta = 3\alpha$ for an isotropic substance.

2. An air bubble originating from a deep-sea diver has a radius of 2 mm at some depth h. When the bubble reaches the surface of the water, it has a radius of 3 mm. Assuming the temperature of the air in the bubble remains constant, determine (a) the depth h of the deep-sea diver and (b) the absolute pressure at this depth.

†3. Show that the volume coefficient of thermal expansion for an isotropic substance is related to the linear coefficient of thermal expansion through the relation $\beta = 3\alpha$. *Hint:* Let $V = L_x L_y L_z$, and expand $\frac{1}{V}\frac{dV}{dT}$, which is the definition of β.

4. Show that the volume coefficient of thermal expansion for an ideal gas at constant pressure is given by $\beta = 1/T$. Recall that the equation of state for an ideal gas is $pV = \mu RT$, and use the definition of β.

5. A fluid has a desnity, ρ, defined by $\rho = M/V$. (a) Show that the *fractional* change in density is given by $\Delta\rho/\rho = -\bar{\beta}\Delta T$. What does the negative sign signify? (b) Water has a maximum density of 1.000 g/cm³ at 4°C. At 10°C, it has a density of 0.9997 g/cm³. What is $\bar{\beta}$ for water?

6. The concrete sections of a certain superhighway are designed to have a length of 30 m. If several sections are poured and cured at 10°C, what minimum spacing must the engineer leave *between the sections* to eliminate "buckling" if the concrete is to reach a temperature as high as 45°C? Take the average coefficient of linear expansion for concrete to be $1 \times 10^{-5} (°C)^{-1}$.

7. If a substance of density ρ, molecular weight M, occupies a volume V, show that the *total number of molecules* is given by

$$N = \frac{\rho V N_0}{M}$$

while $\mu = N/N_0$ is the number of moles.

8. A 1 kg block of aluminum, initially at 20°C, is dropped in a large vessel of liquid nitrogen, which is boiling at 77°K. Assuming the vessel is thermally insulated from its surroundings, calculate the number of liters of nitrogen which boils off by the time the aluminum reaches 77°K. *Note:* The specific heat capacity of aluminum is 0.21 cal/g°C. Liquid nitrogen has a heat of vaporization of 48 cal/g and a density of 0.8 g/cm³.

9. An aluminum rod, 1 m in length and cross-sectional area 2 cm², is inserted vertically into a thermally insulated vessel containing liquid helium, which boils at 4.2°K. The rod is initially at 300°K. (a) If one half of the rod is inserted in the helium, how many liters of helium boil off by the time this half is cooled to 4.2°K? (Assume the upper half does not get cooled.) (b) If the upper portion of the rod is maintained at 300°K, what is the *approximate* boil-off rate of liquid helium *after* the lower half has reached 4.2°K? *Note:* Aluminum has a thermal conductivity of 31 J/sec cm°K at 4.2°K, a specific heat capacity of 0.21 cal/g°C and a density of 2.7 g/cm^3. See Example 13.6 for data on helium.

10. A rigid hollow sphere contains one mole of helium gas at 300°K, at a pressure of one atmosphere. If the sphere is cooled to 77°K by immersing it in liquid nitrogen, what is the final pressure in the sphere? An interesting classroom demonstration demonstrating this effect is cooling an inflated rubber balloon. What do you suppose happens in this case?

11. One mole of an ideal gas is compressed isothermally at 300°K from a volume V_0 to a final volume of $V_0/3$. How much work is done?

12. (a) Show that the work done by an ideal gas in an *adiabatic expansion* is given by

$$W = \frac{pV - p_0V_0}{1 - \gamma}$$

where p, V are final values of pressure and volume, while p_0, V_0 are initial values. *Hint:* Note that pV^γ = constant for an adiabatic process. (b) If the initial and final temperatures of an ideal gas are known to be T_0 and T, and its molar heat capacity at constant volume is c_v, show that the work done for an adiabatic process is given by

$$W = \mu c_v (T_0 - T)$$

13. The *equipartition theorem* states that the average energy associated with each *translational* and *rotational* degree of freedom of a molecule is $\frac{1}{2}$kT, while each *vibrational* degree of freedom has an average energy of kT (divided equally between potential and kinetic energy). Use this theorem to show that the *molar* specific heat of a *monatomic* gas is predicted to be 3 cal/mole°K, while for a diatomic gas it is 7 cal/mole°K. Why do you suppose there is a large discrepancy with experimental results for the diatomic gas, while the monatomic gas is in good agreement?

14. A system is taken from a to c along two different paths, as in Figure 13-3. (a) What is the work done along the path abc? (b) What is the work done along the path ac? (c) If ΔQ = 400 kcal along abc, what is the change in the internal energy of the system? (d) What is ΔQ for the process ac?

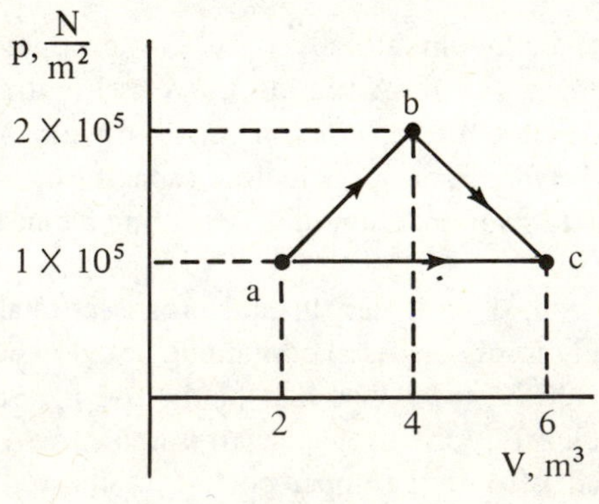

Figure 13-3

15. (a) Show that the rms speed of the molecules of an ideal gas is given by

$$v_{rms} = \sqrt{\frac{3\,RT}{M}}$$

where M is the molecular weight of the gas. (b) Calculate the rms speed of neon molecules at 273°K. Neon has a molecular weight of 20.1 g/mole.

***16.** One technique for containing liquid helium is illustrated in Figure 13-4. The cross-section shows, starting from the outside, a vacuum jacket, a liquid nitrogen jacket, a second vacuum jacket and, finally, the helium vessel. A safety "pop-valve" is provided on the helium vessel to prevent air from condensing and freezing at the neck. Suppose the "pop-valve" releases when the "gauge pressure" is 5 $lb/in^2 \cong 0.34 \times 10^5\ N/m^2$, and a "heat-leak" is present which causes the helium to boil-off at a rate of 0.1 liters/hr. (a) Estimate the number of times per hour the safety valve should release, assuming the cold gas occupies a volume of 5000 cm^3. (b) If a full container has 25 liters of liquid helium, and the vessel accidentally freezes at the neck, to what absolute pressure will the inner vessel rise if all the liquid is converted into gas at 300°K? (Obviously, a very explosive situation which should be avoided.) *Note:* See Example 13.6 for data on helium. You should prove that one liter of liquid helium is equivalent to 700 liters of gas at STP.

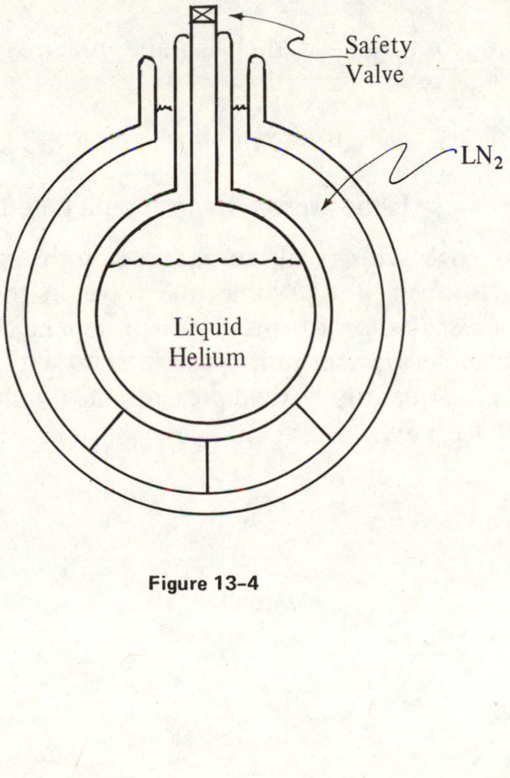

Figure 13-4

14

KINETIC THEORY AND THERMODYNAMICS

14.1 KINETIC THEORY OF GASES

If N molecules occupy a volume V, the absolute pressure is given by

$$p = \frac{2}{3}\frac{N}{V}\langle E \rangle \tag{14.1}$$

where $\langle E \rangle = \frac{1}{2} m \langle v^2 \rangle$ is the average translational kinetic energy per molecule. In the last chapter, we used this result to obtain the rms speed of the molecules, defined by $\sqrt{\langle v^2 \rangle}$. However, not all molecules travel at the same speed. Maxwell showed that there is a *distribution* of speeds which depends on temperature. If we call N_v the number of molecules per unit speed interval with a speed v, and plot N_v *vs* v as in Figure 14–1, the quantity $N_v dv$ represents the number of molecules whose speeds lie between v and v + dv.

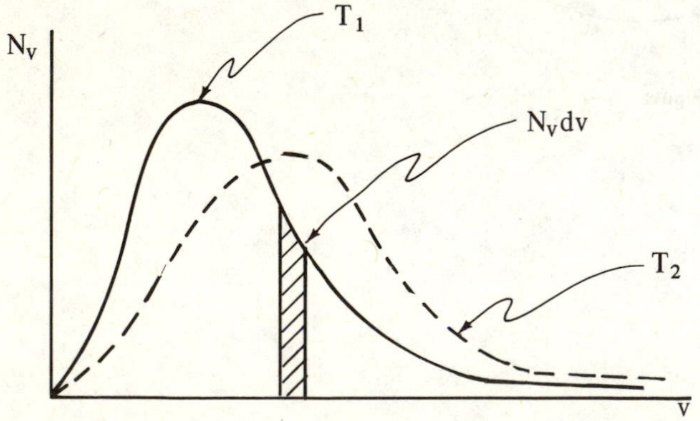

Figure 14–1 The Maxwell distribution of molecular speeds of a gas. *Note:* $T_2 > T_1$.

If we integrate the quantity $N_v dv$, allowing all possible speeds from 0 to ∞, we get the *total number* of particles, N.

$$N = \int_{v=0}^{\infty} N_v dv \tag{14.2}$$

The dotted line in Figure 14–1 shows the speed distribution at a temperature $T_2 > T_1$. Note that at higher temperatures, the curve shifts to the right, corresponding to a higher probable speed for a given molecule. However, the area under both curves is N. Sometimes, the *probability* distribution function $P(v) = N_v/N$ is plotted *vs* v, where P(v) represents the *fraction* of molecules with speeds between v and v + dv. These curves are essentially scaled down versions of the N_v *vs* v curve.

Using the procedures of statistical mechanics, and assuming that all molecules can occupy any region of the vessel with equal probability, one obtains the *Maxwell speed distribution function* given by

$$N_v = \frac{4N}{\sqrt{\pi}} \left(\frac{m}{2kT}\right)^{3/2} v^2 e^{-\frac{mv^2}{2kT}} \tag{14.3}$$

This is a fundamental expression in the development of the kinetic theory of gases. It represents the most probable speed distribution function for the molecules of a gas at a temperature T. The function $N_v \to 0$ as $v \to \infty$ and as $v \to 0$, which signifies that *all* speeds are possible Also, the function obeys the relation $N^{-1} \int_0^\infty N_v dv = 1$, which means that the probability for the molecule having a speed $0 < v < \infty$ is unity.

The *average value* of a quantity w which is any function of v is given by

$$< w > = \frac{1}{N} \int_0^\infty w N_v dv \tag{14.4}$$

For example, if w is taken to be v or v^2, the average values are calculated to be

$$< v > = \frac{1}{N} \int_0^\infty v N_v dv = \sqrt{\frac{8kT}{\pi m}} \tag{14.5}$$

$$< v^2 > = \frac{1}{N} \int_0^\infty v^2 N_v dv = \frac{3kT}{m} \tag{14.6}$$

We see that Equation (14.6) is consistent with previous results in Chap. 13, namely, $v_{rms} = \sqrt{< v^2 >}$. Finally, the *most probable* speed v_p is the speed for which N_v has its *maximum* value. This occurs when $dN_v/dv = 0$, where N_v is given by Equation (14.3). The result is

$$v_p = \sqrt{\frac{2kT}{m}} \tag{14.7}$$

Since the *average kinetic energy* is given by $< E > = \frac{1}{2} m < v^2 >$, where $< v^2 >$ is given by Equation (14.6), direct substitution gives

$$< E > = \frac{3}{2} kT \tag{14.8}$$

Again, this is a familiar result. The *most probable energy* is given by $\frac{1}{2}$kT.

14.2 MEAN FREE PATH

The molecules of a gas make elastic collisions with each other at a rate which depends on the concentration of molecules as well as on their size. The average distance between such collisions, called the *mean free path,* is given by

$$l = \frac{1}{4\pi\sqrt{2}\,na^2} \tag{14.9}$$

where n = N/V, the number of molecules per unit volume, and a is the radius of each molecule, assumed to be hard spheres.

The *average collision frequency,* or the rate at which molecules collide, is given by

$$f = \frac{\langle v \rangle}{l} = 4\pi\sqrt{2}\,na^2 \langle v \rangle \tag{14.10}$$

The inverse of f is the *average time between* collisions for a molecule.

Typical values of l and f at STP for a gas are of the order of 5×10^{-8} m and 10^{10} Hz, respectively. As the gas pressure is reduced, n gets smaller, so l increases while f decreases.

Example 14.1

Calculate the mean free path, the average speed and the collision frequency for nitrogen molecules contained in a vessel at a pressure of 10^{-3} atm and temperature 300°K. Assume the molecular radius is 1.7×10^{-3} m.

Solution

Since 10^{-3} atm = 1.01×10^2 N/m^2, and $pV = \mu RT = \frac{N}{N_0}RT$, we get

$$n = \frac{N}{V} = \frac{pN_0}{RT} = \frac{1.01 \times 10^2 \frac{N}{m^2} \times 6.02 \times 10^{23} \frac{\text{molecules}}{\text{mole}}}{8.3149 \frac{J}{\text{mole}°K} \times 300°K}$$

$$n = 2.4 \times 10^{22} \text{ molecules/m}^3$$

$$\therefore\ l = \frac{1}{4\pi\sqrt{2}\,na^2} = \frac{1}{4\pi\sqrt{2} \times 2.2 \times 10^{22} \times (1.7 \times 10^{-10})^2} \cong 8.8 \times 10^{-5}\,\text{m}$$

To obtain the collision frequency, we first must calculate $\langle v \rangle$ from Equation (14.5), using m = 4.7×10^{-26} kg.

$$\langle v \rangle = \sqrt{\frac{8kT}{\pi m}} = \sqrt{\frac{8 \times 1.38 \times 10^{-23}\ \mathrm{J/^\circ K} \times 300^\circ\mathrm{K}}{\pi \times 4.7 \times 10^{-26}\ \mathrm{kg}}} = 475\ \frac{\mathrm{m}}{\mathrm{sec}}$$

The collision frequency, given by Equation (14.9), is

$$f = \frac{\langle v \rangle}{l} = \frac{475\ \mathrm{m/sec}}{8.8 \times 10^{-5}\ \mathrm{m}} = 5.4 \times 10^6\ \frac{\mathrm{collisions}}{\mathrm{sec}}$$

14.3 FIRST LAW OF THERMODYNAMICS

In chapter 13 we introduced the first law of thermodynamics and stated that it represents a conservation of energy expression, namely,

$$\Delta U = \Delta Q - \Delta W \tag{14.11}$$

where ΔQ is the heat added to the system, ΔW is the work done by the system on its environment and ΔU is the *change* in the internal energy of the system. It is conventional to take ΔQ positive when heat is added to the system, and ΔW positive when the system does work on its surroundings. The initial and final states of the system are *equilibrium* states, that is, the system is in equilibrium with its environment at these times. The intermediate states are in general nonequilibrium states, since the *thermodynamic coordinates* (for example, p, V and T for a gas) undergo finite changes during the thermodynamic process. For an infinitesimal change of the system, we can write the first law as

$$dU = đQ - đW \tag{14.12}$$

The symbolism $đQ$ and $đW$ is used to indicate that these *are not exact differentials,* since both Q and W are not functions of the system's coordinates. That is, both Q and W depend on the *path* taken between the initial and final equilibrium states, during which time the system interacts with its environment. On the other hand, dU is an *exact differential,* therefore the internal energy function U is a *state variable.* The function U is analogous to the potential energy function used in mechanics when dealing with conservative forces. Recall that such a function depends only on the system's coordinates.

14.4 THE SECOND LAW OF THERMODYNAMICS – THE CARNOT CYCLE

Before discussing the second law of thermodynamics, it is useful to first describe a particular *cyclic process* (one which begins and ends with the same state variables). The most efficient cyclic process is called the *Carnot cycle,* described in the pV diagram of Figure 14–2.

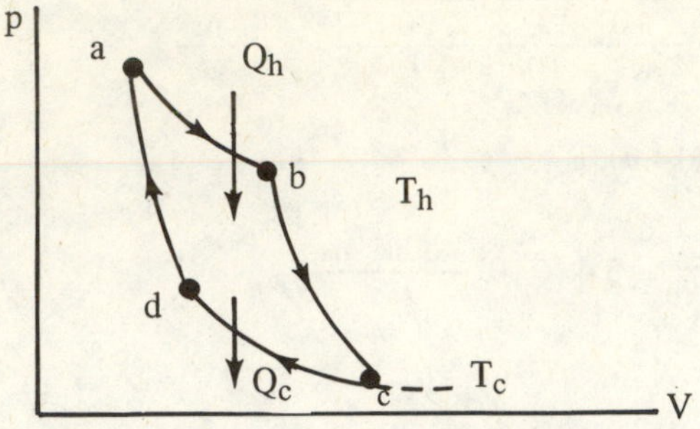

Figure 14-2 The Carnot cycle.

The *Carnot* cycle consists of two adiabatic and two isothermal processes, all being *reversible.* (1) The process ab in Figure 14-2 is an isotherm, during which time the gas expands at a constant temperature T_h, and absorbs heat Q_h from a hot reservoir. (2) The process bc is an adiabatic expansion ($Q = 0$), during which time the gas expands and cools to a temperature T_c. (3) The process cd is a compression stage at constant temperature T_c, whereby the gas gives up heat Q_c to a cold reservoir. (4) Finally, the gas is compressed from d to a adiabatically ($Q = 0$), to a final temperature of T_h. The net work done in this cyclic process is the *area* enclosed by the curve abcda. Since the system returns to its original state, $\Delta U = 0$ for the cycle, and the work done equals the net heat flow into the system. That is,

$$W = Q_h - Q_c \tag{14.13}$$

The *thermal efficiency* η of the Carnot engine, operating in the Carnot cycle, is defined as the work done divided by the heat input ("what you get out for what you put in") or

$$\eta = \frac{Q_h - Q_c}{Q_h} = 1 - \frac{Q_c}{Q_h} \tag{14.14}$$

Since the internal energy of the system remains constant along the isotherms ab and cd, the first law shows that $W_{ab} = Q_h$ and $W_{dc} = Q_c$. By calculating the work done along ab and cd, it is left as an exercise to show that

$$\frac{Q_c}{Q_h} = \frac{T_c}{T_h} \tag{14.15}$$

Thus, it follows from (14.14) and (14.15) that the thermal efficiency of a *Carnot engine* can be written as

$$\eta = 1 - \frac{T_c}{T_h} \tag{14.16}$$

Equation (14.16) is actually valid for *any* system operating in a Carnot cycle. It is not restricted to ideal gases. All engines are *less* efficient than the Carnot engine. We see that the efficiency is increased as T_c is made small compared to T_h. Equation (14.16) is also used to define the *absolute* thermodynamic temperature scale. A 100% efficient engine would require $T_c = 0°K$. However, absolute zero is *impossible* to obtain, so a 100% efficient engine cannot be constructed.

The *second law of thermodynamics* can now be stated in several ways.

Clausius statement: No thermodynamic process can occur whose only result is to transfer heat from a colder to a hotter body—this process is only possible if work is done on the system.

Kelvin-Planck statement: It is impossible for a thermodynamic process to occur whose only final result is the complete conversion of heat extracted from a hot reservoir into work. This is equivalent to stating that it is impossible to construct a *perpetual motion machine of the second kind* (that is, those which violate the second law). Perpetual motion machines of the *first kind* are those which violate energy conservation (that is, they violate the first law). In effect, the second law states that heat will *spontaneously* flow from a hotter to a colder body, but not the reverse.

A schematic of a *heat engine* is illustrated in Figure 14-3(*a*). Again, Q_h is the heat absorbed from the hot reservoir, Q_c is the heat given up to the cold reservoir and $W = Q_h - Q_c$ is the net work done. A *refrigerator* is the reverse of the heat engine, as shown schematically in Figure 14-3(*b*). For the *refrigerator* cycle, heat Q_c is absorbed from the cold reservoir, heat Q_h is given up to the hot reservoir and work $W = Q_h - Q_c$ is done *on* the system. The refrigerator is sometimes referred to as a *heat pump.*

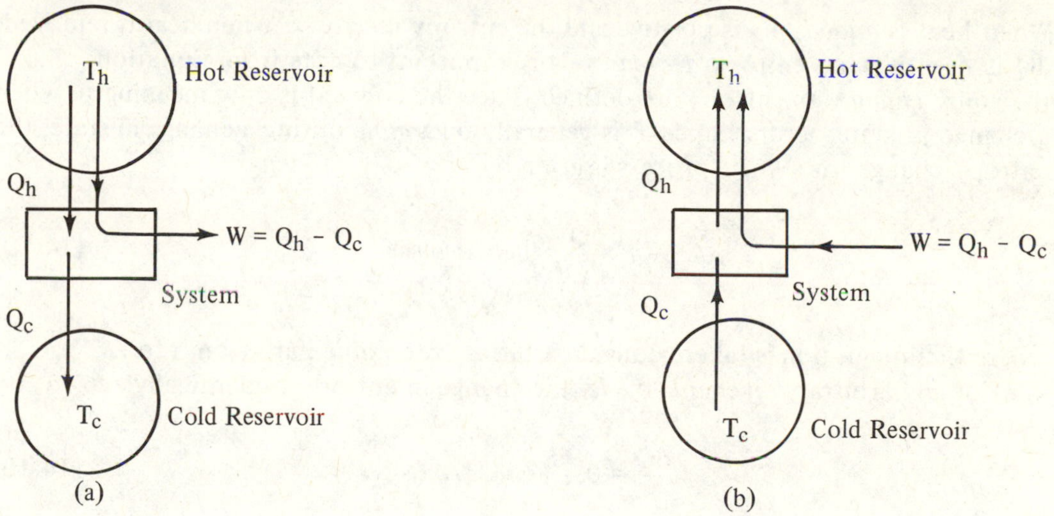

Figure 14-3 (*a*) Schematic diagram of a heat engine. (*b*) Schematic diagram of a refrigerator (or heat pump).

In practice, no working engine is 100% efficient, even when losses due to friction are neglected. One can obtain some theoretical limits on the efficiency of real engines by comparison with the ideal Carnot engine. A *reversible engine* is an engine which will operate with the same efficiency in the forward and reverse directions. The Carnot engine is an example of a reversible engine. Therefore, a *reversible engine* is one which can run forwards or backwards without any change in the magnitudes of Q_h, Q_c and W. *Carnot's theorems,* consistent with the first and second laws of thermodynamics, are as follows:

I. All reversible engines operating between T_h and T_c have the *same* efficiency, given by Equation (14.16).

II. No real (irreversible) engine can have an efficiency greater than that of a reversible engine operating between the same two temperatures.

14.5 ENTROPY AND THE SECOND LAW OF THERMODYNAMICS

We have seen that the concept of *temperature* T is involved in the *zeroth* law of thermodynamics, and the concept of *internal energy* U is involved in the *first law.* The functions, T and U, are thermodynamic variables. Another thermodynamic variable, called the *entropy,* S, is used to describe the second law of thermodynamics. Entropy is not as simple to describe as temperature and internal energy. We can think of entropy as the measure of the *disorder* in a system. For example, the molecules of a gas at a low temperature are in a more highly ordered state (lower entropy) than a gas at a higher temperature. The concept of entropy can be expressed in terms of heat and temperature. If a system changes from one equilibrium state to another, and a quantity of heat $đQ$ is added (or removed) at an *absolute* temperature T, the *change in entropy* is given by

$$dS = \frac{đQ}{T} \tag{14.17}$$

When heat is *added,* $đQ$ is positive and the entropy *increases.* When heat is removed, $đQ$ is negative and entropy *decreases.* It is important to note from Equation (14.17) that only *changes* in entropy are defined, since the concept is only meaningful when a change in state occurs. Since T is generally a *variable* during a change in state, the entropy change for a *finite* process is given by

$$\Delta S = \int_i^f \frac{đQ}{T} \quad \text{Reversible Path} \tag{14.18}$$

where the integration is taken along an arbitrary *reversible* path from i to f.

For any arbitrary *reversible cycle,* the change in entropy is identically *zero.*

$$\oint \frac{đQ}{T} = 0 \quad \text{Reversible cycle} \tag{14.19}$$

The definition of ΔS as given by Equation (14.18) assumes that changes occur between *equilibrium* states. However, entropy *changes* can be calculated from Equation (14.18) even if a process is *irreversible,* since ΔS depends only on the end points, not on the path. Note, however, that an irreversible process cannot be described by a definite path on a pV diagram. In addition, ΔS is *always greater than zero* for an *irreversible* process.

Let us apply the concept of entropy to the *Carnot cycle.* From Equation (14.15), we can write

$$\frac{Q_h}{T_h} - \frac{Q_c}{T_c} = 0$$

Q_h/T_h represents the heat divided by temperature during the isothermal expansion from a to b (see Figure 14-1). Since heat enters the system, Q_h is positive. $-Q_c/T_c$ represents the heat divided by temperature during the isothermal compression from c to d. This term is negative since heat leaves the system. The sum of the two terms is *zero*, since we are dealing with a cycle taken along reversible paths. Therefore, Equation (14.19) is satisfied. Since any *reversible* cycle can be reduced to a series of Carnot cycles, we can write

$$\sum_i \frac{Q_i}{T_i} = 0 \quad \text{Reversible Cycle} \tag{14.20}$$

Example 14.2

A solid substance has a mass m, a melting temperature T_m, a heat of fusion L and a specific heat capacity c. Determine the change in entropy of the substance if it is slowly heated from a temperature $T_0 < T_m$ to a temperature $T = T_m$. Assume that just enough heat is provided to melt all of the substance. Note that this is an *irreversible* process.

Solution

We can solve the problem in steps. First, we calculate the change in entropy of the substance when it is taken from T_0 to its melting point T_m. During this process, T is *not* constant, but $đQ$ can be expressed in terms of T. That is, $đQ = mcdT$; therefore,

$$\Delta S_1 = \int \frac{đQ}{T} = mc \int_{T_0}^{T_m} \frac{dT}{T} = mc\,ln\left(\frac{T_m}{T_0}\right)$$

Next, we must supply a total heat mL to melt the substance completely. Since $T = T_m$ = constant, the change in entropy during this process is

$$\Delta S_2 = \frac{1}{T_m} \int đQ = \frac{mL}{T_m}$$

Therefore, the total change in entropy is given by

$$\Delta S = \Delta S_1 + \Delta S_2 = mc\,ln\left(\frac{T_m}{T_0}\right) + \frac{mL}{T_m}$$

Of course, $\Delta S > 0$ for this irreversible process.

Example 14.3

A substance of mass m_1, specific heat c_1 and temperature T_1 is mixed with a second substance of mass m_2, specific heat c_2 and initial temperature T_2. (These could both be liquids, or one might be a solid, the other a liquid). (a) Calculate the final temperature T of the system after equilibrium is established. (b) Determine the total entropy change for the system.

Solution

(a) Energy conservation requires that the heat lost by one substance equal the heat gained by the other. Since $\Delta Q = mc\Delta T$ for each substance, we have

$$\Delta Q_1 = -\Delta Q_2$$

$$m_1c_1(T - T_1) = -m_2c_2(T_2 - T)$$

or,

$$T = \frac{m_1c_1T_1 + m_2c_2T_2}{m_1c_1 + m_2c_2}$$

Note that $T_1 < T < T_2$ as would be expected.

(b) Since $đQ = mcdT$ for an infinitesimal change, we have for the *total* change in entropy

$$\Delta S = \int \frac{đQ_1}{T} + \int \frac{đQ_2}{T}$$

$$\Delta S = m_1c_1 \int_{T_1}^{T} \frac{dT}{T} + m_2c_2 \int_{T_2}^{T} \frac{dT}{T}$$

$$\Delta S = m_1c_1 ln\left(\frac{T}{T_1}\right) + m_2c_2 ln\left(\frac{T}{T_2}\right)$$

The student should show that ΔS is *always* positive. Why is this the case?

The second law of thermodynamics can be stated in terms of entropy as follows: The *total entropy* of an *isolated* system cannot decrease in time; or, the total entropy of an isolated system always *increases* in time if the process is *irreversible.* For a *reversible process,* the total entropy of an isolated system *remains constant.* We can also think of the increase in entropy of a system as a change to a state of greater *disorder.* Therefore, a highly ordered system has *low* entropy, whereas a disordered system has *high* entropy.

Many events in everyday life involve the second law. Consider the following events: (1) a warm dinner on the table gets cooler; (2) objects with no support fall freely towards the earth; (3) wood or other combustible materials burn when ignited with a match. These are all "one-way" processes, that is, processes which clearly show that entropy is increasing. We can only observe the reverse of these processes on a time-reversed world (or on a film of these events running in reverse. You may have experienced the humor in a slap-stick comedy run in reverse.) To be more precise, we say that the time-reversed events are *highly improbable,* since they represent events in which entropy is decreasing. Therefore, processes in nature follow a *preferred direction in time,* namely the "forward" direction. This statement implies that the *entropy of the universe is always increasing;* or, the entropy of an isolated system *cannot* decrease.

14.6 PROGRAMMED EXERCISES

†1.A

In the kinetic theory of gases, the Maxwell distribution of speeds is the most probable distribution. Sketch this function N_v *vs* v at a temperature T_1 and $T_2 > T_1$.

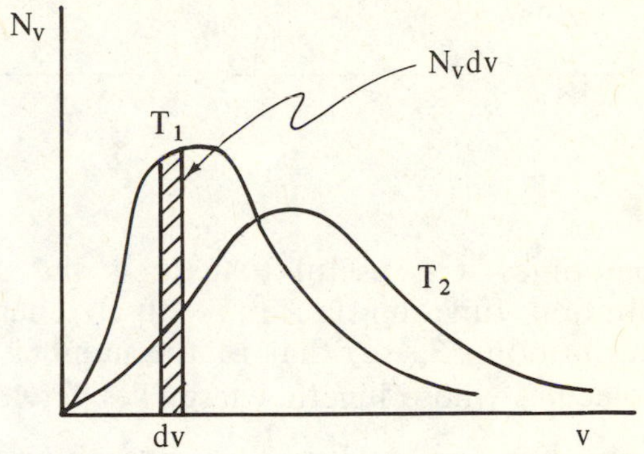

1.B

What does the quantity $N_v\,dv$ represent?

$N_v\,dv$ is the number of molecules whose speeds lie between v and v + dv.

1.C

The function N_v is given by

$$N_v = Av^2 e^{-\frac{mv^2}{2kt}} \qquad (1)$$

where

$$A = \frac{4N}{\sqrt{\pi}}\left(\frac{m}{2kT}\right)^{3/2} \qquad (2)$$

What is $\int_0^\infty N_v\,dv$ equal to?

$$N = \int_0^\infty N_v\,dv \qquad (3)$$

where N is the *total number of molecules* in the sample gas. *Note: All* possible speeds are allowed, and N is the total area under the N_v *vs* v curve.

1.D

Determine the *fraction* of particles whose speeds lie between v_1 and v_2.

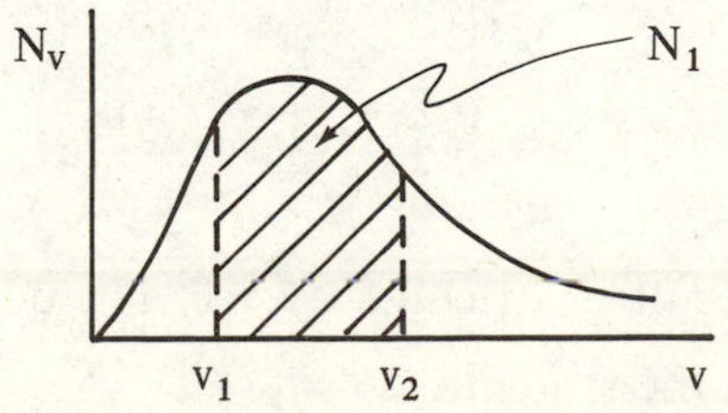

The *total* number of particles in the range v_1 to v_2 is given by

$$N_1 = \int_{v_1}^{v_2} N_v\,dv \qquad (4)$$

Therefore, the *fraction* in this range is

$$f = \frac{N_1}{N} \qquad (5)$$

1.E

What general expression would you use to calculate the *average* value of some quantity w(v)?

$$<w> = \frac{1}{N}\int_0^{\infty} w N_v \, dv \qquad (6)$$

1.F

Sometimes it is useful to express the distribution function in terms of an energy distribution, $N_E\,dE$, that is, the number of molecules whose kinetic energy lies between E and E + dE. Use $E = \frac{1}{2}mv^2$, to obtain an expression for $N_E\,dE$. Note that $N_E\,dE = N_v\,dv$, where N_v is given by (1).

$$E = \frac{1}{2}mv^2$$

$$dE = mv\,dv$$

or,

$$dv = \frac{dE}{\sqrt{2mE}}$$

Substituting this into (1) gives

$$N_E\,dE = A\left(\frac{2E}{m}\right) e^{-\frac{E}{kT}} \frac{dE}{\sqrt{2mE}}$$

or,

$$N_E\,dE = BE^{1/2} e^{-\frac{E}{kT}}\,dE \qquad (7)$$

where

$$B = \frac{2N}{\sqrt{\pi}}\left(\frac{1}{kT}\right)^{3/2} \qquad (8)$$

1.G

How would you now proceed to calculate the average kinetic energy per molecule? Simply *outline* the procedure and make use of (6) and (7).

Using the same form as (6), with $N_v\,dv$ replaced by $N_E\,dE$, we get

$$<E> = \frac{B}{N}\int_0^{\infty} E^{3/2} e^{-\frac{E}{kT}}\,dE \qquad (9)$$

The result is, of course, $\frac{3}{2}kT$. The details are left to the student.

1.H

Compare the coefficients of (9) and (1). In what way does the energy distribution differ from the speed distribution? [*Note:* The plot of N_E *vs* E will have the same general form as N_v *vs* v, where the peak occurs at the *most probable energy*, $E_p = \frac{1}{2}kT$.]

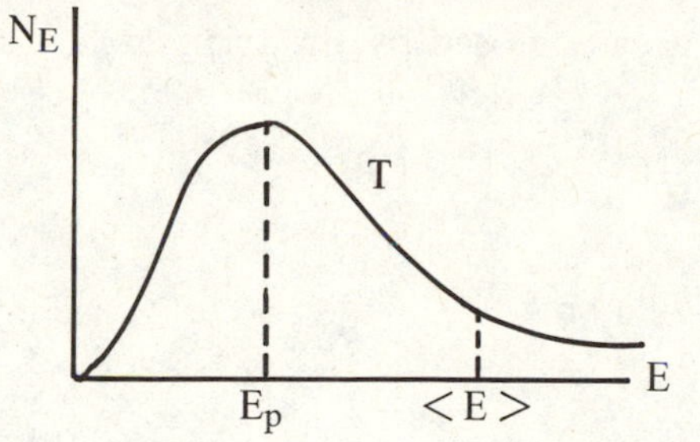

The speed distribution function depends on both m and T, whereas the energy distribution depends *only* on T. Therefore, a gas with a small m value will have a higher proportion of molecules with high speeds, compared to a gas with large m. On the other hand, at a given temperature T, the energy distribution for the molecules of *any* gas is the *same.*

2.A

The *first law of thermodynamics* is a statement of *conservation of energy.* For a finite process, we write the law as

$$\Delta U = \Delta Q - W$$

Apply this expression to a system which undergoes a *differential* change, identifying all terms.

$$dU = \text{đ}Q - \text{đ}W$$

but

$$\text{đ}W = pdV$$

$$\therefore \quad dU = \text{đ}Q - pdV \qquad (1)$$

where dU is the change in internal energy, $\text{đ}Q$ is the heat gained (or lost) by the system and pdV is the work done by (or on) the system.

2.B

Is (1) restricted to gases?

No. The first law is applicable to *any* process involving solids, liquids or gases.

2.C

In the arbitrary *cyclic process* shown, what is the net work done by the system?

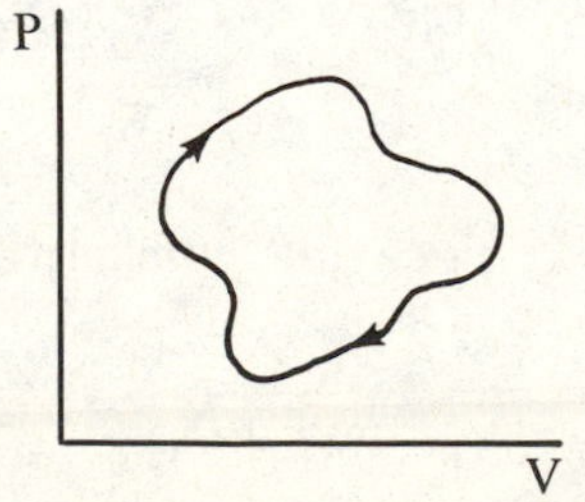

The net work done is the *area* enclosed by the curve representing the process in the pV diagram.

2.D

If the *net* heat flowing into the cyclic engine is Q, what is the net work done?

For a cyclic process, $\Delta U = 0$, ∴ from the first law,

$$W = Q$$

2.E

Determine the net work done by a gas taken through the cyclic process shown below.

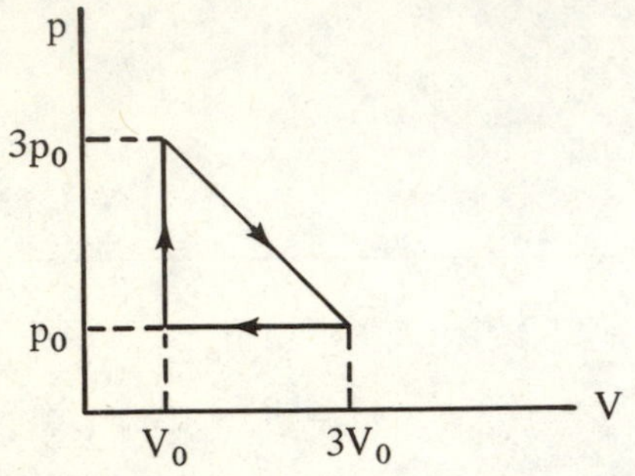

W = area enclosed by the triangle

$$W = \frac{1}{2}(3V_0 - V_0)(3P_0 - P_0)$$

$$W = 2P_0V_0$$

3.A

The figure below is a pV diagram of the *Carnot cycle.* Identify the processes ab, bc, cd, and da. Assume the system is an ideal gas.

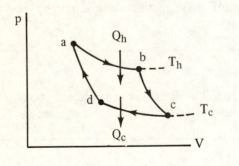

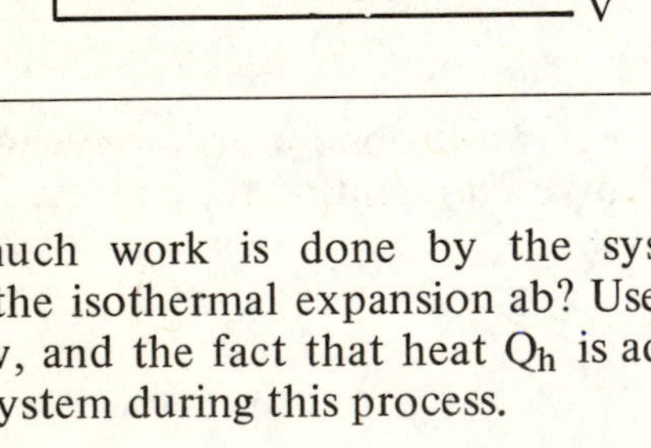

ab is an isothermal expansion.

bc is an adiabatic expansion.

cd is an isothermal compression.

da is an adiabatic compression.

3.B

How much work is done by the system during the isothermal expansion ab? Use the first law, and the fact that heat Q_h is added to the system during this process.

From the first law,

$$\Delta U = \Delta Q - W$$

but $\Delta U = 0$, since T = constant for an ideal gas,

$$\therefore \quad W_{ab} = Q_h \qquad (1)$$

3.C

How much heat is added (or removed) from the system during the adiabatic processes bc and da?

Zero. By definition, an adiabatic process is one for which no heat is lost or gained by the system.

3.D

Determine the work done by the system during the isothermal compression cd, noting that the system gives up heat Q_c to the cold reservoir.

Again, since T = constant, and $\Delta U = 0$, the first law gives

$$W_{cd} = -Q_c \tag{2}$$

The negative sign signifies that heat is removed from the system, so work is done *on* the system.

3.E

What is the *net* work done by the system? Use (1) and (2) to obtain this result. Note also that $\Delta U = 0$ for the complete cycle.

Since the process is cyclic, that is, the system returns to its original state, $\Delta U = 0$; ∴ from the first law we get

$$W_{net} = W_{ab} + W_{cd} = Q_h - Q_c \tag{3}$$

3.F

Draw a schematic of the engine which operates in a cyclic fashion between the temperature T_h and T_c.

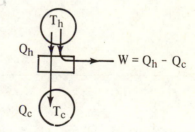

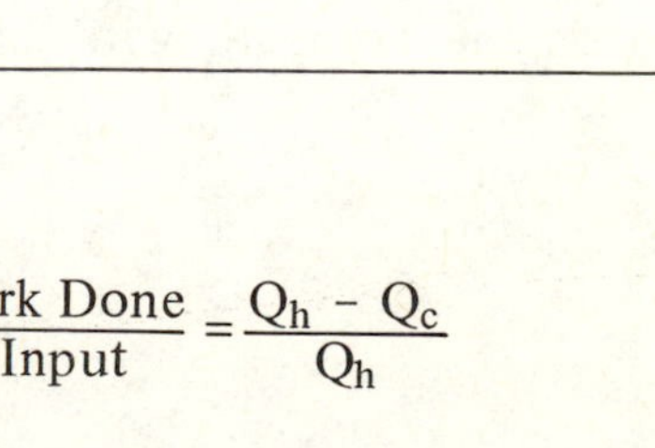

3.G

Determine the thermal efficiency, η, of a cyclic engine. Recall that efficiency is the ratio of the work done to the heat input per cycle.

$$\eta = \frac{\text{Net Work Done}}{\text{Heat Input}} = \frac{Q_h - Q_c}{Q_h}$$

or,

$$\eta = 1 - \frac{Q_c}{Q_h} \tag{4}$$

3.H

The student should show that

$$\frac{Q_c}{Q_h} = \frac{T_c}{T_h}$$

Use this fact and (4) to determine another expression for η. Can η ever be 1, corresponding to 100% efficiency? Explain.

$$\eta = 1 - \frac{T_c}{T_h} \qquad (5)$$

η will be 1 only if $T_c = 0°K$. This is *impossible* to obtain; $\therefore \eta$ is always <1. Also, *the second law of thermodynamics* states that a 100% efficient cyclic engine is impossible.

3.I

Is (5) restricted only to ideal gases?

No. This expression is valid for *any* system operating in a Carnot cycle.

3.J

Can *real* engines, such as gasoline or diesel, have a thermal efficiency given by (5)? Explain.

No. Gasoline and diesel engines do not operate in a Carnot cycle, hence their efficiencies are less than η. The *Carnot engine* is the *most* efficient engine.

3.K

Determine the *upper limit* on the efficiency of a steam engine which operates between 320°K and 373°K (the boiling point of water).

The upper limit of η is given by (5).

$$\eta = 1 - \frac{320}{373} = 0.14$$

or 14%. The actual engine would have a lower efficiency.

3.L

In theory, how does the efficiency of a gasoline engine compare with that of the steam engine?

T_h can be made much higher than 373°K, so the gasoline engine should have the higher efficiency. The compression ratio limits the efficiency. Of course, friction, heat losses, and so forth, reduce the efficiencies of both devices. Also, the processes involved in the cycles differ.

4 One mole of an ideal gas is carried through a reversible, cyclic process described by the pV diagram in frame 4.A. We will analyze the thermal efficiency of an engine operating in this cycle, and compare it with a Carnot engine operating between T_1 and T_2.

4.A

Processes ab and cd are isothermal, while bc and da are *isochoric* (constant volume). What is the work done by the gas in the process ab?

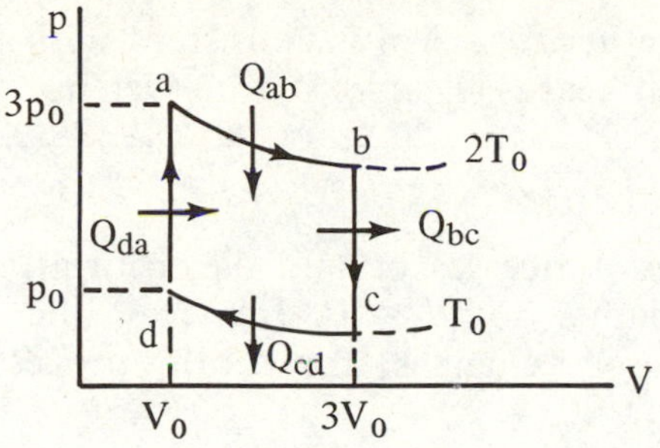

$$W = \int p dV$$

but $p = \frac{2\mu R T_0}{V}$ from the equation of state, where $2T_0$ is a constant. Since $\mu = 1$,

$$W_{ab} = 2RT_0 \int_{V_0}^{3V_0} \frac{dV}{V}$$

$$W_{ab} = 2RT_0 \ln 3 \qquad (1)$$

4.B

Determine the heat Q_h added to the system (extracted from the hot reservoir) during the process ab. Use the *first law* and (1).

Since we are dealing with an ideal gas, $\Delta U = 0$ for an *isothermal* process, and the first law gives

$$Q_{ab} = W_{ab} = 2RT_0 \ln 3 \qquad (2)$$

4.C

What is the work done during the processes bc and da?

$W_{bc} = W_{da} = 0$, since $V =$ constant, $dV = 0$; hence, $W = 0$ along both paths.

4.D

Calculate the work done by the gas during the isothermal compression cd. Note that $T = T_0$ along this path.

[Recall that $\ln(1/x) = -\ln x$.]

Along cd, $p = \frac{RT_0}{V}$, so

$$W_{cd} = \int p dV = RT_0 \int_{3V_0}^{V_0} \frac{dV}{V}$$

$$W_{cd} = RT_0 \ln\left(\frac{1}{3}\right) = -RT_0 \ln 3 \qquad (3)$$

4.E

The negative sign in (3) implies that work is done *on* the system. Now use (1) and (3) to calculate the *net* work done by the system.

$$W_{net} = W_{ab} + W_{bc} + W_{cd} + W_{da}$$

But $W_{bc} = W_{da} = 0$, so

$$W_{net} = 2RT_0 \ln 3 - RT_0 \ln 3$$

$$W_{net} = RT_0 \ln 3 \qquad (4)$$

4.F

Recall that the net work done is the area *enclosed* by the cyclic path. How much heat, Q_{cd}, is removed from the system (or given up to the cold reservoir) during the process cd?

Again, following 4.B, $\Delta U = 0$ for this isothermal process, so

$$Q_{cd} = W_{cd} = -RT_0 \ln 3 \qquad (5)$$

The *negative* sign is consistent with the fact that heat is *given up* by the system.

4.G

Is there any heat lost or gained by the system during the processes bc and da? Explain.

Yes. Processes bc and da occur at constant volume, so $W = 0$. However, the *internal energy* changes for both processes, so $\Delta Q \neq 0$.

4.H

Determine the heat *given up* by the system during the process bc. Note that T changes from $2T_0$ to T_0 along this path. Again, recall that $W_{bc} = 0$.

From the first law,

$$\Delta U = Q_{bc} - \not{W}_{bc}$$

But $\Delta U = \frac{3}{2}\mu R \Delta T$ for an ideal gas,

$$\therefore \quad Q_{bc} = \frac{3R}{2}(T_0 - 2T_0)$$

$$Q_{bc} = -\frac{3RT_0}{2} \qquad (6)$$

4.I

What is the heat *added* to the system during the process da, where T goes from T_0 to $2T_0$?

Following frame 4.H,

$$Q_{da} = \frac{3R}{2}(2T_0 - T_0)$$

$$Q_{bc} = -\frac{3RT_0}{2} \qquad (7)$$

4.J

Now calculate the *thermal efficiency* of the engine operating in this cycle, using (2) and (4) and (7). Note that $Q_{in} = Q_{ab} + Q_{da}$ is the *total heat put into the system* for one cycle.

$$\eta = \frac{W_{net}}{Q_{in}} = \frac{W_{net}}{Q_{ab} + Q_{da}}$$

$$\eta = \frac{RT_0 \ln 3}{2RT_0 \ln 3 + \frac{3RT_0}{2}}$$

$$\eta = \frac{\ln 3}{2 \ln 3 + 1.5} \cong 0.30 \qquad (8)$$

This corresponds to an efficiency of ~30%.

4.K

Determine the thermal efficiency of a *Carnot* engine operating between the two temperatures $2T_0$ and T_0, and compare your result with (8).

$$\eta_c = 1 - \frac{T_c}{T_h} = 1 - \frac{T_0}{2T_0}$$

$\eta_c = 0.5$ *or* 50%.

$\therefore \eta < \eta_c$

5.A

The *second law* of thermodynamics can be stated in various ways. The Clausius statement says that processes *do not* occur whose sole result is the transfer of heat from a ______ to a ______ body. What is the Kelvin-Planck statement of the second law?

colder to hotter

The Kelvin-Planck statement says that processes do not occur whose sole result is the conversion of heat extracted from a hot reservoir into work.

5.B

The thermodynamic variable *entropy* is a measure of the ______ in a system.

disorder

5.C

If an infinitesimal process occurs such that a system gains or loses heat $\bar{d}Q$ while at a temperature T, the *change* in entropy of the system, dS, is given by ______.

$$dS = \frac{\bar{d}Q}{T} \qquad (1)$$

5.D

If the system is taken by i to f along the path shown, what is the change in entropy (symbolic)?

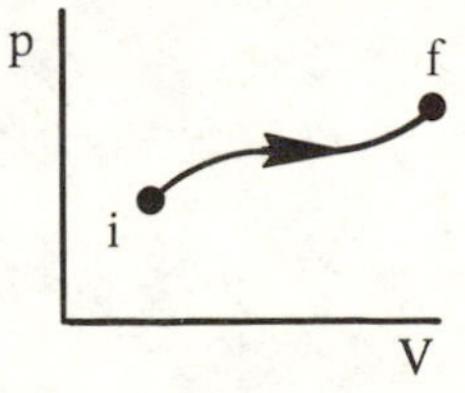

From (1), we have

$$\int_{S_i}^{S_i} dS = \int_i^f \frac{\bar{d}Q}{T}$$

or,

$$\Delta S = S_f - S_i = \int_i^f \frac{\bar{d}Q}{T} \qquad (2)$$

5.E

What kind of path must i to f be in order to calculate ΔS?

The path must be *reversible.* That is, the process occurs between a series of *equilibrium* states of the system, infinitesimally close together.

5.F

What is ΔS for a reversible, cyclic path?

From (2), we see that for any *reversible cycle,*

$$\oint \frac{đQ}{T} = 0 \qquad (3)$$

5.G

Make use of (3) for the Carnot cycle described in Figure 14–2, and verify that

$$\frac{Q_c}{Q_h} = \frac{T_c}{T_h}$$

Note that heat Q_h *enters* the system along ab (positive heat), while heat $-Q_c$ leaves the system along cd (negative heat).

Since heat is only exchanged between the system and its surroundings along the *isotherms,* ab and cd, we have

$$\oint \frac{đQ}{T} = \int_a^b \frac{đQ}{T} + \int_c^d \frac{đQ}{T} = 0$$

But $T = T_h$ along ab, while $T = T_c$ along cd.

$$\therefore \quad \oint \frac{đQ}{T} = \frac{Q_h}{T_h} - \frac{Q_c}{T_c} = 0 \quad \text{or} \quad \frac{Q_c}{Q_h} = \frac{T_c}{T_h}$$

5.H

Consider the *reversible, isothermal* expansion of μ moles of an ideal gas from V_0 to $3V_0$. What is the change in entropy, ΔS?

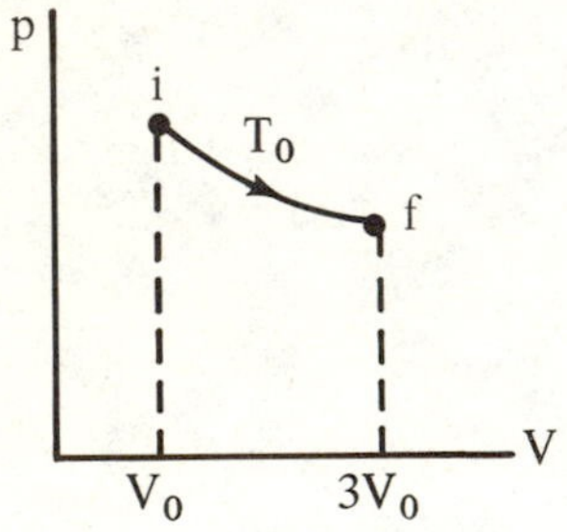

Since $\Delta U = 0$ for an isothermal process for an ideal gas, the first law gives $Q_{if} = W_{if}$, where

$$Q_{if} = W_{if} = \mu R T_0 \int_{V_0}^{3V_0} \frac{dV}{V} = \mu R T_0 \ln 3$$

Since $T = T_0$ = constant, (2) gives

$$\Delta S = \frac{1}{T_0} \int đQ = \frac{Q_{if}}{T_0} = \mu R \ln 3 \qquad (4)$$

5.I

What is the change in entropy of μ moles of an ideal gas which undergoes *reversible, adiabatic* expansion from V_1 to V_2?

By definition, $đQ = 0$ for any adiabatic process, $\therefore$ $\Delta S = 0$. That is, the entropy function is *constant* along a reversible, adiabatic path.

5.J

How can one, in principle, calculate the entropy change in an irreversible process between the states i and f?

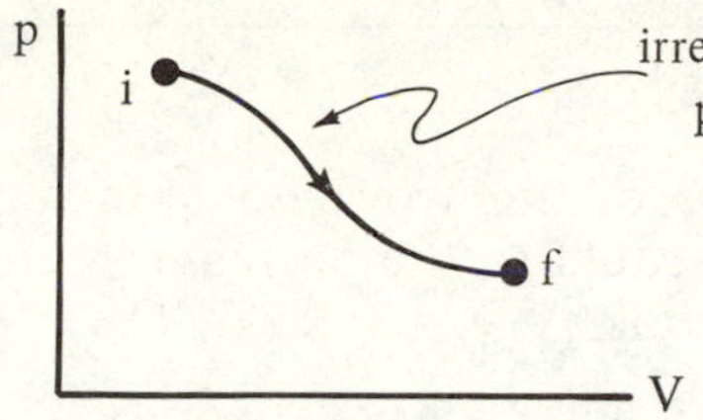

Devise a *reversible* path (which could be one or more reversible processes) between i and f. The entropy change is then equivalent to the entropy change calculated for this reversible path.

6 This is an exercise which demonstrates that the entropy change of the universe for an *irreversible* process is *greater* than zero. An iron bar of mass 5 kg is heated to 300°C, and is then submerged in a very large vat of water which is at 27°C.

6.A

What is the heat lost by the iron bar for an infinitesimal change in temperature, dT?

$$đQ = C_I dT \tag{1}$$

where C_I is the heat capacity of iron.

6.B

Calculate the entropy change of the iron for this change in temperature, dT.

$$dS = \frac{đQ}{T} = C_I \frac{dT}{T} \tag{2}$$

6.C

Assume that the water reservoir is large, so its temperature remains at 20°C. Calculate the *total* change in entropy of the iron bar. *Note:* The specific heat of iron is 0.113 cal/g°C.

The heat capacity of iron is

$$C_I = Mc_I$$

$$C_I = 5 \times 10^3 \text{ g} \times 0.113 \text{ cal/g°C}$$

$$C_I = 565 \text{ cal/°C} \tag{3}$$

Using (2), with T_0 = 300°C = 573°K and T = 27°C = 300°K gives

$$\Delta S_I = C_I \int_{573}^{300} \frac{dT}{T} = C_I ln\left(\frac{300}{573}\right)$$

$$\Delta S_I = 565\,(-0.648) = -366 \text{ cal/°C} \tag{4}$$

6.D

What is the physical significance of the negative sign in the result for ΔS_I?

This indicates that the entropy of the iron bar *decreases,* which corresponds to a higher degree of order for the bar in the final state.

6.E

If the cooling process were reversible, what would be the change in entropy of the water? What would be the entropy change of the universe? *Note:* The iron bar and water form the *isolated* system.

The entropy of the water would increase by an amount equal to the decrease in entropy of iron. That is, $\Delta S_W = 366$ cal/°C. ∴

$$\Delta S_{universe} = \Delta S_W + \Delta S_I = 0$$

6.F

The real process being considered is actually *irreversible.* What is the change in entropy of the water for this case? *Note:* The water temperature is assumed to remain at 300°K.

$$\Delta S_W = \frac{\Delta Q}{T_w} = \frac{C_I \Delta T}{T_w}$$

$$\Delta S_W = 565 \text{ cal/°C} \times \left(\frac{573 - 300}{300}\right)$$

$$\Delta S_W \cong 510 \text{ cal/°C} \qquad (5)$$

6.G

Now calculate the change in entropy of the universe. What is the physical significance of the result?

$$\Delta S_{universe} = \Delta S_W + \Delta S_I$$

$$\Delta S_{universe} = 510 - 366 = 144 \text{ cal/°C}$$

$\Delta S_{universe} > 0$, which it must be

for any irreversible process. Therefore, the entropy of the universe is always increasing.

14.7 SUMMARY

The Maxwell distribution function N_v [the most probable speed distribution for molecules defined by Equation (14.3)], provides a means for calculating the average value of a quantity w, using the relation

$$< w > = \frac{1}{N} \int_0^{\infty} w N_v \, dv \qquad (14.4)$$

where N is the total number of molecules making up the gas. Using Equation (14.4), one can calculate such quantities as the *average speed* $< v >$, the *rms speed* $\sqrt{< v^2 >}$, the *most probable* speed v_p and the *average kinetic energy* $< E >$.

The *mean free path* is the average distance a molecule travels between collisions. The *collision frequency* is the rate at which the molecules collide.

The *first law of thermodynamics* applied to a system undergoing an infinitesimal change in state is given by

$$dU = đQ - đW \qquad (14.12)$$

where dU is the change in internal energy, đQ is the heat added (or removed) and đW is the work done on (or by) the system. The internal energy function U, like temperature T, is a *state variable,* hence dU is an *exact differential.* Q and W depend on the path of a process, hence they are *not* state variables.

A *Carnot cycle* represents the most efficient cyclic process. It consists of an isothermal expansion, an adiabatic expansion, an isothermal compression and, finally, an adiabatic compression. The *efficiency* of a Carnot engine operating between a hot reservoir of temperature T_h and cold reservoir or temperature T_c is given by

$$\eta = 1 - \frac{T_h}{T_c} \qquad (14.16)$$

The *second law of thermodynamics* can be stated in various ways: (1) Heat must flow spontaneously from a hotter to a colder body. (2) A thermodynamic process, whose sole result is the transfer of heat from a colder to a hotter body, is impossible. (3) A thermodynmaic process, whose sole result is the complete transformation of heat into work, is impossible. (4) Perpetual motion machines are impossible. (5) The entropy of an isolated system either increases or remains constant.

The change in entropy of a system is defined as

$$\Delta S = \int_i^f \frac{đQ}{T} \qquad (14.18)$$

$\Delta S = 0$ for any *reversible cycle,* such as the Carnot cycle, whereas $\Delta S > 0$ for any *irreversible* process. Entropy is a state variable which is a measure of the degree of disorder in a system. Whenever spontaneous changes occur, the system transforms to a state of greater disorder.

14.8 PROBLEMS

1. A vessel contains oxygen gas at a temperature of 300°K. Calculate (a) the rms speed, (b) the average speed and (c) the most probable speed for an oxygen molecule. (The mass of O_2 is 53.5×10^{-27} kg)

2. Verify Equations (14.5) and (14.6) for the average and rms speeds of the molecules of a gas at a temperature T. Use the following integrals:

$$\int_0^\infty x^3 e^{-ax^2}\,dx = \frac{1}{4a}\sqrt{\frac{\pi}{a}};\qquad \int_0^\infty x^4 e^{-ax^2}\,dx = \frac{3}{8a^2}\sqrt{\frac{\pi}{a}}$$

3. Show that the most probable speed of the molecules of a gas as obtained from the Maxwell distribution is given by Equation (14.7). Note that $v = v_p$ corresponds to the *peak* of the N_v *vs* v curve, where $dN_v/dv = 0$.

4. A vessel contains one kg-mole of helium gas at a temperature of 300°K. Calculate the approximate number of molecules whose speeds lie in the range 400 to 450 m/sec. *Hint:* The number of molecules with speeds in the range v to v + Δv is approximately ($N_v\Delta v$).

5. Twenty particles have the following speeds: two have speeds v_0, three have speeds $2v_0$, five have speeds $3v_0$, four have speeds $4v_0$, three have speeds $5v_0$, two have speeds $6v_0$ and one has a speed $7v_0$. Calculate (a) the average speed, (b) the root mean square speed and (c) the most probable speed.

6. If the twenty particles described in problem 5 each have a mass m and are confined to a volume V, (a) what pressure do they exert on the walls of the vessel, and (b) what is the average kinetic energy per particle?

7. (a) Show that the mean free path of the molecules of a gas is related to the absolute pressure and temperature according to the relation

$$l = \frac{kT}{4\pi\sqrt{2}a^2}\left(\frac{1}{p}\right)$$

(b) Use this expression to find the mean free path of hydrogen molecules ($a \cong 5 \times 10^{-11}$ m) at T = 300°K and p = 1 atm.

8. In deriving the efficiency of an engine operating in the Carnot cycle, Equation (14.15) was assumed to be true. Verify this expression explicitly, noting that $W_{ab} = Q_h$ and $W_{dc} = Q_c$.

9. The efficiency of a Carnot engine is 30%. The engine extracts 200 cal of heat per cycle from a high temperature reservoir which is at 500°K. Determine (a) the heat given up per cycle to the cold reservoir and (b) the temperature of the cold reservoir.

10. The pV diagram of an ideal gasoline engine cycle is shown in Figure 14-4. The processes ab and cd are *adiabatic,* while bc and da are *isochoric* (constant volume). Assuming the gas is ideal, with constant specific heat, show that the thermal efficiency is given by

$$\eta = 1 - \left(\frac{V_2}{V_1}\right)^{\gamma-1}$$

The ratio V_1/V_2 is sometimes referred to as the *compression ratio. Hint:* Recall that $Q = mC_V\Delta T$ for a constant volume process, and that PV^{γ} = constant [or $TV^{\gamma-1}$ = constant] in an *adiabatic* process.

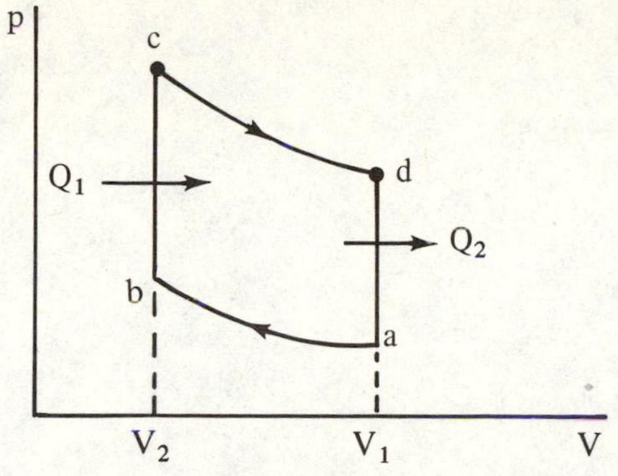

Figure 14-4 pV diagram for an ideal gasoline engine.

11. A 1 kg sample of mercury is initially at a temperature of −100°C. What is its change in entropy if enough heat is added to raise it to a temperature of 100°C? Mercury has a melting temperature of −39°C, a heat of fusion of 2.8 cal/g and a heat capacity of 0.033 cal/g°C.

12. Calculate the entropy change $\Delta S = S_b - S_a$ for the reversible *isobaric* process shown in Figure 14-5, where the system is μ moles of an ideal gas. *Hint:* Imagine that the system goes from a to b, first along an isotherm, then along an adiabatic curve, for which there is no entropy change.

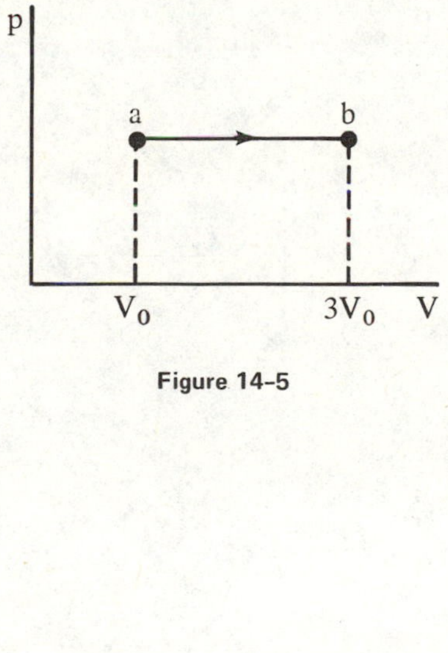

Figure 14-5

APPENDIX 1

TABLE OF TRIGONOMETRIC FUNCTIONS

Angle θ		sin θ	cos θ	tan θ	Angle θ		sin θ	cos θ	tan θ
Degree	Radian				Degree	Radian			
0	0.0000	0.0000	1.0000	0.0000					
1	0.0175	0.0175	0.9998	0.0175	46	0.8029	0.7193	0.6947	1.0355
2	0.0349	0.0349	0.9994	0.0349	47	0.8203	0.7314	0.6820	1.0724
3	0.0524	0.0523	0.9986	0.0524	48	0.8378	0.7431	0.6691	1.1106
4	0.0698	0.0698	0.9976	0.0699	49	0.8552	0.7547	0.6561	1.1504
5	0.0873	0.0872	0.9962	0.0875	50	0.8727	0.7660	0.6428	1.1918
6	0.1047	0.1045	0.9945	0.1051	51	0.8901	0.7771	0.6293	1.2349
7	0.1222	0.1219	0.9925	0.1228	52	0.9076	0.7880	0.6157	1.2799
8	0.1396	0.1392	0.9903	0.1405	53	0.9250	0.7986	0.6018	1.3270
9	0.1571	0.1564	0.9877	0.1584	54	0.9425	0.8090	0.5878	1.3764
10	0.1745	0.1736	0.9848	0.1763	55	0.9599	0.8192	0.5736	1.4281
11	0.1920	0.1908	0.9816	0.1944	56	0.9774	0.8290	0.5592	1.4826
12	0.2094	0.2079	0.9781	0.2126	57	0.9948	0.8387	0.5446	1.5399
13	0.2269	0.2250	0.9744	0.2309	58	1.0123	0.8480	0.5299	1.6003
14	0.2443	0.2419	0.9703	0.2493	59	1.0297	0.8572	0.5150	1.6643
15	0.2618	0.2588	0.9659	0.2679	60	1.0472	0.8660	0.5000	1.7321
16	0.2793	0.2756	0.9613	0.2867	61	1.0647	0.8746	0.4848	1.8040
17	0.2967	0.2924	0.9563	0.3057	62	1.0821	0.8829	0.4695	1.8807
18	0.3142	0.3090	0.9511	0.3249	63	1.0996	0.8910	0.4540	1.9626
19	0.3316	0.3256	0.9455	0.3443	64	1.1170	0.8988	0.4384	2.0503
20	0.3491	0.3420	0.9397	0.3640	65	1.1345	0.9063	0.4226	2.1445
21	0.3665	0.3584	0.9336	0.3839	66	1.1519	0.9135	0.4067	2.2460
22	0.3840	0.3746	0.9272	0.4040	67	1.1694	0.9205	0.3907	2.3559
23	0.4014	0.3907	0.9205	0.4245	68	1.1868	0.9272	0.3746	2.4751
24	0.4189	0.4067	0.9135	0.4452	69	1.2043	0.9336	0.3584	2.6051
25	0.4363	0.4226	0.9063	0.4663	70	1.2217	0.9397	0.3420	2.7475
26	0.4538	0.4384	0.8988	0.4877	71	1.2392	0.9455	0.3256	2.9042
27	0.4712	0.4540	0.8910	0.5095	72	1.2566	0.9511	0.3090	3.0777
28	0.4887	0.4695	0.8829	0.5317	73	1.2741	0.9563	0.2924	3.2709
29	0.5061	0.4848	0.8746	0.5543	74	1.2915	0.9613	0.2756	3.4874
30	0.5236	0.5000	0.8660	0.5774	75	1.3090	0.9659	0.2588	3.7321

Angle θ		sin θ	cos θ	tan θ	Angle θ		sin θ	cos θ	tan θ
Degree	Radian				Degree	Radian			
31	0.5411	0.5150	0.8572	0.6009	76	1.3265	0.9703	0.2419	4.0108
32	0.5585	0.5299	0.8480	0.6249	77	1.3439	0.9744	0.2250	4.3315
33	0.5760	0.5446	0.8387	0.6494	78	1.3614	0.9781	0.2079	4.7046
34	0.5934	0.5592	0.8290	0.6745	79	1.3788	0.9816	0.1908	5.1446
35	0.6109	0.5736	0.8192	0.7002	80	1.3963	0.9848	0.1736	5.6713
36	0.6283	0.5878	0.8090	0.7265	81	1.4137	0.9877	0.1564	6.314
37	0.6458	0.6018	0.7986	0.7536	82	1.4312	0.9903	0.1392	7.115
38	0.6632	0.6157	0.7880	0.7813	83	1.4486	0.9925	0.1219	8.144
39	0.6807	0.6293	0.7771	0.8098	84	1.4661	0.9945	0.1045	9.514
40	0.6981	0.6428	0.7660	0.8391	85	1.4835	0.9962	0.0872	11.430
41	0.7156	0.6561	0.7547	0.8693	86	1.5010	0.9976	0.0698	14.301
42	0.7330	0.6691	0.7431	0.9004	87	1.5184	0.9986	0.0523	19.081
43	0.7505	0.6820	0.7314	0.9325	88	1.5359	0.9994	0.0349	28.636
44	0.7679	0.6947	0.7193	0.9657	89	1.5533	0.9998	0.0175	57.290
45	0.7854	0.7071	0.7071	1.0000	90	1.5708	1.0000	0.0000	∞

APPENDIX 2

TABLE OF NATURAL LOGARITHMS (BASE e)

x	ln x	x	ln x	x	ln x	x	ln x
10^{-9}	-20.723	1.05	0.0488	4.2	1.4351	16	2.773
10^{-6}	-13.816	1.10	0.0953	4.4	1.4816	17	2.833
10^{-5}	-11.513	1.15	0.1398	4.6	1.5261	18	2.890
10^{-4}	-9.210	1.20	0.1823	4.8	1.5686	19	2.944
10^{-3}	-6.908	1.25	0.2231	5.0	1.6094	20	2.996
0.01	-4.6052	1.30	0.2624	5.2	1.6487	22	3.091
0.02	-3.9120	1.35	0.3001	5.4	1.6864	24	3.178
0.03	-3.5066	1.40	0.3365	5.6	1.7228	26	3.258
0.04	-3.2189	1.45	0.3716	5.8	1.7579	28	3.332
0.05	-2.9957	1.50	0.4055	6.0	1.7918	30	3.401
0.06	-2.8134	1.55	0.4383	6.2	1.8245	32	3.466
0.07	-2.6593	1.60	0.4700	6.4	1.8563	34	3.526
0.08	-2.5257	1.65	0.5008	6.6	1.8871	36	3.584
0.09	-2.4079	1.70	0.5306	6.8	1.9169	38	3.638
0.10	-2.30259	1.75	0.5596	7.0	1.9459	40	3.689
0.12	-2.1203	1.80	0.5878	7.2	1.9741	42	3.738
0.14	-1.9661	1.85	0.6152	7.4	2.0015	44	3.784
0.16	-1.8326	1.90	0.6419	7.6	2.0281	46	3.829
0.18	-1.7148	1.95	0.6678	7.8	2.0541	48	3.871
0.20	-1.6094	2.00	0.69315	8.0	2.0794	50	3.912
0.22	-1.5141	2.1	0.7419	8.2	2.1041	55	4.007
0.24	-1.4271	2.2	0.7885	8.4	2.1282	60	4.094
0.26	-1.3471	2.3	0.8329	8.6	2.1518	65	4.174
0.28	-1.2730	2.4	0.8755	8.8	2.1748	70	4.248
0.30	-1.2040	2.5	0.9163	9.0	2.1972	75	4.317
0.35	-1.0498	2.6	0.9555	9.2	2.2192	80	4.382
0.40	-0.9163	2.7	0.9933	9.4	2.2407	85	4.443
0.45	-0.7985	2.8	1.0296	9.6	2.2618	90	4.500
0.50	-0.6931	2.9	1.0647	9.8	2.2824	95	4.554
		3.0	1.0986	10.0	2.30259	100	4.605

x	ln x	x	ln x	x	ln x	x	ln x
0.55	−0.5978	3.1	1.1314	10.5	2.3514	200	5.298
0.60	−0.5108	3.2	1.1632	11.0	2.3979	300	5.704
0.65	−0.4308	3.3	1.1939	11.5	2.4423	400	5.991
0.70	−0.3567	3.4	1.2238	12.0	2.4849	500	6.215
0.75	−0.2877	3.5	1.2528	12.5	2.5257	600	6.397
0.80	−0.2231	3.6	1.2809	13.0	2.5649	10^3	6.908
0.85	−0.1625	3.7	1.3083	13.5	2.6027	10^4	9.210
0.90	−0.1054	3.8	1.3350	14.0	2.6391	10^5	11.513
0.95	−0.0513	3.9	1.3610	14.5	2.6741	10^6	13.816
1.00	0.0000	4.0	1.3863	15.0	2.7081	10^9	20.723

APPENDIX 3

USEFUL MATHEMATICAL RELATIONS

Roots of a Quadratic Equation:

$$ax^2 + bx + c = 0 \qquad x = \frac{-b \pm \sqrt{b^2 - 4ac}}{2a}$$

Trigonometric Relations:

$$\sin\theta = \frac{y}{r} \qquad \cos\theta = \frac{1}{\sin\theta} = \frac{r}{y}$$

$$\cos\theta = \frac{x}{r} \qquad \sec\theta = \frac{1}{\cos\theta} = \frac{r}{x}$$

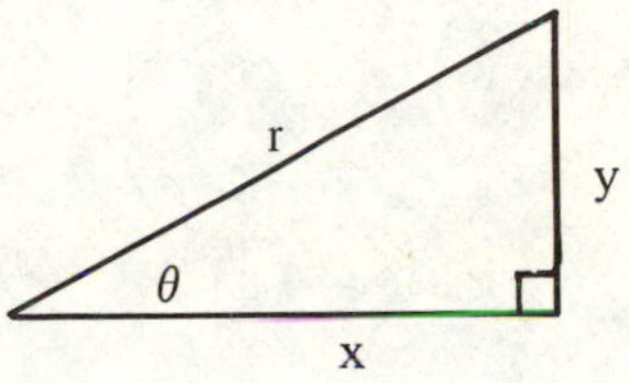

$$\tan\theta = \frac{y}{x} \qquad \cot\theta = \frac{1}{\tan\theta} = \frac{x}{y}$$

$$\sin^2\theta + \cos^2\theta = 1 \qquad x^2 + y^2 = r^2$$

$$\sin 2\theta = 2\sin\theta\cos\theta \qquad \sin(-\theta) = -\sin\theta$$

$$\cos 2\theta = \cos^2\theta - \sin^2\theta = 1 - 2\sin^2\theta \qquad \cos(-\theta) = \cos\theta$$

$$\sin(\theta \pm \psi) = \sin\theta\cos\psi \pm \cos\theta\sin\psi \qquad \tan(-\theta) = -\tan\theta$$

$$\cos(\theta \pm \psi) = \cos\theta\cos\psi \mp \sin\theta\sin\psi$$

Law of Cosines:

$$c^2 = a^2 + b^2 - 2ab\cos\gamma$$

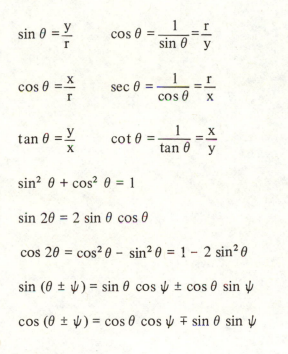

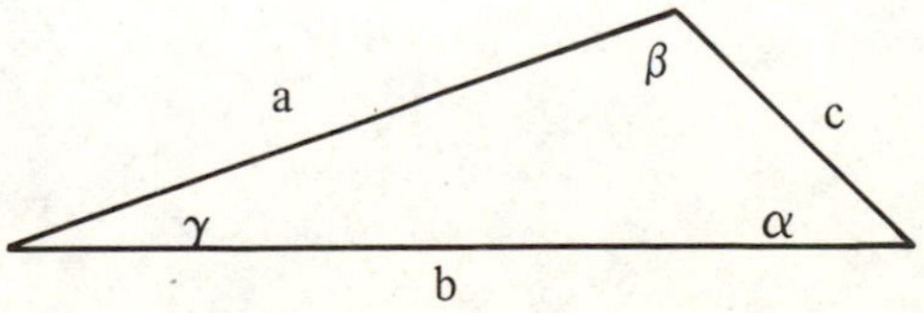

Law of Sines:

$$\frac{\sin\alpha}{a} = \frac{\sin\beta}{b} = \frac{\sin\gamma}{c}$$

Series Expansions:

$$e^x = 1 + x + \frac{x^2}{2!} + \frac{x^3}{3!} + \ldots$$

$$\sin x = x - \frac{x^3}{3!} + \frac{x^5}{5!} - \frac{x^7}{7!} + \ldots$$

$$\cos x = 1 - \frac{x^2}{2!} + \frac{x^4}{4!} - \frac{x^6}{6!} + \ldots$$

$$ln\,(1 \pm x) = \pm x - \frac{1}{2}x^2 \pm \frac{1}{3}x^3 - \frac{1}{4}x^4 + \ldots$$

$$(1 \pm x)^n = 1 \pm nx + \frac{n\,(n-1)\,x^2}{2!} \mp \frac{n\,(n-1)\,(n-2)\,x^3}{3!} + \ldots$$

APPENDIX 4

SOME CONVERSION FACTORS

Length:	1 m = 39.37 in = 3.28 ft = 100 cm
	1 ft = 30.48 cm 1 in = 2.54 cm
	1 mi = 5280 ft = 1.609 km
	1A = 10^{-8} cm = 10^{-10} m
	1μ (micron) = 10^{-4} cm = 10^{-6} m
Mass:	1 slug = 14.59 kg = 32.2 lb (mass)
	1 g = 10^{-3} kg = 6.85×10^{-5} slug
Time:	1 year = 365 days = 3.16×10^{7} sec
	1 day = 24 hrs = 1.44×10^{3} min = 8.64×10^{4} sec
Area:	1 cm^2 = 0.155 in^2 1 m^2 = 10.76 ft^2
	1 in^2 = 6.452 cm^2 1 ft^2 = 144 in^2 = 0.0929 m^2
Volume:	1 m^3 = 10^6 cm^3 = 10^3 liters
	1 m^3 = 35.3 ft^3 = 6.1×10^4 in^3
	1 ft^3 = 2.83×10^{-2} m^3 = 28.32 liters
Velocity:	1 mi/hr = 1.47 ft/sec = 0.447 m/sec
	1 m/sec = 100 cm/sec = 3.281 ft/sec
	1 mi/min = 60 mi/hr = 88 ft/sec
Acceleration:	1 m/sec^2 = 3.28 ft/sec^2 = 100 cm/sec^2
	1 ft/sec^2 = 0.3048 m/sec^2 = 30.48 cm/sec^2
Force:	1 N = 10^5 dyne = 0.2247 lb
Pressure:	$1 \frac{N}{m^2} = 10 \frac{dyne}{cm^2} = 1.45 \times 10^{-4}$ lb/in^2
Energy and Power:	1 J = 10^7 ergs = 0.239 cal = 0.738 ft lb
	1 eV = 1.6×10^{-19} J = 1.6×10^{-12} ergs
	1 cal = 4.18 J
	1 hp = 745 W = 550 ft lb/sec

APPENDIX 5

SOME FUNDAMENTAL CONSTANTS*

Universal gravitational constant	G	6.67×10^{-11} Nm^2/kg^2
Speed of light in a vacuum	c	2.9979×10^8 m/sec
Avogadro's number	N_0	6.02×10^{23} $\frac{\text{molecules}}{\text{mole}}$
Universal gas constant	R	8.314×10^3 J/mole °K
Boltzmann's constant	$k = R/N_0$	1.38×10^{-23} J/°K
Charge of electron	e	-1.6×10^{-19} C
Rest mass of electron	m_e	9.1×10^{-31} kg
Rest mass of proton	m_p	1.67×10^{-27} kg
Planck's constant	h	6.626×10^{-34} J sec
Permittivity of space	ϵ_0	8.85×10^{-12} C^2/Nm^2
Permeability constant	μ_0	$4\pi \times 10^{-7}$ N/A^2
Mass of earth		5.98×10^{24} kg
Mean radius of earth		6.37×10^6 m

*See B. N. Taylor, et al.: Fundamental Constants and Quantum Electrodynamics. In Reviews of Modern Physics Monographs, *41:* 375 (1969) for more complete list of constants.

APPENDIX 6

ANSWERS TO PROBLEMS

Chapter 1

1. 24 ft^2
2. 464 m/sec
3. 6.71×10^8 mi/hr
4. 1.013×10^5 N/m^2
5. (179 ± 3) cm^3
6. (157.50 ± 0.39) in
7. (0.196 ± 0.008) kg m^2
8. (a) 5.3×10^5 (b) 2.5×10^{-4} (c) 5×10^6 (d) 1.3×10^{10} (e) 1.492×10^3
9. (a) 2.1×10^{14} (b) 1.2×10^{-4} (c) 4.2×10^{-9} (d) 3.3×10^{12}

Chapter 2

1. (a) $A_x = 4 \cos\theta$ $A_y = 4 \sin\theta$ (b) $\mathbf{A} = 4 \cos\theta\,\mathbf{i} + 4 \sin\theta\,\mathbf{j}$
2. (a) $x = -14$ m, $y = 14$ m (b) $\mathbf{r} = (-14\mathbf{i} + 14\mathbf{j})$ m
3. (a) $\mathbf{i} + 2\mathbf{j}$ (b) $5\mathbf{i} + 6\mathbf{j}$ (c) $\sqrt{5}$, 63.4° from +x axis (d) −14 (e) $14\mathbf{k}$
4. $R = 6.4$ m, $\theta = 231.4°$, $\mathbf{R} = (-4\mathbf{i} - 5\mathbf{j})$ m
6. $\alpha = 60.8°$, $\beta = 35.8°$, $\gamma = 71.1°$
10. 12.3 mi, 72.5° N of E
11. $\mathbf{F} = (15.2\mathbf{i} + 4.4\mathbf{j})$ N
13. (a) $\mathbf{F}_1 = 6.4\mathbf{i} + 4.8\mathbf{j}$, $\mathbf{F}_2 = -7.2\mathbf{i} - 9.6\mathbf{j}$
    $\mathbf{F}_1 + \mathbf{F}_2 = -0.8\mathbf{i} - 4.8\mathbf{j}$
    $\mathbf{F}_1 - \mathbf{F}_2 = 13.6\mathbf{i} + 14.4\mathbf{j}$
    (b) $\mathbf{F} = 0.8\mathbf{i} + 4.8\mathbf{j}$
14. (a) $\mathbf{r}_1 = a\mathbf{i}$ $\mathbf{r}_2 = a(\mathbf{i} + \mathbf{j})$ $\mathbf{r}_3 = a(\mathbf{i} + \mathbf{j} + \mathbf{k})$
    (b) $\theta_1 = 45°$, $\theta_2 = 35.3°$, $\theta_3 = 54.8°$
16. (b) 8 m^2
17. $\mathbf{a} = \mathbf{i} \cos\theta + \mathbf{j} \sin\theta$ $\mathbf{b} = \mathbf{i} \cos\phi + \mathbf{j} \sin\phi$

Chapter 3

1. (a) 37.5 ft/sec^2 (b) 2.67 sec
2. (a) 2 m/sec^2 (b) 4 m/sec (c) $\sqrt{2}$ sec (d) 2.8 m/sec
3. (a) -1.50×10^6 ft/sec^2 (b) 3.33×10^{-4} sec
4. (a) 80 ft/sec (b) 5 sec
5. (a) 1320 ft (b) $V_A = 230$ ft/sec, $V_B = 252$ ft/sec
6. (a) 83 ft/sec (b) 3.1 sec (c) 39 ft
7. (b) 11.2 sec (c) 4930 ft from observer (d) 567 ft/sec, 39° below the horizontal
8. (a) 20 m/sec, 5 sec (b) 31 m/sec, 59° below the horizontal (c) 6.5 sec (d) 24 m

10. (a) 1528 m (b) 36.2 sec (c) 4041 m
12. (b) $\mathbf{r} = 32\,\mathbf{j}$ m, $\mathbf{v} = (2\mathbf{i} + 40\mathbf{j})$ m/sec, $\mathbf{a} = (6\mathbf{i} + 32\mathbf{j})$ m/sec^2
14. $v = -t^2 + t^3,\ x = x_0 - \frac{t^3}{3} + \frac{t^4}{4}$
15. (a) 35 ft/sec (b) 3.4 ft

Chapter 4

1. (a) 0.22 rad/sec (b) $a_r = 58$ ft/sec^2
2. (a) 0.11 rad/sec (b) 7.3×10^{-3} rad/sec^2 (c) 10.9 ft/sec^2 (d) 3.98×10^4 ft
3. (a) $40\,\pi$ rad/sec (b) $800\,\pi^2$ ft/sec^2 (c) $200\,\pi$ ft
4. (a) 4 rad/sec, 8 rad/sec (b) $\mathbf{v}_1 = 8\hat{\theta}$ ft/sec, $\mathbf{v}_2 = 160\hat{\theta}$ ft/sec, $\mathbf{a}_1 = (8\hat{\theta} - 32\hat{r})$ ft/sec^2, $\mathbf{a}_2 = (8\hat{\theta} - 128\hat{r})$ ft/sec^2 (c) $\theta_1 = 3$ rad, $\theta_2 = 9$ rad
5. (a) 2.60×10^{-6} rad/sec (b) 996 m/sec (c) 2.60×10^{-3} rad/sec^2
6. (a) 1.99×10^{-7} rad/sec (b) 2.97×10^4 m/sec (c) 5.91×10^{-3} m/sec^2
7. (a) (4a + 8b) rad (b) (4a + 12b) rad/sec (c) (2a + 12b) rad/sec^2

Chapter 5

1. (a) 0.17 m/sec^2 (b) 4.9 sec (c) 0.83 m/sec
2. 3.6 m/sec^2
3. (a) $\frac{F}{m_1 + m_2} - \mu g$ (b) $\frac{m_1 F}{m_1 + m_2}$
4. (b) 11.3 lb (c) 0.18
5. (b) $a = \frac{(m_1 - m_2)\,g \sin\theta - (m_1 + m_2)\,\mu g \cos\theta}{m_1 + m_2}$,

   $T = \frac{2\,m_1 m_2 g \sin\theta}{m_1 + m_2}$
6. (b) 24.4 N (c) 0.31
7. (b) 16 ft/sec^2 (c) $T_1 = 1.6$ lb, $T_2 = 5$ lb
8. (c) 8 ft/sec^2 (d) 20 lb
9. (a) 20 lb (b) 12 lb (c) 18 ft/sec^2
10. (b) 17.5°
11. (a) $g\,(\mu \cos\theta - \sin\theta)$ (b) $v_0^2/2\,g\,(\mu\cos\theta - \sin\theta)$ (c) $g\,(\mu - \theta)$; $v_0^2/2\,g\,(\mu - \theta)$ (d) 13 ft/sec^2; 296 ft
12. (a) $a = \frac{\sin\theta + \mu\cos\theta}{\cos\theta - \mu\cos\theta}$ (b) $F = (M + m)\,g\left[\frac{\sin\theta + \mu\cos\theta}{\cos\theta - \mu\cos\theta}\right]$
13. (a) $\omega = 3.8$ rad/sec (b) 76 cm/sec, 294 cm/sec^2
14. 8.1 rad/sec

Chapter 6

1. 5×10^3 ft lb
2. (a) 80 ft lb (b) −20 ft lb (c) 60 ft lb
3. (a) 0.9 J (b) 2.5 J (c) 1.6 J
4. (a) 9×10^3 J (b) 300 N
5. (a) 51 J (b) 69 J
6. (a) $\mathbf{F}_{ext} = 4k\mathbf{i}$ N (b) 6.3k J
7. (a) 31.3 m/sec (b) 147 J (c) 4

8. (a) 9 J, no (b) −9 J
9. (a) $\frac{1}{2}mR^2\omega^2$ (b) 2.5 J
10. (a) 5.88×10^6 ergs (b) 2.25×10^6 ergs (c) 3.63×10^6 ergs
11. (a) 2.7 m/sec (b) 4.7 m/sec (c) 5.5 m/sec
12. (a) 2A/b (b) $\sqrt{v_0^2 + \frac{4A}{mb}}$
13. (a) 16 J (b) 4 J (c) 7.3 J (d) nonconservative
14. (a) $(2gmh/k)^{1/2}$ (b) 0.28 m
15. (a) 2 m (b) 2.25 m (c) 1.25 m
16. (a) 350 J, 680 J, 740 J (b) 175 N, 340 N, 370 N (c) yes
17. 590 N/m
18. 0.11
19. (a) $v = \sqrt{\frac{g}{L}(L^2 - y_0^2)}$ (b) $t = \sqrt{\frac{L}{g}}\, ln\left[\frac{L + \sqrt{L^2 - y_0^2}}{y_0}\right]$
20. (a) $v = \sqrt{2gl(1 + \sin\theta)}$ (b) $T = mg\,(3 + 2\sin\theta)$ (c) 5.9 m/sec, 9.0 N
21. (a) 334 N (b) 0.71 m/sec
22. (a) $U(x) = -\frac{Ax^2}{2} + \frac{Bx^3}{3}$ (b) $U = -\frac{5}{2}A + \frac{19}{3}B$
23. (a) $F(x) = -3 - 8x + 3x^2$
26. (a) GMm/3r (b) GMm/r
27. (a) 2.37×10^3 m/sec (b) 6.17×10^5 m/sec
28. 2.2 hp
29. (a) 2.1×10^5 J (b) 3×10^3 W
30. (a) 3400 lb (b) 8 ft/sec (c) 3200 ft lb (d) 50 hp
31. 2.94×10^6 dynes/cm

Chapter 7

1. (a) 1.25 kg m/sec (b) 31 J (c) 1.25 kg m/sec
2. (a) 6.2×10^3 lb sec in the NE direction (b) 2070 lb
3. (a) 7.0 kg m/sec to the left (b) 700 N (c) 17.5 J
4. (a) 25 mi/hr to the East (b) 8.9×10^4 ft lb
5. 1.3×10^{-20} m/sec
6. 24 cm/sec
7. (a) It doesn't change (b) 38 ft/sec in the Southerly direction (c) 22 ft/sec at an angle 24° W of S
8. 100 m/sec
11. (a) $\mathbf{r}_c = 3\,\mathbf{j}$ m/sec (b) $\mathbf{v}_c = (4\mathbf{i} + 2\mathbf{j})$ m/sec (c) $\mathbf{a}_c = (3\mathbf{i} - \mathbf{j})$ m/sec^2
12. 2.7 m/sec to the left
13. (a) 3.8 ft from the shore (b) No. The turtle will be 6.2 ft from the tip of the boat, so the boy can't reach it.
14. 4.67×10^6 m
15. (a) $\mathbf{r}_c = -\frac{1}{3}(4\mathbf{i} + 10\mathbf{j})$ cm. (b) $\mathbf{v}_1 = (3\mathbf{i} + 11\mathbf{j})$ cm/sec, $\mathbf{v}_2 = -(8\mathbf{i} + 6\mathbf{j})$ cm/sec (c) $\mathbf{p} = -(13\mathbf{i} + \mathbf{j})$ g cm/sec (d) $\mathbf{v}_c = -\frac{1}{3}(13\mathbf{i} + \mathbf{j})$ cm/sec (e) $\mathbf{a}_1 = 4\mathbf{j}$ cm/sec^2, $\mathbf{a}_2 = -4\mathbf{i}$ cm/sec^2 (f) $\mathbf{a}_c = \frac{4}{3}(-2\mathbf{i} + \mathbf{j})$ cm/sec^2
16. (a) 6.9 m/sec (b) 1.35m
17. 1320 m/sec

18. (a) 106 cm/sec, 30° below the +x axis (b) 0.32
19. (a) 6 cm/sec, 34 cm/sec (b) 0.25
20. (a) $v_p = -1.2 \times 10^6$ m/sec, $v_{He} = 0.8 \times 10^6$ m/sec (b) 0.64
21. (a) 9×10^6 N (b) 3 m/sec^2 (c) 2.1×10^5 m/sec
22. (c) 1.97×10^5 kg
23. 200 m/sec
24. (a) 5 rad/sec (b) 132 J
25. (a) 0.5 m/sec (b) 11.25 J

Chapter 8

1. (a) 4 mvd directed out of the plane (b) zero (c) 2mvd directed into the plane
2. (a) 2.7×10^{40} kg m^2/sec (b) 2.7×10^{33} J (c) 2.6×10^{29} J (d) 8.5×10^{16}yr
3. (a) 6k kg m^2/sec (b) (2i + 5j) m (c) 7j m/sec (d) 14k kg m^2/sec
(e) 8k kg m^2/sec (f) 8k Nm
5. (a) $2(m_1 a^2 + m_2 b^2)$ (b) $(m_1 a^2 + m_2 b^2)$ (c) $m_1 a^2 \omega^2$ (d) $(m_1 a^2 + m_2 b^2)\omega$
6. (a) $v_c = 2\sqrt{\dfrac{Rg}{3}}$ (b) $v_c = \sqrt{Rg}$
8. (a) $\dfrac{7}{12} Ml^2$ (b) $\dfrac{4}{3} Ml^2$ (c) $\dfrac{1}{6} Ml^2$ (d) $\dfrac{11}{12} Ml^2$
9. (a) $\mathbf{L} = md(v_0 + gt)\,\mathbf{k}$ (b) $\boldsymbol{\tau} = mgd\mathbf{k}$
10. (a) 2 rad/sec^2 (b) 6 rad/sec (c) 30 kg m^2/sec (d) 90 J (e) 4.5 m
11. (a) 9.3 rad/sec (b) 9.3 rad/sec
12. (a) 0.62 rad/sec (b) 0.12 J
13. (a) $a = \dfrac{m_1 g}{m_1 + m_2 + \dfrac{I}{R^2}}$ (b) $T_1 = m_1 g \left\{\dfrac{m_2 + I/R^2}{m_1 + m_2 + I/R^2}\right\}$, $T_2 = \dfrac{m_1 m_2 g}{m_1 + m_2 + I/R^2}$
(c) $a = 1.2$ m/sec^2, $T_1 = 43$ N, $T_2 = 12$ N
14. (a) 0.3 rad/sec in the counterclockwise direction (b) 45 ft/lb
15. (a) 12 kg m^2 (b) 2.4 Nm (c) 44 rev
16. (a) $T = \dfrac{mMg}{M + 4m}$ (b) $a = \dfrac{4mg}{M + 4m}$
17. (a) $\omega = r_0^2 \omega_0 / r^2$ (b) $W = \dfrac{1}{2} M r_0^2 \omega_0^2 \left[\left(\dfrac{r_0}{r}\right)^2 - 1\right]$ (c) 45 rad/sec, 0.45 J
18. (a) $\omega = \sqrt{\dfrac{3g}{l}}$ (b) $\alpha = \dfrac{3g}{2l}$ (c) $a_x = \dfrac{3}{2} g$, $a_y = \dfrac{3}{4} g$ (d) $R_x = \dfrac{3}{2} Mg$, $R_y = \dfrac{1}{4} Mg$
19. (a) $\dfrac{\sqrt{2}}{2} R$ (b) $\sqrt{\dfrac{2}{5}} R$ (c) R (d) $\dfrac{\sqrt{3}}{3} l$ (e) $\dfrac{\sqrt{3}}{6} l$ (f) $\sqrt{\dfrac{2}{3}} R$
20.

Object	*Acceleration, a*	*Speed, v*	*Coefficient of friction, μ*
(1) Disc	$\dfrac{2}{3} g \sin\theta$	$\sqrt{\dfrac{4}{3} gh}$	$\dfrac{1}{3} \tan\theta$
(2) Hoop	$\dfrac{1}{2} g \sin\theta$	$\sqrt{gh}$	$\dfrac{1}{2} \tan\theta$
(3) Hollow sphere	$\dfrac{3}{5} g \sin\theta$	$\sqrt{\dfrac{6}{5} gh}$	$\dfrac{2}{5} \tan\theta$

21. (a) $v = \sqrt{\dfrac{4}{3} g \left(\dfrac{R^3 - r^3}{r^2}\right)}$ (b) $v \cong 5.3 \times 10^4$ m/sec (or Mach #160!)
22. (a) $\sqrt{3l/g \sin\theta}$ (b) $\dfrac{2}{3}\sqrt{3gl \sin\theta}$
23. (a) $\sqrt{\dfrac{kd^2 + 2mgd \sin\theta}{I + mR^2}}$ (b) 1.8 rad/sec

24. $\omega = \sqrt{\frac{10}{7}\frac{g}{R^2}(1 - \cos\theta)(a - R)}$

25. (a) $t = \frac{v_0}{\mu g(1 + R^2/k^2)}$ (b) $v_c = \frac{v_0}{1 + k^2/R^2}$ (c) $x = \frac{v_0^2}{2\mu g}\frac{(1 + 2R^2/k^2)}{(1 + R^2/k^2)^2}$

Chapter 9

1. $T_1 = 98$ N, $T_2 = 59$ N, $T_3 = 79$ N
2. (a) $F_x = 12.4$ lb, $F_y = 15$ lb (b) $F = 19.5$ lb, $\theta = 50.4°$ from x (c) $F' = 19.5$ lb opposite to F
3. (c) $\mu_s = 0.5$, $f_s = 6$ lb
4. 75 cm mark
6. (b) $T = 204$ N, $R_x = 176$ N, $R_y = 192$ N
7. (b) $T = 17.5$ lb (c) $d = 0.76\,l$
8. (a) $\mu_k = 0.57, \frac{6}{7}$ ft from the lower right corner (b) $h = \frac{5}{3}$ ft
9. $W_x = \frac{W(\mu_s\cos\phi + \sin\phi)}{\sin\theta - \mu_s\cos\theta}$
10. $T = 520$ lb, $R_x = 415$ lb, $R_y = 490$ lb
11. $\mu_s = 0.27$
12. (a) $W = \frac{w}{2}\left[\frac{2\mu_s\sin\theta - \cos\theta}{\cos\theta - \mu_s\sin\theta}\right]$ (b) $R = (w + W)\sqrt{1 + \mu_s^2}$ (c) $F = \sqrt{W^2 + \mu_s^2(w + W)^2}$

Chapter 10

1. $0.2\,\pi$ sec
2. $0.4\pi^2$ N/m
3. (a) 4.3 cm (b) -5 cm/sec (c) -17.2 cm/sec^2 (d) π sec
4. (a) $\delta \cong 44° = 0.77$ rad (b) $x = 2\cos(4\pi t + 0.77)$ cm
   (c) $v = -8\pi\sin(4\pi t + 0.77)$ cm/sec (d) $a = -32\pi^2\cos(4\pi t + 0.77)$ cm/sec^2
8. (a) 4 sec (b) $\Delta T \cong 3 \times 10^{-2}$ sec
9. $T = 2\pi\sqrt{\frac{2l}{3g}} \cong 1.6$ sec
10. (a) $\theta_0 = 0.088$ rad, $\omega = 2.2$ rad/sec, $\delta = 0$ (b) $|v_{max}| = 0.39$ m/sec (c) $T = 30$ N
12. (a) $K = \frac{1}{2}\left(\frac{m}{3}\right)v^2$ (b) $T = 2\pi\sqrt{\frac{M + m/3}{k}}$
13. (a) 0.36 J (b) (0.16 J (c) 0.20 J (d) 8.5 cm
14. (a) $x = 16\cos\left(\frac{\pi}{2}t\right)$ cm, $x(t = 0.5) = 8\sqrt{2}$ cm (b) $F = -100\sqrt{2}\pi^2$ dynes (to the left)
   (c) $\frac{2}{3}$ sec (d) $v = -8\pi\sin\left(\frac{\pi}{2}t\right)\frac{\text{cm}}{\text{sec}}$, $v(x = 8) = -4\sqrt{3}\,\pi\frac{\text{cm}}{\text{sec}}$
   (e) $E = 1.6\pi^2 \times 10^3$ ergs; $k \overset{2}{=} 12.5\pi^2\frac{\text{dynes}}{\text{cm}}$
15. (a) $T_p = Mg\left(1 + \frac{y}{l}\right)$ (b) $T_{top} = 2Mg$ (c) period $= T = \frac{8\pi}{3}\sqrt{\frac{l}{g}} = 3.8$ sec

Chapter 11

1. (a) 1.5×10^{11} m (b) 3.5×10^{28} N

2. The point lies along the axis joining the particles at a distance d/3 from M, or $\frac{2}{3}$d from the mass 4M.
3. $F_x = G\left[\frac{2\,m^2}{b^2} + \frac{3\,m^2 b}{(a^2+b^2)^{3/2}}\right]$, $F_y = G\left[\frac{2\,m^2}{a^2} + \frac{3\,m^2 a}{(a^2+b^2)^{3/2}}\right]$
4. $U = -Gm^2\left[\frac{2}{a} + \frac{2}{b} + \sqrt{\frac{3}{a^2+b^2}}\right]$
5. (a) 9.60 m/sec (b) 6.6×10^4 m
6. $F = \frac{4\sqrt{2}\,GMm}{\pi R^2}$ in the +y direction
7. (a) $\rho_0 = \frac{5M}{4\pi R^5}$ (b) $\frac{GMm}{r^2}$ (c) $\frac{GMm}{R^5}r^3$
8. 1.3×10^6 m from the earth's surface.
9. $\left(\frac{GM}{4R^3}\right)^{1/2}$
10. The space vehicle will accelerate towards the planet until it reaches the hole, where it will have some maximum velocity v. It will then remain moving with this velocity inside the planet (where F = 0) and continue moving until it collides with the inner surface of the planet.
11. $\frac{g}{4} = 2.45$ m/sec^2
12. 86 min
13. 6.0×10^7 J
14. (a) $3.0 \times 10^{-19}\ \frac{kg}{Nm^2}$ (b) 7.6×10^6 sec
15. 5.03×10^3 m/sec
16. 5.8×10^4 m/sec
19. (a) $-\frac{GMm}{r}$ (b) $-\frac{GMm}{R}$ (c) $\frac{GMm}{2R}\left[\frac{r^2 - 3R^2}{R^2}\right]$
21. (a) $U_{total} = -\frac{3GM}{R}\left[M + \frac{\sqrt{3}}{3}m\right]$ (b) $v = \left(\frac{\sqrt{3}\,GM}{3R} + \frac{GM}{R}\right)^{1/2}$

Chapter 12

1. 1.96×10^3 N/m
2. 513 g
3. (a) 8×10^5 lb (b) 4.1×10^3 ft^3/sec, or 3×10^4 gal/sec
6. (a) 11.7 N (b) $F_t = 1.0196 \times 10^3$ N; $F_b = 1.0374 \times 10^3$ N
8. (a) 1.02 ft/sec (b) 13.3 lb/in^2
10. (a) $v_1 = 2.42$ m/sec, $v_2 = 3.43$ m/sec (b) $x_1 = 0.91$ m, $x_2 = 0.98$ m (c) 8.27 A

Chapter 13

1. 1.02×10^{-5}
2. (a) 24 m (b) 3.4×10^5 N/m^2
5. (b) 5×10^{-5} (°C)$^{-1}$
6. 1.05 cm
8. 11.7 *l*
9. (a) 27 *l* (b) 8.2 *l*/min
10. 0.257 atm
11. -2.74×10^3 J

14. (a) 6×10^5 J (b) 4×10^5 J (c) 10.7×10^5 J (d) 14.7×10^5 J
15. (b) 584 m/sec
16. (a) About 14 times per hour. (b) 7.90×10^7 N/m^2 $\cong$ 790 atm

Chapter 14

1. (a) 482 m/sec (b) 444 m/sec (c) 393 m/sec
4. 7.68×10^{24} molecules
5. (a) $5.65v_0$ (b) $4.0v_0$ (c) $3v_0$
6. (a) $\dfrac{107mv_0{}^2}{V}$ (b) $8mv_0{}^2$
7. (b) 9.3×10^{-7} m
9. (a) 140 cal (b) 350°K
11. 37 cal/°C
12. $\mu R\left(\dfrac{\gamma}{\gamma - 1}\right) ln3$

INDEX

Page numbers in *italic* type refer to illustrations.
Page numbers followed by (t) refer to tables.